Dudley G. Moon, Ph.D.

D1472798

Immunolabelling for Electron Microscopy

Immunolabelling for Electron Microscopy

edited by

Julia M. Polak and Ian M. Varndell

Department of Histochemistry
Royal Postgraduate Medical School
Hammersmith Hospital
Du Cane Road
London W12 0HS
U.K.

ELSEVIER
AMSTERDAM · NEW YORK · OXFORD

ISBN 0-444-80563-X
1st edition 1984
2nd printing 1985
3rd printing 1985
4th printing 1987

PUBLISHED BY:

Elsevier Science Publishers B.V.
P.O. Box 211
1000 AE Amsterdam
The Netherlands

SOLE DISTRIBUTORS FOR THE U.S.A. AND CANADA:

Elsevier Science Publishing Company, Inc.
52 Vanderbilt Avenue,
New York, N.Y. 10017
U.S.A.

Library of Congress Cataloging in Publication Data
Main entry under title:

Immunolabelling for electron microscopy.

Bibliography: p.
Includes index.
1. Immunochemistry—Technique. 2. Tracers (Biology)
3. Tracers (Chemistry.) 4. Electron microscopy—
Technique. I. Polak, Julia M. II. Varndell, Ian M.
QR183.6.I464 1984 616.07'9 84-10308
ISBN 0-444-80563-X

Printed in The Netherlands

Preface

In 1959, almost 20 years after Albert H. Coons [1] introduced his revolutionary concept now widely referred to as 'Immunohistochemistry', a short paper by S.J. Singer [2] was published in *Nature* which heralded the birth of electron microscopical immunocytochemistry. A quarter of a century has elapsed since his description of horse spleen ferritin covalently bound to immunoglobulin molecules and in that time a new generation of techniques and applications has given many fields of biology and medicine a new technology—immunolocalisation. It would serve no useful purpose to list all the scientists who have made significant contributions to the growth of this subject, instead we have attempted to invite as many of them as possible to contribute chapters to this book. Graciously, and almost without exception, they agreed to do so. To ensure that *Immunolabelling for Electron Microscopy* would be current at the time of publication, the Editors imposed severe deadlines for the receipt of manuscripts. With professional courtesy and inconceivable tolerance all the authors responded and it is to them and their achievements that we dedicate this book.

The format of this book documents the current state of ultrastructural antigen detection, reviews all the major well-established techniques and introduces new areas of sub-cellular research.

In the opening chapters three of the most eminent electron immunocytochemists, Edigio and Martha Romano and Marc Horisberger, present comprehensive reviews on the background to ultrastructural immunolocalisation, the techniques and markers available and draw on their own extensive research to illustrate the applications of immunolabelling at the electron microscopical level. Each well-established technique is then treated in more detail. John Priestley (Chapter 4) has reviewed and described the use of immunoenzyme reagents in pre-embedding procedures and has clearly demonstrated the advantages of this approach in the study of central nervous system arborisation. In his contribution he introduced and illustrated the use of radiolabelled immunoreagents—a theme that is expanded upon by Georges Pelletier and Gerard Morel (Chapter 7) and Virginia Pickel and Alain Beaudet (Chapter 18). Post-embedding immunoenzyme techniques are also utilised by these authors and the flexibility of this methodology is highlighted by Geoffrey Newman and Bharat Jasani (Chapter 5). The latter authors also describe in some detail their innovative work with the use of DNP-labelled ligands in combination with a newly introduced acrylic resin. This begs the question, "Which embedding medium to use for electron microscopical immunocytochemistry?" All too frequently the reply is biased and proferred without consideration of all the alternatives and parameters. In Chapter 3, Brian Causton, a polymer chemist with considerable experience of embedding media for electron microscopy, has presented a comprehensive resumé of embedding media in current usage and points to their advantages and disadvantages for immunocytochemistry.

Once the ultrathin sections have been cut the electron immunocytochemist is faced with a wealth of post-embedding techniques in addition to the immunoenzyme procedures already mentioned. The majority of these techniques involve the use of colloidal gold as the electron-dense marker. Updated views of the well-known techniques are presented by the people who established them—Jurgen Roth (Chapter 9), Jan Slot and Hans Geuze (Chapter 11), Moïse Bendayan and Heather Stephens (Chapter 12) on protein A-gold; Lars-Inge Larsson (Chapter 10) on GLAD and other *g*old-*l*abelled *a*ntigen *d*etection methods. For our part we have presented an overview of single and double immunostaining methods utilising immunoglobulin-gold (Chapter 13).

Two relatively recent introductions to the range of electron immunocytochemical techniques have been those of avidin-biotin-marker complexes and cryoultramicrotomy. We are honoured that the leading exponents of both technologies, Jean-Pierre Kraehenbuhl, David Papermaster and Claude Bonnard (Chapter 8) and K.T. Tokuyasu (Chapter 6) have contributed detailed chapters. The use of frozen tissue for electron immunocytochemistry has gained credibility in recent years, largely through the excellence of workers such as Professors Tokuyasu, Slot and Geuze, however, the concept of rapid freezing is not new. Over 40 years ago experiments were conducted to evaluate the use of frozen biological tissues for electron microscopy. With the introduction of reliable freezing devices and sophisticated freeze driers electron microscopists have been able to achieve good morphological preservation. In Chapter 16, Ronald Dudek and co-workers have investigated the combination of fixation by rapid freezing and freeze-drying with immunogold staining procedures and have been able to demonstrate a new and highly effective technique for improved immunostaining.

The section on Techniques is completed by overviews of three powerful and rapidly developing areas—freeze-fracture immunocytochemistry by Pedro Pinto da Silva (Chapter 14), scanning immunoelectron microscopy by Gisele Hodges, Marie Smolira and D. Campbell Livingston (Chapter 15) and lectin cytochemistry by Marc Horisberger (Chapter 17). It is with deep sadness, however, that we must record the death of D.C. Livingston, one of the contributors to this book, which occurred shortly after receipt of his manuscript.

In the final section of *Immunolabelling for Electron Microscopy* we have attempted to cover some applications of immunocytochemistry which involve specialist or less well-known aspects of electron immunocytochemistry.

Isolated cells, unicellular organisms, bacteria and viruses all present peculiar problems to the electron immunocytochemist. In Chapter 19, Marc De Waele has demonstrated the application of several immunostaining procedures to the investigation of haematological cell surface antigens. The problems of 'comparative immunocytochemistry' are raised by Derek Le Roith (Chapter 21) who has discussed in considerable detail the phylogenetic conservation of peptide molecules and the means by which unicellular organisms may be harvested and investigated. Descending the phylogenetic arborisation still further, the unique problems of microbiological identification and investigation are beautifully presented by Julian Beesley (Chapter 20) and Daphne Wright (Chapter 22). The final chapter, ably written by Nancy Hutchison, describes a new technology—the study of genetic elements by the hybridisation to cellular DNA or RNA of labelled nucleotide probes. This technology, young but metamorphosing, is surely set to reach maturity in the next few years and in so doing will direct the emphasis of basic sub-cellular morphological research in the decade to come.

References

1. Coons, A.H., Creech, H.G. and Jones, R.N. (1941) Proc. Soc. Exp. Biol. Med. 47, 200–202.
2. Singer, S.J. (1959) Nature (London), 183, 1523–1524.

Julia M. Polak
Ian M. Varndell
London, February 1984

List of contributors

A. Beaudet
Department of Neurology and Neurosurgery, Montréal Neurological Institute, 3801 University St., Montréal, Québec, Canada.

J. E. Beesley
Wellcome Research Laboratories, Langley Court, Beckenham, Kent BR3 3BS, U.K.

M. Bendayan
Department of Anatomy, Université de Montréal, Montréal, Québec, Canada.

C. Bonnard
Swiss Institute for Experimental Cancer Research and the Institute for Biochemistry, University of Lausanne, CH-1066 Epalinges, Switzerland.

B. E. Causton
Department of Materials Science in Dentistry, The London Hospital Medical College, Turner Street, London E1 2AD, U.K.

M. De Waele
Department of Haematology, University Hospital of the Free University Brussels (V.U.B.), B-1090 Brussels, Belgium.

R. W. Dudek
Department of Anatomy, University of East Carolina, Greenville, NC, U.S.A.

H. J. Geuze
Centre for Electron Microscopy, School of Medicine, University of Utrecht, Nicolaas Beetsstraat 22, 3511 HG Utrecht, The Netherlands.

G. M. Hodges
Tissue Interaction Laboratory, Imperial Cancer Research Fund, Lincoln's Inn Fields, London WC2, U.K.

M. Horisberger
Research Department, Nestlé Products Technical Assistance Co. Ltd., P.O. Box 88, CH-1814 La Tour de Peilz, Switzerland.

N. J. Hutchison
Division of Genetics, Fred Hutchinson Cancer Research Center, 1124 Columbia Street, Seattle, WA 98104, U.S.A.

B. JASANI — Department of Pathology, Welsh National School of Medicine, Heath Park, Cardiff, U.K.

J.-P. KRAEHENBUHL — Swiss Institute for Experimental Cancer Research and the Institute for Biochemistry, University of Lausanne, CH-1066 Epalinges, Switzerland.

L.-I. LARSSON — Unit of Histochemistry, Frederik den V's vej 11, DK-2100 Copenhagen O, Denmark.

D. LE ROITH — University of Cincinnati, College of Medicine, Division of Endocrinology and Metabolism, Mail Location 547, Cincinnati, OH 45267, U.S.A.

D. C. LIVINGSTON† — Chemistry Laboratory, Imperial Cancer Research Fund, Lincoln's Inn Fields, London WC2, U.K.

G. MOREL — Laboratoire d'Histologie-Embryologie, CNRS-ERA981, Faculté de Médecine, Lyon-Sud, Oullins, France.

G. R. NEWMAN — Department of Pathology, University Hospital of Wales, Heath Park, Cardiff, U.K.

D. S. PAPERMASTER — Department of Pathology, Yale Medical School, New Haven, CT 06510, U.S.A.

G. PELLETIER — MRC Group in Molecular Endocrinology, Le Centre Hospitalier de l'Université Laval, Québec G1V 4G2, Canada.

V. M. PICKEL — Laboratory of Neurobiology, Cornell University Medical College, New York, NY 10021, U.S.A.

P. PINTO DA SILVA — Section of Membrane Biology, National Cancer Institute, Frederick Cancer Research Facility, Frederick, MD 21701, U.S.A.

J. M. POLAK — Department of Histochemistry, Royal Postgraduate Medical School, Hammersmith Hospital, Du Cane Road, London W12 0HS, U.K.

J. V. PRIESTLEY — Neuroanatomy-Neuropharmacology Group, University Departments of Pharmacology and Human Anatomy, South Parks Road, Oxford, U.K.

E. L. Romano — Centro de Medicina Experimental, Instituto Venezolano de Investigaciones Cientificas (IVIC), Caracas, Venezuela.

M. Romano — Centro de Microbiologia y Biologia Celular, Instituto Venezolano de Investigaciones Cientificas (IVIC), Caracas, Venezuela.

J. Roth — Biocenter, Department of Electron Microscopy, University of Basel, CH-4056 Basel, Switzerland.

J. W. Slot — Centre for Electron Microscopy, School of Medicine, University of Utrecht, Nicolaas Beetsstraat 22, 3511 HG Utrecht, The Netherlands.

M. A. Smolira — Tissue Interaction Laboratory, Imperial Cancer Research Fund, Lincoln's Inn Fields, London WC2, U.K.

H. Stephens — Department of Anatomy, Université de Montréal, Montréal, Québec, Canada.

K. T. Tokuyasu — Department of Biology, University of California at San Diego, La Jolla, CA 92093, U.S.A.

I. M. Varndell — Department of Histochemistry, Royal Postgraduate Medical School, Hammersmith Hospital, Du Cane Road, London W12 0HS, U.K.

D. M. Wright — Plant Virus Unit, Ministry of Agriculture, Fisheries and Food, Agricultural Development and Advisory Service, Block C, Brooklands Avenue, Cambridge, U.K.

Contents

Introduction

Immunolabelling for Electron Microscopy (Polak/Varndell, eds)

CHAPTER 1

Historical aspects

Edigio L. Romano[1] and Mirtha Romano[2]

[1]*Centro de Medicina Experimental and* [2]*Centro de Microbiologia y Biologia Celular, Instituto Venezolano de Investigaciones Cientificas (IVIC), Caracas, Venezuela*

CONTENTS

1. Introduction

A major development in cytochemical studies of cells and tissues at the light microscope level was the introduction by Coons [1] of the use of fluorescent antibodies to identify sites of antigen-antibody reaction. To take advantage of the much greater resolving power of the electron microscope, it was necessary to develop markers which render antibodies visible at the electron microscope level so as to identify antigens within or on the surface of cells.

The introduction by Singer [2] of the iron-containing protein ferritin as an electron-dense marker which could be conjugated to other proteins such as antibodies opened a new field, immunolabelling for electron microscopy, and thereby made a significant advance in the study of the molecular structure of cells and tissues. Antibodies so labelled can be individually located and identified at or near the site of interaction with the antigen. Singer's ferritin-labelled antibody technique was immediately accepted and quickly became widely used. An excellent review of the use of ferritin-conjugated antibodies in electron microscopy was published by Morgan [3].

In 1966, Nakane and Pierce [4] applied the reaction of horseradish peroxidase (HRP) and diaminobenzidine (DAB) previously introduced into histochemistry by Graham and Karnovsky [5] to localise antigens by conjugating antibody to HRP. A few months earlier Ram and co-workers [6] had demonstrated that the approach of conjugating enzyme to antibody could be useful for ultrastructural studies; however, they used the enzyme acid phosphatase. Antibodies conjugated with this enzyme were found to give variable results and this led Nakane and Pierce to try HRP which proved to be a better enzyme for that purpose. Thus, the introduction of the enzyme-labelled antibody technique became a second major landmark in immunolabelling for electron microscopy.

Feldherr and Marshall [7] introduced colloidal gold particles as a tracer for electron microscopy in 1962 but it was 9 years before colloidal gold was applied as a specific marker for antisera by Faulk and Taylor [8] and for isolated antibodies—3 years later by Romano and colleagues [9]. The immunogold staining technique (IGS) has become increasingly popular during the last 10 years and its use has also been extended to the scanning electron microscope by Horisberger et al. [10]. Reviews on the use of the IGS technique have been published by Horisberger [11], Goodman et al. [12] and Horisberger [13].

2. The immunoferritin technique

Ferritin is a good choice as a marker for electron microscopy because its iron-containing core is very electron dense and large enough, about 70 Å, to permit visualisation at relatively low magnification. The apoprotein coat permits chemical conjugation to other proteins such as antibodies by means of bivalent reagents. Ferritin from horse spleen can be obtained in large quantities and it is relatively easy to prepare in a highly purified form through several steps of crystallisation with cadmium sulphate. Being a particulate marker it also allows for quantitative analysis provided that the ratio of antibody to ferritin is known.

An important drawback of the immunoferritin technique is that it depends on the chemical coupling of ferritin to antibodies. This is very crucial and several problems arise from it:

1) loss of antibody activity;
2) heterogeneity of the products, the active ferritin-antibody conjugates must be isolated;
3) inefficiency of the coupling procedures.

In his original publication, Singer [2] employed metaxylylene diisocyanate and later suggested toluene 2,4-diisocyanate [14], however, other bivalent reagents have been used for the conjugation reaction [14–18]. According to Isliker et al. [16], the yields of active conjugates with metaxylylene diisocyanate are higher and the procedure gives more satisfactory results, although a two step cross-linking procedure with monomeric glutaraldehyde seems to be simple and adequate [18].

In order to obtain an effective specific labelling of antigen sites in cells and tissues, high titre antisera, or preferably purified antibodies are required for conjugation to ferritin. This is because the cross-linking reaction does not distinguish between specific antibody, other proteins and ferritin, and thus a mixture of conjugates is obtained. For this reason highly purified antibody preparations, such as those obtained using techniques of affinity chromatography, or immunoadsorbance, are highly desirable. When highly purified antibody preparations are available, then the ferritin-antibody conjugate may be applied directly to the sample or specimen, thus the labelling is obtained by a direct one layer method. Otherwise, it is preferable to use an indirect technique in which the specimen is first incubated with unlabelled specific antibody or unfractionated antiserum, the excess washed off and then the sections are treated with ferritin-conjugated antibody to the γ-globulin of the animal species from which the unlabelled antibody has been prepared. Ferritin-conjugated antiglobulin to γ-globulin of several species of animal are commercially available.

An elegant technique devised to circumvent the undesirable aspects of the chemical coupling of ferritin to antibody was published in 1968 by Hämmerling et al. [19]. They employed hybrid antibodies prepared as described by Nisonoff [20]. The hybrid antibodies contain two different specificities united into one molecule. In theory a hybrid antibody, in which one specificity is directed to ferritin and the other to the antigen to be marked, is an excellent reagent for the purpose. In practise hybrid antibodies with the dual specificity antiferritin/anti-immunoglobulin of a given animal species have been used in an indirect, sandwich technique to locate cell surface antigens [19, 21]. Hybrid antibodies can also be used in conjunction with any other marker which is antigenic; for example, with enzymes such as horseradish peroxidase and viruses such as the southern bean mosaic virus [22].

The immunoferritin technique has been applied to a wide variety of biological problems. Some examples of these studies are: the arrangement of fibrin in clots; the composition of amyloid; identification of viruses and of viral antigens in cells and tissues; identification of components of bacterial, fungal and rickettsial origin; studies of the surface of many types of cells: red blood cell antigens, HLA antigens, tumour cell associated antigens, components of cell nuclei; studies regarding the production of proteins by plasma cell myelomas, pancreatic and carcinomatous cells, and glomerular basement membranes; studies regarding the mechanisms of experimental nephritis, the pathology of human acute glomerulonephritis, the deposition of antigen-antibody complexes and the pathology associated with human renal transplants.

The treatment of specimens with ferritin-conjugated antibody merits some consideration. If it is assumed that the antibody has a suitable specificity, that binding potency has been preserved and there has been adequate conjugation, then the basic problems that remain to be solved are:

1) the specimen should be treated in such a way as to allow the penetration of the

large ferritin-antibody conjugate to bring it into intimate contact with the antigenic grouping of cells or tissues;
2) the antigenic specificity should be preserved by such treatment;
3) the integrity of the fine structure of cells and tissue should be preserved as much as possible.

Several situations may be considered. One is the labelling of surface antigens of cells or of microorganisms. This is normally achieved by direct or indirect methods without or with only slight initial fixation followed by conventional processing. Another situation is the labelling of inter- or intracellular antigens within tissues. In this regard many efforts have been made to design a treatment that fulfills the conditions of preserving the fine structure whilst also maintaining intact the antigenic specificity. The fixatives used and the embedding and processing methods are central to this problem and will be discussed elsewhere. Various methods of immunolabelling thin sections of specimens with good preservation of the tissue, cell structure and antigenic reactivity are in use and are the subject of chapters later in this book.

3. Immunoperoxidase labelling for electron microscopy

One of the advantages of horseradish peroxidase over ferritin is its small molecular weight of 40,000 which is about one-tenth that of ferritin, thus there is less of a problem of penetration of the conjugate. An important difference between ferritin and colloidal gold on one hand and enzymes on the other, is that the former are particulate markers while the catalytic activity of enzymes results in the accumulation of reaction products which diffuse away from the labelled site. This diffusion may be a positive feature because it aids detection of the antigenic site [23] but may also be a drawback when fine localisation of antigenic sites is required. Another important feature of horseradish peroxidase-antibody conjugates is that they may be used in light microscopy in much the same way as fluorescent antibodies with some advantages: no special microscopy is needed, the reaction products are permanent and retrospective studies are possible.

Peroxidase may be conjugated to antibodies by methods similar to those used for the coupling of ferritin. Nakane and Pierce [4] in their original publication concluded that *p*, *p'*-difluoro-*m*-*m'*-dinitrophenyl sulphone (FNPS) was an adequate bifunctional reagent, an alternative being the water soluble carbodiimides. Avrameas [24] in 1969 introduced a procedure in which enzymes were coupled to γ-globulins with glutaraldehyde, claiming that the conjugates retained a substantial part of their immunological and enzymatic activity and could be employed for the detection of intracellular antigens. Both FNPS and glutaraldehyde are currently widely used.

A number of immunoperoxidase labelling procedures has been described after the original work of Nakane and Pierce [4]. The technique was soon employed for the detection of cell surface and intracellular antigens. The pioneering work by

Nakane and Pierce [25] and Bouteille and Avrameas [26] deserves special mention. Direct and indirect or sandwich techniques were used and enzymes were also used as antigens. The indirect methods have the advantage of being more sensitive and more versatile thanks to the extensive use of peroxidase conjugated with antiglobulin antibodies. There is no need to purify the first antibody, whole antisera can be used, with the condition that additional specificity controls should be carried out particularly by substitution of the primary antisera with non-specific sera from the same species.

A major contribution to immunolabelling was the introduction of immunoperoxidase non-conjugate procedures. The enzyme bridge method and the peroxidase-antiperoxidase method (PAP technique), as well as the hybrid antibody method [19], were designed to circumvent some of the disadvantages inherent in the conjugation procedures in which there may be antibody denaturation, inactivation of enzymes and residual uncoupled antibody and enzymes. Mason et al. [27] described in 1967 the enzyme bridge method in which after the binding of primary antibody an excess of antiglobulin to the γ-globulin of the first species is added to ensure binding through only one site. A third layer is then formed by the addition of antibody to peroxidase of the same origin as the primary antibody, which will bind to the free arm of the antiglobulin. Finally, a layer of free peroxidase is added which is bound by the antiperoxidase antibody. The PAP method, which was initially introduced by Sternberger et al. [28], is based on the same principle as the enzyme bridge method, however, in the last step peroxidase-antiperoxidase soluble complex is used in the place of free peroxidase. The unlabelled antibody peroxidase method provided a high sensitivity for the histochemical localisation of a variety of antigens and quickly became very popular both in light and electron microscopy.

Horseradish peroxidase was introduced as a tracer for light microscopy studies by Straus [29] in 1957, and by Graham and Karnovsky [5] for electron microscopy in 1966. Since the introduction of peroxidase-antibody conjugates for ultrastructural studies [4], modifications of this immunolabelling technique have been extensively applied both in biology and medicine. Many viruses, enzymes, hormones and immunoglobulins have been localised in tissue sections by these methods and it is beyond the scope of this chapter to review such studies. It must suffice to mention only a few examples of recent studies such as the localisation of human platelet factor 4 in tissue mast cells [30]; the localisation of hormones [31–33], the demonstration of prolactin binding to human breast cancer cells in tissue culture [34] and other studies on immunocytochemical methodology [35, 36].

As with other labelling methods, the sensitivity of the immunoperoxidase procedures depends on the conditions of fixation and tissue processing. A compromise has to be made between the requirements for the preservation of the tissue and cell fine structure, good fixation, and the retention of antigenic activity—a theme that will be mentioned in later chapters.

Some particular problems inherent in the immunoperoxidase procedures are related to the need to inhibit endogenous peroxidase activity which may be achieved by the use of phenylhydrazine [37] and to the purity and toxicity of diaminobenzidine [38], the chromogen commonly used.

4. Immunogold labelling for electron microscopy

4.1. General aspects

Colloidal gold offers many advantages as a tracer for electron microscopy. Gold granules are particulate and very distinct, they allow a fine localisation of the marked sites. A great advantage is that due to their high electron density, gold particles are easily detected in the electron microscope, in contrast to the lower opacity of the iron core of ferritin. Because of the granular nature of gold particles, quantification of the degree of labelling can be made by direct counting of the number of particles in a given area, which cannot be done with the amorphous reaction product floccules of the immunoperoxidase techniques. Colloidal gold can be easily prepared by several methods. Its binding to macromolecules like immunoglobulins, protein A, lectins, although poorly understood, is a simple matter of adsorption at the correct conditions of pH, reagent concentration, and ionic strength; no chemical conjugation procedure is involved. Thus the method is easy, quick, and cheap because it requires only small amounts of specific macromolecules. Another advantage is the low non-specific adsorption of the coated gold particles. Gold labelling can also be used in scanning electron microscopy as originally introduced by Horisberger et al. [10] (see Chapter 15, pp. 189–233) and in light microscopy as introduced by Geoghegan et al. [39].

As colloidal gold can be produced in a variety of sizes ranging from 5 nm to 150 nm, multiple labelling can be achieved by adsorption to different antibodies or macromolecules. Gold labelling with small particles, 5–20 nm, has the further advantage of being less susceptible to steric hindrance due to the small size of the label. A disadvantage of immunogold reagents is the relatively low stability of the label when the colloid is not properly coated with macromolecules and is thus unprotected from aggregation by electrolytes (see Chapter 15, pp. 189–233).

4.2. Immunogold procedures

The preparation of a gold marker is based on the adsorption of macromolecules onto the surface of the gold particles. Fortunately, upon adsorption, full biological activity of the macromolecules is preserved. They remain globular and little change in their native state occurs [40]. As pointed out by Horisberger and Rosset [41], catalase is the only known exception to this rule, losing its activity when bound to gold particles. Gold particles have been labelled with a great variety of macromolecules, for a review see Horisberger [13]. As a result of the adsorption of macromolecules onto gold particles, the hydrophobic negatively charged colloidal gold is stabilised, 'protected', against subsequent aggregation by electrolytes, becoming a lyophilic colloid.

Colloidal gold can be prepared by several methods. In our hands, a convenient and reproducible method is the one described by Frens [42] based on the reduction of a dilute solution of tetrachloroauric acid by sodium citrate although in the initial works of Feldherr and Marshall [7], Faulk and Taylor [8] and Romano et al. [9], the procedure used was that outlined by Weiser and described by Feldherr and Marshall [7], namely reduction of tetrachloroauric acid by a solution of phosphorus in ether.

In the method described by Faulk and Taylor [8] in 1971, whole serum which contained anti-*Salmonella* antibody was labelled with colloidal gold. This reagent was demonstrated to be useful as a means of investigating the distribution of antigens on cell surfaces by electron microscopy. Romano et al. [9] in 1974, described an indirect method in which affinity-purified horse antibodies to human IgG were labelled with colloidal gold and used to map the distribution of Rh antigen sites on human erythrocyte ghosts. A similar reagent was utilised by Romano et al. [43] to map the distribution of the A, D and C antigens of human red cell membranes under different conditions. The gold-labelled reagent allowed the study of the distribution of the sites and also permitted a quantitative analysis. These studies confirmed and expanded the results obtained by Nicolson et al. [44] using a ferritin-antibody conjugate on the distribution of the D antigens as single entities dispersed in the erythrocyte membrane. The clustering of A sites was found to be proportional to the amount of antibody bound to the cells. Increased mobility of the antigen sites after enzymic treatment of the red cells was also observed in these studies. Since 1975, many studies have been reported which used colloidal gold markers for labelling surface antigens of cells, viruses, bacteria, protoplasts, yeasts and also for marking intracellular antigens either with pre-embedding or with post-embedding techniques on thin sections. For these studies gold particles have been coated with immunoglobulins [8, 9, 39, 45–49], antigens [50], lectins [46, 51–57] and with staphylococcal protein A, as described originally by Romano and Romano [58] in 1977 and by Roth et al. [59] in 1978. During the last 5 years the protein A-gold technique has become increasingly popular and it is now widely used for the localisation of both surface and intracellular antigens. For this reason special consideration will be given to this technique.

4.3. The staphylococcal protein A-gold technique: basic aspects of protein A immunochemistry

This method is based on the ability of protein A to bind to the Fc portion of IgG molecules of many animal species [60–62]. It is fortunate that the binding is very strong to most IgG subclasses from human, rabbit, guinea pig, swine, and dog [63], because the primary antibodies commonly used originate in these species.

It must be mentioned that for IgG from cow, mouse and horse the binding of protein A is less strong and is considerably weaker with IgG from sheep, goat, rat and chicken [64]. This is an important consideration because it is not uncommon to use antibodies produced in sheep or goat; protein A-based immunoassay methods should not depend on these antibodies. It is also of interest that human IgG subclasses 1, 2 and 4, which comprise about 95% of the normal human IgG bind protein A [65]. Binding to IgG3 cannot be shown using myeloma proteins and is very weak with polyclonal IgG. It has been suggested that a proportion of human IgG, IgA and IgE but not IgD can also bind to protein A (see review of Langone

[64]); thus it is not advisable to add protein A-gold to whole antisera because complexes with various immunoglobulin classes may be formed which in some circumstances may not be desirable. As shown by Endresen and Grov [66] and others, binding of human IgG to protein A requires an intact Fc region. Isolated CH_2 and CH_3 fragments are not reactive. Thus, immunolabelling with protein A-gold using F(ab) or $F(ab)_2$ fragments is not possible, but may serve as a basis for specificity controls.

The strength of binding of protein A to human and rabbit IgG is very high, of the order of 2 to 5×10^{-8} M, which is similar to the binding affinity observed for the antigen-antibody reaction. The kinetics of binding are very fast, equilibrium being reached in less than 30 min at temperatures ranging from 4 to 37°C, most of the protein A-IgG complexes being formed within the first few minutes (see Langone [64] for a review). The antigen-binding activity of the antibody complexed to protein A is fully preserved and in some cases an increase in affinity for antigen has been observed. On the other hand, antigen-antibody complexes of IgG show an increased affinity for binding with protein A. For all the above reasons it seems appropriate to allow an incubation time of 30–60 min with protein A-gold or with antibody-protein A-gold labels at temperatures in the range of 4° to 37°C.

Some important physico-chemical properties of protein A-gold immunolabelling (see Langone [64] for a review) follow:

a) Molecular weight of 42,000 is generally accepted as the average from several determinations.
b) Protein A probably has an elongated shape as suggested by a high frictional ratio, and intrinsic viscosity in comparison with globular proteins.
c) It has an extinction coefficient of 1.65 at 275 nm for a 1% solution.
d) The isoelectric point of staphylococcal protein A is 5.1, therefore to produce stable complexes the pH of the protein A solution to be mixed with colloidal gold should be in the range of 4.6 to 5.6 and not 6.9 as suggested by Romano and Romano [58], and Roth et al. [59], since according to the work of Geoghegan and Ackerman [46] the adsorption of proteins to colloidal gold is maximal at a pH close to the isoelectric point.
e) According to Sjödahl [67], protein A from intact *Staphylococcus aureus* possesses 4 Fc binding regions with homologous amino acid sequence; however in protein A released from the bacterial wall, only two of them are expressed, therefore protein A is functionally bivalent [64].
f) The reason why protein A forms precipitating complexes with human IgG, but not with rabbit IgG, is not fully understood. Binding of protein A to Fab regions outside the antigen binding site may be involved [64]. Tyrosine and histidine groups of protein A appear to be important for binding to Fc [68], thus procedures which modify tyrosine or histidine residues, for example, iodination with chloramine T, iodine monochloride or with enzymatic methods, may decrease the reactivity of protein A. However, the structural basis for IgG binding to protein A is not known yet.

After the protein A-gold reagent is prepared, it is desirable to assess its biological activity prior to application in labelling studies. There are several methods; one is based on the use of normal human IgG to absorb the reagent, another method that we follow and find very useful, is to test the ability of the reagent to agglutinate human Rh positive red cells coated with IgG anti-Rh as in the Coombs reaction.

4.4. Use of Immunogold labelling

Some recent examples of the use of the protein A-gold technique in immunocytochemical studies are the localisation of regulatory peptides and neurotransmitters [69], of kallikrein in the rat pancreas [70], of mitochondrial proteins in the rat hepatocyte [71], of actin in muscle, epithelial and secretory cells [72], of viral proteins [73, 74] and of pea seed storage proteins [75]. For descriptions of methodology see Batten and Hopkins [76], Roth et al. [77] and Bendayan [78].

4.5. Double and multiple labelling

Colloidal gold of different sizes may be easily distinguished in the electron microscope and therefore can be used for double [78–80] or multiple labelling (see Chapters 12 and 13). Gold markers can also be used in combination with ferritin or with peroxidase [56] since gold particles may be distinguished from ferritin or from peroxidase.

5. References

1. Coons, A.H., Creech, H.J. and Jones, R.N. (1941) Immunological properties of an antibody containing a fluorescent group. Proc. Soc. Exp. Biol. 47, 200–202.
2. Singer, S.J. (1959) Preparation of an electron dense antibody conjugate. Nature (London) 183, 1523–1524.
3. Morgan, C. (1972) The use of ferritin-conjugated antibodies in electron microscopy. Int. Rev. Cytol. 32, 291–326.
4. Nakane, P.K. and Pierce, G.B. (1966) Enzymes-labeled antibodies: preparation and application for the localization of antigens. J. Histochem. Cytochem. 14, 929–931.
5. Graham, R.C. and Karnovsky, M.J. (1966) The early stages of absorption of injected horse radish peroxidase in the proximal tubules of mouse kidney: ultrastructural cytochemistry by a new technique. J. Histochem. Cytochem. 14, 291–302.
6. Ram, J.S., Nakane, P.K., Rawlins, D.G. and Pierce, G.B. (1966) Enzyme labeled antibody for ultrastructural studies. Fed. Proc. 25, 732–741.
7. Feldherr, C.M. and Marshall, J.M. (1962) The use of colloidal gold for studies of intracellular exchange in amoeba *Chaos chaos*. J. Cell. Biol. 12, 640–645.
8. Faulk, W.P. and Taylor, G.M. (1971) An immunocolloid method for the electron microscope. Immunochemistry 8, 1081–1083.
9. Romano, E.L., Stolinski, C. and Hughes-Jones, N.C. (1974) An antiglobulin reagent labelled with colloidal gold for use in electron microscopy. Immunochemistry 11, 521–522.
10. Horisberger, M., Rosset, J. and Bauer, H. (1975) Colloidal gold granules as markers for cell surface receptors in the scanning electron microscope. Experientia 31, 1147–1149.
11. Horisberger, M. (1979) Evaluation of colloidal gold as a cytochemical marker for transmission and scanning electron microscope. Biol. Cell. 36, 253–258.
12. Goodman, S.L., Hodges, G.M. and Livingston, D.C. (1980) A review of the colloidal gold marker

system. In: O. Johari (Ed.) Scanning Electron Microscopy, Vol. 2. SEM Inc., AMF O'Hare (Chicago) IL., pp. 133–145.

13. Horisberger, M. (1981) Colloidal gold a cytochemical marker for light and fluorescent microscopy and for transmission and scanning electron microscopy. In: O. Johari (Ed.) Scanning Electron Microscopy, Vol. 2. SEM Inc., AMF O'Hare (Chicago) IL., pp. 9–31.
14. Singer, S.J. and Schick, A.F. (1961) The properties of specific stains for electron microscopy prepared by the conjugation of antibody molecules with ferritin. J. Biophys. Biochem. Cytol. 9, 519–537.
15. Sri Ram, J., Tawde, S.S., Pierce, G.B. and Midgley, A.R. (1963) Preparation of antibody-ferritin conjugation for immuno-electron microscopy. J. Cell. Biol. 17, 673–675.
16. Isliker, H., Le Maire, B. and Morgan, C. (1964) The use of ferritin-conjugated antibody fragments in electron microscopic studies of viruses. Pathol. Microbiol. 27, 521–531.
17. Rifkind, R.H., Hsu, K.C. and Morgan, C. (1964) Immunochemical staining for electron microscopy. J. Histochem. Cytochem. 12, 131–136.
18. Otto, H., Takamiya, H. and Vogt, H. (1973) A two stage method for cross linking antibody globulin to ferritin by glutaraldehyde. Comparison between the one-stage and the two-stage method. J. Immunol. Methods 3, 137–146.
19. Hämmerling, U., Aoki, T., De Harven, E., Boyse, E.A. and Old, L.J. (1968) Use of hybrid antibody with anti-G and anti-ferritin specificities in locating cell surface antigens by electron microscopy. J. Exp. Med. 128, 1461–1469.
20. Nisonoff, A. and Rivers, M.M. (1961) Recombination of a mixture of univalent antibody fragments of different specificity. Arch. Biochem. Biophys. 93, 460–462.
21. Aoki, T., Hämmerling, U., De Harven, E., Boyse, E.A. and Old, L.J. (1969) Antigenic structure of cell surfaces. An immunoferritin study of the occurrence and topography of H-2′, θ, and TL alloantigens of mouse cells. J. Exp. Med. 130, 979–1002.
22. Hämmerling, U., Polliack, A. and De Harven, E. (1974) Hybrid antibodies in immuno-electron microscopy, Vol 2 Eighth International Congress on Electromicroscopy. Canberra, Australia, pp. 110–111.
23. Bretton, R. (1970) Comparison of peroxidase and ferritin labeling for localization of specific cell surface antigens. Seventh International Congress of Electron microscopy. Grenoble, pp. 527–528.
24. Avrameas, S. (1969) Coupling of enzymes to proteins with glutaraldehyde. Use of the conjugates for the detection of antigens and antibodies. Immunochemistry 6, 43–52.
25. Nakane, P.K. and Pierce, G.B. (1967) Enzyme labelled antibodies for the light and electron microscopic localization of tissue antigens. J. Cell. Biol. 33, 307–318.
26. Bouteille, M. and Avrameas, S. (1967) Study with the electronmicroscope of antibody formation with the aid of antigens labeled with peroxidase. C.R. Acad. Sci. (D) (Paris) 265, 2097–2103.
27. Mason, T.E., Phifer, R.F., Spicer, S.S., Swallow, R.A. and Dreskin, R.B. (1969) An immunoglobulin-enzyme bridge method for localizing tissue antigens. J. Histochem. Cytochem. 17, 563–569.
28. Sternberger, L.A., Hardy, Jr., P.H., Cuculis, J. and Meyer, H.G. (1970) The unlabeled antibody enzyme method of immunochemistry. Preparation and properties of soluble antigen-antibody complex (Horseradish peroxidase antihorseradish peroxidase) and its use in the identification of spirochetes. J. Histochem. Cytochem. 18, 315–333.
29. Straus, W. (1957) Segregation of an intravenously injected protein by droplet cells of rat kidney. J. Biophys. Biochem. Cytol. 3, 1037–1041.
30. McLaren, K.M. and Pepper, D.S. (1983) The immunoelectron-microscopic localization of human platelet factor 4. In: Tissue Mast Cells. Histochem. J. 15, 795–800.
31. Tougard, C., Picart, R. and Tixier-Vidal, A. (1980) Immunocytochemical localization of glycoprotein hormones in the rat anterior pituitary. A light and electron microscope study using antisera against β-subunits: a comparison between the preembedding and the postembedding methods. J. Histochem. Cytochem. 28, 101–114.
32. Larsson, L. (1980) Immunocytochemical characterization of ACTH-like immunoreactivity in cerebral nerves and in endocrine cells of the pituitary and gastrointestinal tract by using region-specific antisera. J. Histochem. Cytochem. 28, 133–141.

33. PIEKUT, D.T. and KNIGGE, K.M. (1982) Immunocytochemical analysis of the rat pineal gland using antisera generated against analogs of luteinizing hormone-releasing hormone (LHRH). J. Histochem. Cytochem. 30, 106–110.
34. HSU, S. and RAINE, L. (1982) Immunocytochemical and auto-radiographic demonstration of prolactin binding to human breast cancer cells in tissue culture. J. Histochem. Cytochem. 30, 157–161.
35. BUSSOLATI, G., ALFANI, V., WEBER, K. and OSBORN, M. (1980) Immunocytochemical detection of actin on fixed and embedded tissues: Its potential use in routine pathology. J. Histochem. Cytochem. 28, 169–173.
36. DUDEK, R.W., CHILDS, G.V. and BOYNE, A.F. (1982) Quick freezing and freeze-drying preparation for high quality morphology and immunocytochemistry at the ultrastructural level: application to pancreatic beta cells. J. Histochem. Cytochem. 30, 129–138.
37. STRAUS, W. (1972) Phenylhydrazine as inhibitor of horse-radish peroxidase for use in immunoperoxidase procedures. J. Histochem. Cytochem. 20, 949–951.
38. PELLINIEMI, L.J., DYM, M. and KARNOVSKY, M.J. (1980) Peroxidase histochemistry using diaminobenzidine tetrahydrochloride stored as a frozen solution. J. Histochem. Cytochem. 28, 191–192.
39. GEOGHEGAN, W.D., SCILLIAN, J.J. and ACKERMAN, G.A. (1978) The detection of human B lymphocytes by both light and electron microscopy utilizing colloidal gold labelled-antiimmunoglobulin. Immunol. Commun. 7, 1–12.
40. BAUER, H., GERBER, H. and HORISBERGER, M. (1975) Morphology of colloidal gold, ferritin and anti-ferritin antibody complexes. Experientia 31, 1149–1151.
41. HORISBERGER, M. and ROSSET, J. (1977) Colloidal gold a useful marker for transmission and scanning electron microscopy. J. Histochem. Cytochem. 25, 295–305.
42. FRENS, G. (1973) Controlled nucleation for the regulation of particle size in monodisperse gold suspensions. Nature Phys. Sci. 241, 20–22.
43. ROMANO, E.L., STOLINSKI, C. and HUGHES-JONES, N.C. (1975) Distribution and mobility of the A, D and c antigens on human red cell membranes: studies with a gold-labelled antiglobulin reagent. Br. J. Haematol. 30, 507–516.
44. NICOLSON, G.L., MASOUREDIS, S.P. and SINGER, S.J. (1971) Quantitative two-dimensional ultrastructural distribution of Rho(D) antigenic sites on human erythrocyte membranes. Proc. Natl. Acad. Sci. 68, 1416–1420.
45. HORISBERGER, M. and VONLANTHEN, M. (1977) Localization of mannan and chitin on thin sections of budding yeasts with gold markers. Arch. Microbiol. 115, 1–7.
46. GEOGHEGAN, W.D. and ACKERMAN, G.A. (1977) Adsorption of horseradish peroxidase, ovomucoid and antiimmunoglobulin to colloidal gold for the indirect detection of Concanavalin A, wheat germ agglutinin and goat anti-human immunoglobulin-G on cell surfaces at the electron microscopic level: a new method, theory and application. J. Histochem. Cytochem. 25, 1187–1200.
47. ROMANO, E.L., LAYRISSE, M., ROMANO, M., SOYANO, A. and LAYRISSE, Z. (1980) Electron microscopic demonstration of IgG antibodies directed to erythroblast in primary acquired pure red cell aplasia. Clin. Immunol. Immunopathol. 17, 330–334.
48. GOODMAN, S.L., HODGES, G.M., TREJDO, J., SIEWICZ, L. and LIVINGSTON, D.C. (1981) Colloidal gold markers and probes for routine application in microscopy. J. Microsc. 123, 201–203.
49. HOLGATE, C.S., JACKSON, P., COWEN, P.N. and BIRD, C.C. (1983) Immunogold-silver staining: new method of immunostaining with enhanced sensitivity. J. Histochem. Cytochem. 31, 938–944.
50. LARSSON, L.-I. (1979) Simultaneous ultrastructural demonstration of multiple peptides in endocrine cells by a novel immunocytochemical method. Nature (London) 282, 743–746.
51. GARLAND, J.M. (1971) Preparation and performance of gold-labelled Concanavalin A for the location of specifically reactive sites in walls of *S. faecalis* 8191. In: E. Wisse, W. Th. Daems, I. Molenaar and P. van Duijn (Eds.) Electron Microscopy and Cytochemistry. Elsevier North-Holland, Amsterdam, pp. 303–307.
52. BAUER, H., HORISBERGER, M., BUSH, D.A. and SIGARKAKIE (1972) Mannan as a major component of the bud scars of *Saccharomyces cerevisae*. Arch. Mikrobiol. 85, 202–208.
53. HORISBERGER, M. and ROSSET, J. (1976) Localization of wheat germ agglutinin receptor sites on yeast cells by scanning electron microscopy. Experientia 32, 998–1000.

54. Wagner, M., Roth, J. and Wagner, B. (1976) Gold-labeled protectin from *Helix pomatia* for the localization of blood group A antigen of human erythrocytes by immuno freeze-etching. Exp. Pathol. Bd. 12, 277–281.
55. Roth, J. and Wagner, M. (1977) Peroxidase and gold complexes of lectins for double labeling of cell surface-binding sites by electron microscopy. J. Histochem. Cytochem. 25, 1181–1186.
56. Roth, J. and Binder, M. (1978) Colloidal gold, ferritin and peroxidase as markers for electron microscopic double labeling lectin. J. Histochem. Cytochem. 26, 163–169.
57. Roth, J. (1983) Application of lectin-gold complexes for electron microscopic localization of glyco-conjugates on thin sections. J. Histochem. Cytochem. 31, 987–999.
58. Romano, E.L. and Romano, M. (1977) Staphylococcal protein A bound to colloidal gold: a useful reagent to label antigen-antibody sites in electron microscopy. Immunochemistry 14, 711–715.
59. Roth, J., Bendayan, M. and Orci, L. (1978) Ultrastructural localization of intracellular antigens by the use of protein-A gold complex. J. Histochem. Cytochem. 26, 1074–1081.
60. Forsgren, A. and Sjöquist, J. (1966) 'Protein A' from *S. aureus*. I. Pseudoimmune reaction with human γ globulin. J. Immunol. 97, 822–827.
61. Kronvall, G., Seal, V.S., Finstad, J. and Williams, Jr., R.C. (1970) Phylogenetic insight into evolution of mammalian Fc Fragment of γG globulin using staphylococcal protein A. J. Immunol. 104, 140–147.
62. Kronvall, G., Seal, V.S., Svensson, S. and Williams, Jr., R.C. (1974) Phylogenetic aspects of protein A-reactive serum globulins in birds and mammals. Acta Pathol. Microbiol. Scand. Sect. B 82B, 12–18.
63. Langone, J.J. (1978) Protein-A: A tracer for general use in immunoscience. J. Immunol. Methods 24, 269–285.
64. Langone, J.J. (1982) Protein A of *Staphylococcus aureus* and related immunoglobulin receptors produced by streptococci and pneumococci. Adv. Immunol. 32, 157–252.
65. Ankerst, J., Christensen, P., Kjellen, L. and Kronvall, G. (1974) A routine diagnostic test for IgA and IgM antibodies to rubella virus: absorption of IgG with *Staphylococcus aureus*. J. Infect. Dis. 130, 268–273.
66. Endresen, C. and Grov, A. (1976) Further characterization of protein A reactive and non reactive subfragments of Fc from human IgG. Acta Pathol. Microbiol. Scand. Sect. C 84C, 397–402.
67. Sjödahl, J. (1977) Structural studies in 4 repetitive Fc-binding regions in protein A from *Staphylococcus aureus*. Eur. J. Biochem. 78, 471–490.
68. Sjöholm, J., Bjerkén, A. and Sjöquist, J. (1973) Protein A from *Staphylococcus aureus*. XIV. The effect of nitration of protein A with tetranitromethane and subsequent reduction. J. Immunol. 110, 1562–1569.
69. Varndell, I.M., Tapia, F.J., Probert, L., Buchan, A.M.J., Gu, J., De Mey, J., Bloom, S.R. and Polak, J.M. (1982) Immunogold staining procedure for the localization of regulatory peptides. Peptides 3, 259–272.
70. Bendayan, M. and Ørstavik, T.B. (1982) Immunocytochemical localization of Kallikrein in the rat exocrine pancreas. J. Histochem. Cytochem. 30, 58–66.
71. Bendayan, M. and Shore, G.C. (1982) Immunocytochemical localization of mitochondrial proteins in the rat hepatocyte. J. Histochem. Cytochem. 30, 139–147.
72. Bendayan, M. (1983) Ultrastructural localization of actin in muscle, epithelial and secretory cells by applying the protein A-gold immunocytochemical technique. Histochem. J. 15, 39–58.
73. Garzon, S., Bendayan, M. and Kurstak, E. (1982) Ultrastructural localization of viral antigens using the protein A-gold technique. J. Virolog. Methods 5, 67–73.
74. Pares, R.D. and Whitecross, M.J. (1982) Gold-labelled antibody decoration (GLAD) in the diagnosis of plant viruses by immunoelectron microscopy. J. Immunol. Methods 51, 23–28.
75. Craig, S. and Millerd, A. (1981) Pea seed storage proteins-immunocytochemical localization with protein-A-gold by electron microscopy. Protoplasma 105, 333–339.
76. Batten, T.F.C. and Hopkins, C.R. (1979) Use of protein A-coated colloidal gold particles for immunoelectromicroscopic localization of ACTH on ultrathin sections. Histochemistry 60, 317–320.
77. Roth, J., Bendayan, M., Carlemalm, E., Villiger, W. and Garavito, J. (1981) Enhancement of

structural preservation and immunochemical staining in low temperature embedded pancreatic tissue. J. Histochem. Cytochem. 29, 663–671.

78. Bendayan, M. (1982) Double immunocytochemical labeling applying the protein A-gold technique. J. Histochem. Cytochem. 30, 81–85.
79. Tapia, F.J., Varndell, I.M., Probert, L., De Mey, J. and Polak, J.M. (1983) Double immunogold staining method for the simultaneous ultrastructural localisation of regulatory peptides. J. Histochem. Cytochem. 31, 977–981.
80. Varndell, I.M., Harris, A., Tapia, F.J., Yanaihara, N., De Mey, J., Bloom, S.R. and Polak, J.M. (1983) Intracellular topography of immunoreactive gastrin demonstrated using electron immunocytochemistry. Experientia 39, 713–717.

Immunolabelling for Electron Microscopy (Polak/Varndell, eds)

CHAPTER 2

Electron-opaque markers: A review

Marc Horisberger

Research Department, Nestlé Products Technical Assistance Co. Ltd., P.O. Box 88, CH-1814 La Tour de Peilz, Switzerland

CONTENTS

1. Introduction

Marking of cell surface and intracellular components for microscopical observation is an extremely important technique for studying molecular organisation and cell function. Indeed biochemical analysis alone would be incomplete since the data obtained represent average values and give little information regarding the distribution of cellular components. A large number of cytochemical techniques has therefore been developed in order to identify, localise, quantify and understand the dynamics of cell components at the ultrastructural level.

Besides freeze-etching, X-ray microanalysis and the occasional use of markers such as haemocyanin, all other cytochemical methods for transmission electron microscopy including autoradiography depend upon reactive products opaque to the electrons or on the use of electron-opaque particulate markers.

These methods can be classified into different categories:

a) They have a large specificity (e.g. cationic ferritin) or a narrow specificity (e.g. colloidal gold, ferritin, peroxidase conjugates).
b) The methods are general (e.g. colloidal gold, ferritin, labelled and unlabelled peroxidase conjugates, avidin-biotin complexes) or restricted to particular cases (e.g. glycosylated ferritin and peroxidase).
c) Markers are diffuse (e.g. peroxidase conjugates) or particulate (e.g. ferritin, colloidal gold).
d) Conjugates are obtained via non-covalent bonds (e.g. colloidal gold), covalent bonds (e.g. ferritin and peroxidase conjugates) or specific interactions (e.g. peroxidase-antiperoxidase [PAP], avidin-biotin complexes).
e) Marking is achieved in one step or in multiple steps.
f) Some methods can be quantified (e.g. colloidal gold, ferritin). Although most of these markers have found application in transmission electron microscopy (pre-embedding and post-embedding techniques), only a few have been developed for scanning electron microscopy.

In transmission electron microscopy, the localisation of cellular components requires a balanced fixation of the cell under study. Whilst antigenic determinants must not be denatured during fixation, the ultrastructural morphology of the cell should be well preserved. These two requirements tend to be mutually exclusive. However, the recently introduced low temperature embedding procedure has resulted in enhancement of structural preservation and immunocytochemical labelling, especially with the protein A-gold method [1, 2] (see Chapter 9, pp. 113–121). Alternatively, cryoultramicrotomy allows optimal preservation of ultrastructure and immunoreactivity due to mild fixation and the possibility of avoiding dehydration steps [3] (see also Chapter 6, pp. 71–82).

In view of the number of methods available and the variety of possible combinations, immunocytochemists experience difficulties in making the proper choice to solve a specific problem. Useful criteria of selection are: specificity, sensitivity, easiness and cost of techniques (time and money-wise). Ideally, markers should be easily recognised and quantified. They should also be readily prepared from a variety of molecules recognising ligands with a narrow specificity. Although diffuse markers are generally sensitive since they are based on amplification effects, particulate markers are especially convenient for precise localisation and quantification.

2. Particulate markers

2.1. Colloidal gold

A considerable advance in the understanding of the behaviour of colloidal systems was made by the DLVO theory. This acronym derives from the initial letters of the workers who developed it (Deryagin and Landau [4] and Verwey and Overbeek [5]). In this theory, the particle is considered to consist of two components: one arises from the overlap of the electrical double layer and leads to repulsion, the other from electromagnetic effects which leads to Van der Waal attraction. When a colloidal particle approaches a surface, the potential energy of interaction is as follows: At some finite distance, when the surface does not come into molecular contact, an equilibrium is reached between attractive and repulsive forces (secondary minimum, reversible adhesion). At a smaller distance, a net energy barrier occurs. Once this is overcome, the theory predicts, another minimum occurs (primary minimum, irreversible adhesion). Both the height of the energy barrier and secondary minimum depend on ionic strength and electrostatic charge. Although it is still generally believed that the coagulation concentration of metal colloids in the presence of electrolytes is independent of particle size, Frens [6] has shown that the stability of metal colloids depends largely on size. For instance, small gold particles are more stable against electrolyte coagulation than coarser suspensions. These results are explained by the diminished Van der Waal attraction between smaller particles.

Gold particles carry a net negative charge which causes mutual repulsion and as a consequence, stability of the colloid. The addition of electrolytes results in a compression of the ionic double layer surrounding the particles. As a result, the colloid will coagulate. However, coagulation can be prevented when a protective coat is added onto the particles by merely mixing the colloid with a solution of macromolecules. This is the simple principle on which the preparation of all gold markers is based.

The first application of colloidal gold as a transmission electron microscope specific immunocytochemical marker was described in 1971 [7]. The method was then introduced by Horisberger et al. [8] for scanning electron microscopy in 1975 (for reviews, see [9, 10]).

For transmission and scanning electron microscopical applications, monodisperse gold colloids are prepared essentially by three procedures: the smaller particles, Au_5—the subscript indicates the mean diameter of the particles in nm—are obtained by reducing gold chloride (chloroauric acid, $HAuCl_4$) with yellow or white phosphorus [7, 11]. Au_{12} particles are produced in the presence of sodium ascorbate as the reducing agent [12]. For Au_{16} to Au_{150} particles, the reducing agent is sodium citrate [9, 11] (the smaller the particle size, the higher the concentration of the reducing agent).

Colloidal gold has been labelled with a variety of molecules such as toxins, hormones, polysaccharides, glycoproteins, proteins such as protein A, enzymes, lectins, immunoglobulins, lipoproteins (cited in reference [10]). The list is still expanding. With small molecules, difficulties have been experienced in stabilising colloidal gold [10]. In this case, stable markers can be obtained when the labelling molecules are cross-linked to a carrier molecule such as bovine serum albumin [11]. Detailed information on the preparation of gold markers has been published by Horisberger and Rosset [11] and by Geoghegan and Ackerman [13].

Much evidence indicates that macromolecules, adsorbed onto gold particles as a monolayer presumably through a non-covalent binding process, remain firmly attached and keep their bio-activities for months or even years [10, 14].

In transmission electron microscopy, Au_5 to Au_{20} particles are commonly used. The selection of the size is based on a compromise between the degree of magnification necessary and the density of the marking obtained. As a rule, the density of marking increases when the particle size decreases. Due to their electron opacity and characteristic shapes, gold particles are easily recognised, even on post-stained sections. Both the direct and indirect methods are used. The protein A-gold method popularised by Roth [2] and other groups has found numerous applications and gives excellent results especially in combination with embedding at low temperature [1] to retain as much as possible the antigenicity of intracellular proteins.

When post-embedding techniques were compared [15], the density of marking observed was direct lectin < direct antibodies < indirect antibodies < indirect protein A (Fig. 1).

The colloidal gold method is also suitable for multiple marking of cell surface and intracellular components since the colloids are available in different sizes [9, 10]. Finally, precise quantification is possible with the gold markers.

2.2. Ferritin

A vast number of investigations has made use of ferritin as an electron-dense marker for transmission electron microscopy [16–18]. Ferritin (molecular weight 750,000) has a protein shell of about 12 nm outer diameter surrounding an inner core of ferric hydroxide micelle (5.5–6.0 nm in diameter) containing more than 2,000 iron atoms per molecule. The first conjugate of ferritin was made by the

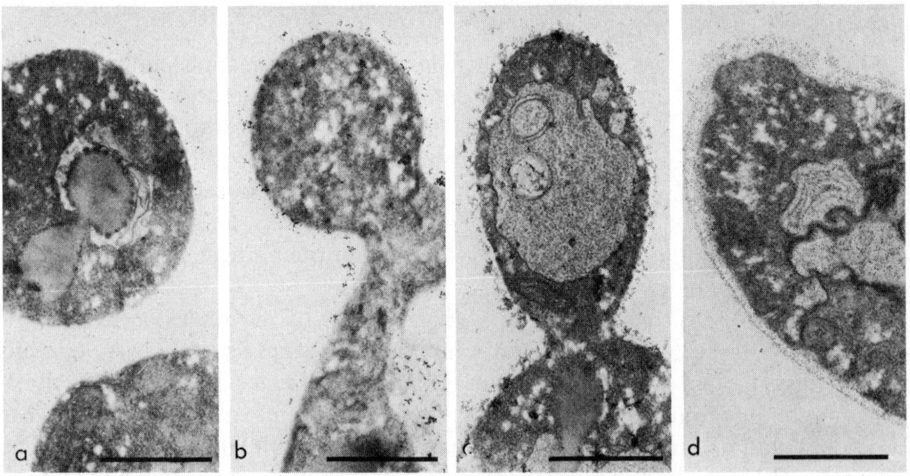

Fig. 1. In an example of the direct method of using gold markers, thin sections of *Candia utilis* were marked for mannan with Concanavalin A-Au_5 (a) and anti-mannan antibodies Au_5 (b). The indirect method is illustrated by sections which were incubated with anti-mannan antiserum and then with goat anti-rabbit IgG-Au_5 (c) and protein A-Au_5 (d). Mannan was found in the cell walls and in some vacuoles. Although the results were similar in all cases, the highest density of marking was achieved by the protein A-gold method. The bars represent 1 μm (From Horisberger [15]).

covalent coupling of antibodies [19, 20]. Ferritin-antibody conjugates are widely used and permit the localisation of specific components in biological specimens with a resolution of about 30 nm [21].

The method is general for transmission electron microscopical application. Both the direct and the indirect techniques are used. Indirect techniques are based on the use of conjugates such as ferritin-protein A [22], ferritin-avidin [23], and biotinylated ferritin [24].

Conjugation of protein to ferritin is carried out by using glutaraldehyde, meta-xylene diisocyanate or toluene 2,4-diisocyanate as coupling agents. The preparation of the conjugates has been described step by step by Hsu [25]. The conjugates must be freed of unconjugated protein and ferritin. $F(ab')_2$ fragments which exhibit a similar affinity to that of IgG can be used for the preparation of the conjugates. The removal of the F_c region of the molecule reduces non-specific binding and decreases steric hindrance [26].

Several difficulties have been encountered in the use of ferritin conjugated to antibodies. Conjugates generally exhibit reduced antibody activity relative to the original antiserum sample (the use of highly purified ferritin and specific antisera of high titre is therefore essential). Ferritin particles are not easily recognised on negatively stained thin sections. In post-embedding techniques, non-specific adsorption is often difficult to eliminate. This is due to the fact that ferritin, which is negatively charged at neutral pH, tends to bind tightly to several internal structures. Several approaches have been used to improve marking specificity [27].

2.3. Iron-dextran complexes

A series of iron-dextran complexes have been developed for use as haematinics for the treatment of iron deficiency anaemia. For instance, Imposil® consists of particles having an electron-opaque core of a ferric oxide-hydroxide micelle complexed to an electron-lucent shell of alkali-modified dextran. The average dimensions of the core are 3×11 nm and those of the shell 12×21 nm. These particles have a narrow distribution size [28, 29]. Preparations of iron-dextran with a size both larger [28] and smaller [30] are available. At present such particles have found only a limited use as markers in transmission electron microscopy.

Iron-dextran [29, 31–33] and iron-mannan particles [34] have been used in a two-step procedure to mark Concanavalin A binding sites on cell surfaces owing to the affinity of the lectin for these polysaccharides. Procedures are available to obtain iron-dextran covalently conjugated with immunoglobulins [29, 31].

Owing to the oblong shape, iron-dextran complexes are readily distinguished from the isometric core of ferritin and have been shown to be suitable for double marking experiments on ultrathin frozen sections [31].

3. Enzymic markers

Among the enzymic tracers proposed (acid phosphatase, glucose oxidase, cytochrome), horseradish peroxidase is now widely used either chemically coupled to ligands such as antibodies or as the unlabelled peroxidase-antiperoxidase (PAP) complex. The highly sensitive PAP method is out of the scope of this article (see Chapters 4, pp. 37–52 and 7, pp. 83–93).

The preparation of antibodies conjugated to peroxidase was developed independently by Nakane [35] and Avrameas [36]. Antibodies or other specific ligands are covalently linked to the enzyme either by the periodate method, glutaraldehyde or other bifunctional coupling agents (reviewed by Sternberger [18]). The marker is then reacted with the tissues (both pre- and post-embedding techniques can be used) and the site of the enzyme is visualised by the formation of osmiophilic reaction products of the enzyme using 3,3′-diaminobenzidine or 4-chloro-1-naphthol. The latter produces highly specific staining and less background staining than the former.

Conjugation of immunoglobulins to peroxidase with glutaraldehyde seems to impair the antigenicity more than the antibody site in multiple step marking [37].

4. Scanning electron microscopy

Although the working resolution is still less than that obtained by transmission electron microscopy, scanning electron microscopy used with the secondary electron detection mode has the advantage of giving a three-dimensional image of complex cell surfaces. Different types of markers have been developed for scanning electron microscopy either from biological origin (e.g. haemocyanin, viruses) or from synthetic origin (e.g. silica spheres, various polymer and co-polymer microspheres, colloidal gold). Information on this topic is given in recent reviews on cell surface marking [38] and on the use of colloidal gold for scanning electron microscopy [9, 10]. These topics are also covered in more detail by Hodges et al. (Chapter 15, pp 189–233).

Markers for scanning electron immunocytochemistry must have several characteristics: they must be large enough to be clearly identified; they should exhibit a uniform and distinct shape and bind tightly to cell surfaces. Most of the marking procedures used in scanning electron microscopy are similar to those developed for transmission electron microscopy. However, ferritin molecules cannot yet be resolved by most scanners. Microsphere markers are generally prepared by covalent coupling of the ligand [38]. An unlabelled antibody haemocyanin technique has been proposed [39] but it is cumbersome.

Contrary to other conventional scanning electron microscope markers, gold particles are good emitters of secondary electrons. The particles are thus easily detected on cell surfaces not coated with a metal. At present with the type of scanner used, the most convenient size of the gold particles is approximately 50 nm although particles as small as Au_{22} have found application [10]. The choice of the particle size depends upon the magnification desired and consideration of steric inaccessibility of binding sites.

As relatively large areas can be examined by scanning electron microscopy, the technique affords information which would be more difficult to obtain by transmis-

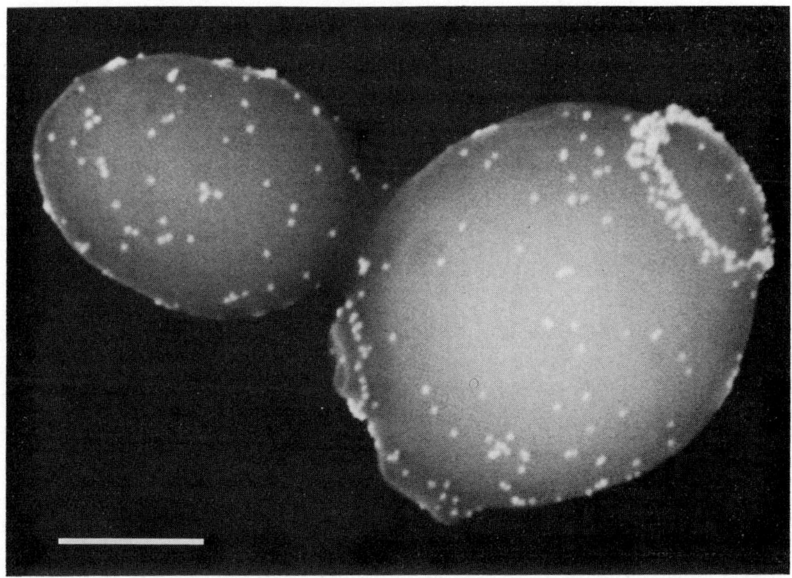

Fig. 2. The osmotolerant yeast *Saccharomyces rouxii* was grown in the presence of 18% NaCl. Glucan, a major cell wall polysaccharide, was marked on the cell surface by the indirect procedure using anti-laminaribiose antibodies and protein A-Au_{60} particles. Glucan was localised as a ring on the bud scars but was only very sparsely distributed over the cell wall. The bar represents 1 μm (unpublished document by Horisberger and Rouvet-Vauthey).

sion studies (Fig. 2). However, as a rule, scanning electron microscopical observations should be corroborated by transmission electron immunocytochemistry with smaller markers. Indeed, some cell surface receptors may not be accessible to large particles especially when the direct procedure is used.

5. What procedure to choose?

It is often difficult to select the correct method to solve a specific problem since few comparative studies are available.

As a rule, the density of marking achieved is superior with indirect procedures since, in principle, several markers can label a single site and hence provide an amplification effect. However, indirect methods require several incubation steps which may adversely affect cell surface morphology. The first bound ligands may also be lost to a certain extent during washing. Occasionally, indirect methods have been found to be less sensitive than direct procedures [40]. When shedding is unavoidable even after fixation, indirect methods are to be preferred since the first ligand applied generally binds rapidly to the cell surface and stabilises the structure [41].

A high density of marking with particulate markers is not always beneficial since the fine organisation of the cell structure may be obscured.

Among the commonly used markers, ferritin and gold particles seem to provide the highest resolution. However, colloidal gold stains may not identify all reactive sites in the post-embedding technique [42]. With frozen thin sections, the protein A-gold method (two steps) was found to be superior for localising antigens when compared to ferritin and peroxidase conjugates [43] (one step). When the sensitivity of ferritin conjugated to goat anti-rabbit immunoglobulin (two steps) was compared to the protein A-Au_5 method (two steps), the number of ferritin particles per unit area was more than three times that of the gold particles. However, using a three-step gold method, staining increased to the level of the two-step ferritin procedure [44].

Besides density of marking and sensitivity, steric hindrance is another factor to consider. Although enzyme conjugates are not devoid of steric hindrance, particulate markers are subject to it especially when pre-embedding procedures are used. Unless the size of particulate markers is varied, false-negative results are difficult to detect.

Contrary to diffuse markers, particulate markers are easily amenable to quantification either by counting the bound particles (ferritin, colloidal gold) or in pre-embedding procedures by spectrophotometric measurements (colloidal gold). Quantitative data should be critically examined since the density of marking depends on several factors such as embedding medium, fixative used, and size of the markers.

Finally, if one wishes to achieve double marking, a number of possibilities exist. Pairs such as gold particles and peroxidase conjugates, gold particles and ferritin conjugates, iron-dextran and ferritin conjugates are compatible. The gold method is particularly well suited for multiple marking in pre- and post-embedding procedures since particles of different sizes are available [9, 11]. Double immunogold procedures are considered in more detail by Varndell and Polak (Chapter 13, pp. 155–177).

6. Acknowledgement

The author thanks Mrs. M. Rouvet for the artwork.

7. References

1. Roth, J., Bendayan, M., Carlemalm, E., Villiger, W. and Garavito, M. (1981) Enhancement of structural preservation and immunocytochemical staining in low temperature embedded pancreatic tissue. J. Histochem. Cytochem. 29, 663–671.
2. Roth, J. (1982) The protein A-gold (pAg) technique. A qualitative and quantitative approach for antigen localization on thin sections. In: G.R. Bullock and P. Petrusz (Eds.) Techniques in Immunocytochemistry, Vol. 1. Academic Press, London and New York, pp. 107–133.
3. Tokuyasu, K.T. (1980) Immunochemistry on ultrathin frozen sections. Histochem. J. 12, 381–403.
4. Deryagin, B.V. and Landau, L. (1941) Theory of the stability of strongly charged lyophobic sols and of the adhesion of strongly charged particles in solutions of electrolytes. Acta Physiochim. U.R.S.S. 14, 633–662.

5. VERWEY, E.J.W. and OVERBEEK, J.T.G. (1948) Theory of the Stability of Lyophobic Colloids. Elsevier, New York and London, 216 pp.
6. FRENS, G. (1972) Particle size and sol stability in metal colloids. Kolloid Z. Z. Polym. 250, 736–741.
7. FAULK, W.P. and TAYLOR, G.M. (1971) An immunocolloid method for the electron microscope. Immunochemistry 8, 1081–1083.
8. HORISBERGER, M., ROSSET, J. and BAUER, H. (1975) Colloidal gold granules as markers for cell surface receptors in the scanning electron microscope. Experientia 31, 1147–1149.
9. HORISBERGER, M. (1979) Evaluation of colloidal gold as a cytochemical marker for transmission and scanning electron microscopy. Biol. Cell. 36, 253–258.
10. HORISBERGER, M. (1981) Colloidal gold: a cytochemical marker for light and fluorescent microscopy and for transmission and scanning electron microscopy. In: O. Johari (Ed.) Scanning Electron Microscopy, Vol. 2. SEM Inc., AMF O'Hare (Chicago) IL., pp. 9–31.
11. HORISBERGER, M. and ROSSET, J. (1977) Colloidal gold, a useful marker for transmission and scanning electron microscopy. J. Histochem. Cytochem. 25, 295–305.
12. HORISBERGER, M. and TACCHINI-VONLANTHEN, M. (1983) Ultrastructural localization of Kunitz inhibitor on thin sections of *Glycine max* (Soybean) cv. Maple Arrow by the gold method. Histochemistry 77, 37–50.
13. GEOGHEGAN, W.D. and ACKERMAN, G.A. (1977) Adsorption of horseradish peroxidase, ovomucoid and antiimmunoglobulin to colloidal gold for the indirect detection of Concanavalin A, wheat germ agglutinin and goat anti-human immunoglobulin on cell surfaces at the electron microscopic level: a new method, theory and application. J. Histochem. Cytochem. 25: 1187–1200.
14. HORISBERGER, M. and TACCHINI-VONLANTHEN, M. (1983) Stability and steric hindrance of lectin-labelled gold markers in transmission and scanning electron microscopy. In: T.C. Bog-Hansen and G.A. Spengler (Eds.) Lectins, Vol. 3. Walter de Gruyter, Berlin, pp. 189–197.
15. HORISBERGER, M. (1981) Colloidal gold as a cytochemical marker in electron microscopy. Gold Bull. 14, 90–94.
16. NICOLSON, G.L. and SINGER, S.J. (1971) Ferritin conjugated plant agglutinins as specific saccharide stains for electron microscopy. Application to saccharides bound to cell membranes. Proc. Natl. Acad. Sci. U.S.A. 68, 942–945.
17. MORGAN, C. (1972) Use of ferritin-conjugated antibodies in electron microscopy. Int. Rev. Cytol. 32, 291–326.
18. STERNBERGER, L.A. (1979) Immunocytochemistry. John Wiley and Sons, New York, pp. 59–81.
19. SINGER, S.J. (1959) Preparation of an electron-dense antibody conjugate. Nature (London) 183, 1523–1524.
20. RIFKIND, R.A., HSU, K.C., MORGAN, C., SEEGAL, B.C., KNOX, A.W. and ROSE, H.M. (1960) Ferritin-conjugated antibody to a localized antigen by electron microscopy. Nature (London) 187, 1094–1095.
21. SINGER, S.J. and SCHICK, A.F. (1961) The properties of specific stains for electron microscopy prepared by conjugation of antibody molecule with ferritin. J. Biophys. Biochem. Cytol. 9, 519–537.
22. TEMPLETON, C.L., DOUGLAS, R.J. and VAIL, W.J. (1978) Ferritin-conjugated protein A. A new immunocytochemical reagent for electron microscopy. FEBS. Lett. 85, 95–98.
23. HEITZMANN, H. and RICHARDS, F.M. (1974) Use of the avidin-biotin complex for specific staining of biological membranes in electron microscopy. Proc. Natl. Acad. Sci. U.S.A. 71, 3537–3541.
24. BAYER, E.A., SKUTELSKY, E. and WILCHEK, M. (1979) The avidin-biotin complex in affinity cytochemistry. Methods Enzymol. 62, 308–315.
25. HSU, K.C. (1981) Preparation of ferritin conjugates and antibodies for the localization and identification of antigens in tissues and cells by electron microscopy. In: O. Johari (Ed.) Scanning Electron Microscopy, Vol. 4. SEM Inc., AMF O'Hare (Chicago) Il., pp. 17–26.
26. KRAEHENBUHL, J.P. and JAMIESON, J.D. (1974) Localization of intracellular antigens by immunoelectron microscopy. In: G.W. Richter and M.A. Epstein (Eds.) Int. Rev. Exper. Pathol., Vol. 13. Academic Press, New York, pp. 1–53.
27. PARR, E.L. (1979) Intracellular labelling with ferritin conjugates. A specificity problem due to the

affinity of unconjugated ferritin for selected intracellular sites. J. Histochem. Cytochem. 27, 1095–1102.

28. Marshall, P.R. and Rutherford, D. (1972) Physical investigation on colloidal iron-dextran complexes. J. Colloid Interface Sci. 37, 390–402.
29. Dutton, A.H., Tokuyasu, K.T. and Singer, S.J. (1979) Iron-dextran antibody conjugates: general method for simultaneous staining of two components in high-resolution immunoelectron microscopy. Proc. Natl. Acad. Sci. U.S.A. 76, 3392–3396.
30. Ricketts, C.R., Cox, J.S.G., Fitzmaurice, C. and Moss, G.F. (1965) The iron-dextran complex. Nature (London) 208, 237–239.
31. Geiger, B., Dutton, A.H., Tokuyasu, K.T. and Singer, S.J. (1981) Immunoelectron microscope studies of membrane-microfilament interactions: distributions of α-actinin, tropomyosin, and vinculin in intestinal epithelial brush border and chicken gizzard smooth muscle cells. J. Cell Biol. 91, 614–628.
32. Martin, B.J. and Spicer, S.S. (1974) Concanavalin A-iron dextran technique for staining cell surface mucosubstances. J. Histochem. Cytochem. 22, 206–209.
33. Schiefer, H.G., Krauss, H., Brunner, H. and Gerhardt, U. (1978) Cytochemical localization of surface carbohydrates on mycoplasma membranes. Experientia 34, 1011–1012.
34. Roth, J. and Franz, H. (1975) Ultrastructural detection of lectin receptors by cytochemical affinity reaction using mannan-iron complex. Histochemistry 41, 365–368.
35. Nakane, P.K. and Pierce, G.N. (1966) Enzyme-labelled antibodies: preparation and application for the localization of antigens. J. Histochem. Cytochem. 14, 929–931.
36. Avrameas, S. and Uriel, J. (1966) Méthode de marquage d'antigènes et d'anticorps avec des enzymes et son application en immunodiffusion. C.R. Acad. Sci. (Paris) 262, 2543–2545.
37. Tougard, C., Tixier-Vidal, A. and Avrameas, S. (1979) Comparison between peroxidase-conjugated antigen or antibody and peroxidase-anti-peroxidase complex in a postembedding procedure. J. Histochem. Cytochem. 27, 1630–1633.
38. Molday, R.S. and Maher, P. (1980) A review of cell surface markers and labelling techniques for scanning electron microscopy. Histochem. J. 12, 273–315.
39. Gonda, M.A., Benton, C.V., Massey, R.J. and Schultz, A.M. (1981) Monoclonal antibodies as immunospecific probes for virus and cell surface antigen localization with the unlabelled antibody hemocyanin bridge: a review. In: O. Johari (Ed.) Scanning Electron Microscopy, Vol. 2. SEM Inc., AMF O'Hare (Chicago) IL., pp. 45–62.
40. Briggman, J.V. and Widnell, C.C. (1983) A comparison of direct and indirect techniques using ferritin-conjugated ligands for the localization of Concanavalin A binding sites on isolated hepatocyte plasma membranes. J. Histochem. Cytochem. 31, 579–590.
41. Horisberger, M. and Vonlanthen, M. (1979) Fluorescent colloidal gold: a cytochemical marker for fluorescent and electron microscopy. Histochemistry 64, 115–118.
42. Childs, G.V. (1983) The use of multiple methods to validate immunocytochemical stains. J. Histochem. Cytochem. 31, 168–176.
43. Beesley, J.E., Orpin, A. and Adlam, C. (1982) A comparison of immunoferritin, immunoenzyme and gold-labelled protein A methods for the localization of capsular antigen on frozen thin sections of the bacterium, *Pasteurella haemolytica*. Histochem. J. 14, 803–810.
44. Tokuyasu, K.T. (1983) Present state of immunocryoultramicrotomy. J. Histochem. Cytochem. 31, 164–167.

Techniques

Immunolabelling for Electron Microscopy (Polak/Varndell, eds)

CHAPTER 3

The choice of resins for electron immunocytochemistry

Brian E. Causton

Department of Materials Science in Dentistry, The London Hospital Medical College, Turner Street, London E1 2AD, U.K.

CONTENTS

---

1. Introduction

Of all the staining techniques performed for electron microscopy, immunocytochemistry is probably the most demanding. Problems arise from two main sources:

a) preservation of antigenicity in the tissue and
b) steric hindrance of antibodies, conjugated or not, during the staining cycle.

These problems impose criteria on resin selection which demand not only close attention to the chemical reactivity of the cured resin, but also to its method of curing and the degree of cross-linking achieved during cure. The added criteria that the resin should be beam stable can cause problems when large amplifications are attained by the use of high molecular weight conjugates. The aim of this chapter will be to impart 'rules of thumb', based on physico-chemical theory, which will allow researchers to adjust their embedding techniques to suit their tissues and staining techniques so that minimum process times and optimum degrees of cross-linking can be achieved.

2. The resins available

Because of the added demands of electron microscopy, some of the resins successfully used at the light microscope level are excluded because they lack the necessary beam stability, though the use of carbon-coated support grids and high accelerator voltages may allow lightly cross-linked poly(hydroxyethyl methacrylate) to be used if nothing else is successful.

The resins that give the best results and greatest flexibility of technique fall into two broad groups:

a) the epoxy cross-linked systems and
b) the cross-linked hydrophilic acrylics.

2.1. *The epoxy resins*

The epoxy resins are often referred to by trivial names exclusive to microscopy, i.e. Araldite, Epon and Spurr resins. For identification these trivial names will still be used to describe general sub-groups of epoxy resins.

2.1.1. *Araldites*

Based on the diglycidyl ether of bisphenol A (DGEBA) and originally formulated around the CIBA product CY.212 (75% DGEBA, 25% Dibutyl phthalate; CIBA, Duxford, Cambridgeshire, U.K.) [1]. The curing system is of the very thermally stable aliphatic anhydride/amine type, usually dodecenyl succinic anhydride (DDSA), to impart flexibility and toughness to blocks with varying amounts of methyl nadic anhydride (MNA) added to stiffen blocks when harder tissue is to be embedded [2]. The amine accelerators can be polyfunctional (DMP-30) or monofunctional (BDMA). For electron immunocytochemistry, Araldites offer good beam stability primarily because of their aromatic character; however, a small degree of cross-linking can also be an advantage. DGEBA is a large molecule and hence has a low rate of diffusion through tissue; also unless the supplier carefully monitors hydroxy content in the monomer, this problem can be compounded by the DGEBA agglomerating in the liquid state due to the formation of hydrogen bonded structures [3].

2.1.2. Epons

This group of resins derives its name from the Shell Epoxy Resin Epon 812 (Shell Chemical Co., Carshalton, Surrey, U.K.), no longer available. Epon 812 was a mixture of seven monomers which resulted from the reaction of glycerol and epichlorohydrin. Epon consists of tri-, di- and mono-substituted glycidyl ethers of glycerol. Large differences in performance can be expected because of differences in the relative proportions of the seven components in each batch and in differences arising from the different clean up treatments the resins receive from various manufacturers. The curing system used for 'Epons' is the same as that for 'Araldites' [4]. The advantages of Epon are good beam stability and a reduced viscosity compared to DGEBA. However, the high hydroxy content of the resin can be expected to lead to some hydrogen bonding in the liquid state and the formation of slow diffusing agglomerates. Care should be taken when handling this resin as it contains di-epoxides which may be carcinogenic [5].

2.1.3. Spurr

Spurr-type resins are based on vinyl cyclohexane dioxide (VCD) cross-linked by a long chain aliphatic anhydride accelerated by an ethanolamine [2, 6]. These resins have the highest diffusion rate of all the embedding epoxy resins and, provided the anhydride is stored dry, they also have the greatest consistency. Their major disadvantage is their toxicity; they should be handled in the same way as other carcinogens and control of resin vapour dissipation should be a priority [5].

2.1.4. Epoxy resin in general

The epoxies listed above all exhibit low water adsorption and many hydrophobic groups in their structures. If antibodies are to interact with tissue embedded in such resin, there must be a gel layer created in the surface, swollen by water and of sufficient mean free path to allow at least the primary antibody to enter the gel. To achieve such a gel two modifications to the resin are possible:

a) The resin can be made more hydrophilic, in which case, despite the same cross-link density, Flory Huggin's theory predicts that the resin matrix will swell in water, or
b) Involves a reduction in cross-linking as well as an increase in hydrophilicity, however, this may only be necessary for staining semi-thin sections used in light microscopy.

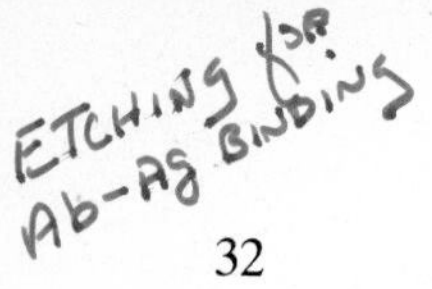

2.1.5. *Treatment with an oxidising agent*

The oxidising agent may be hydrogen peroxide, potassium permanganate or periodic acid. They all have the same effect on the resin, that of oxidising the hydrophobic alkane side chains to alcohols, aldehydes and acids (Fig. 1). This has a

$$RCH_3 + O_2 \rightarrow RCH_2COOH \rightarrow RCHO + H_2O$$

$$\text{alkane} \longrightarrow \text{acid} \longrightarrow \text{aldehyde}$$

Fig. 1. Oxidation of alkane chains by an oxidising agent.

net effect of increasing the hydrophilicity of the resin. It may of course also oxidise essential pendent alkane groups on the antigen and must be seen as a procedure to be avoided if possible.

2.1.6. *Sodium alkoxide 'etching'*

The curing system chosen to cure epoxy embedding resins is of the anhydride/amine type. This results in predominantly ester cross-linking [7]. The result of adding either sodium methoxide or ethoxide to the resin is to cause a transesterification to occur in which the ester cross-link is broken (Fig. 2). The

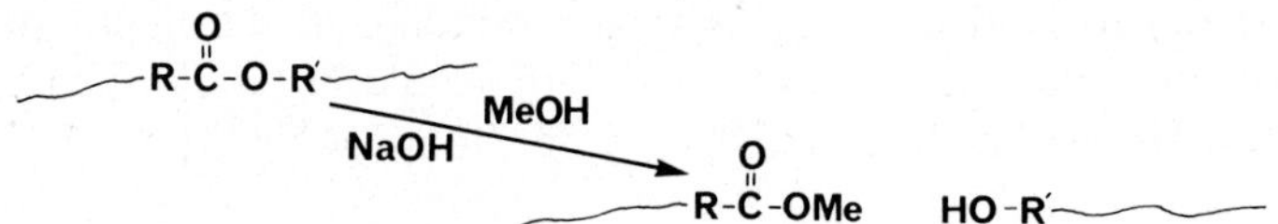

Fig. 2. Transesterification of ester cross-links by sodium methoxide.

resulting reduction in cross-links and net increases in hydrophilic groups result in a swelling of the resin in aqueous media and, if transesterification is allowed to continue, the extraction of low molecular weight resin fragments from the matrix.

2.1.7. *Antigen/epoxy resin interaction*

The epoxy group is a very reactive group capable of ring opening addition reactions to many organic moieties (Fig. 3). The presence of amine accelerators in the resin mixture used for embedding can increase the likelihood of epoxy/peptide group interaction by base catalysis. A list of some of the possible epoxy/peptide interactions is given in Fig. 3 [8]. It is not certain that such reactions occur; however, they are all possible and in some circumstances may be the cause of reduced antigenicity.

2.2. *The acrylics*

The acrylics that are used in electron microscopy are cross-linked to reduce sublimation of the lower acrylic homologues, and preferably aromatically substituted to minimise damage by electron beam bombardment. Acrylics can be

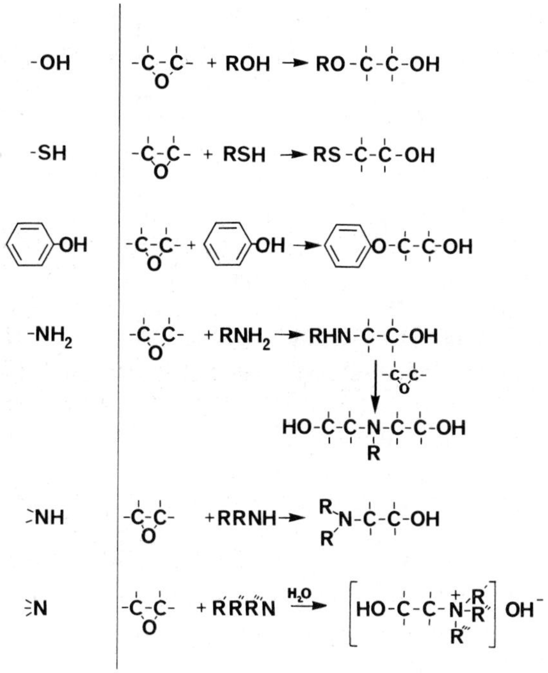

-OH	-C-C-(O) + ROH → RO-C-C-OH
-SH	-C-C-(O) + RSH → RS-C-C-OH
C_6H_5-OH	-C-C-(O) + C_6H_5-OH → C_6H_5-O-C-C-OH
$-NH_2$	-C-C-(O) + RNH_2 → RHN-C-C-OH; + -C-C-(O) → HO-C-C-N(R)-C-C-OH
>NH	-C-C-(O) + RRNH → R(R')N-C-C-OH
≥N	-C-C-(O) + R'R''R'''N $\xrightarrow{H_2O}$ [HO-C-C-N^+R'R''R'''] OH^-

Fig. 3. Some possible reactions between epoxy groups and peptide substituents.

heat-cured or room temperature-cured using an amine accelerator. For low temperature curing radiation curing is necessary either by ultraviolet or visible light, though paradoxically electron beams might be expected to give the best results in thin section; unfortunately such techniques are only available on an industrial scale at the present time.

For immunocytochemistry the hydrophobicity of the acrylic is important. The acrylic can have hydrophilic groups built into them and thus avoid the need for peroxide or 'etch' treatments. This is an advantage when dealing with sensitive antigens or unfixed tissue. The cross-link density of the acrylic can also be controlled without the need for etching. Finally, the mildest curing conditions can be chosen, i.e. low temperatures in heat-cured systems, careful control of exothermy in amine/peroxide room temperature curing systems and the shortest possible irradiation times for light-cured systems.

2.2.1. *Aliphatic cross-linked acrylics*

The commercial acrylic embedding medium Lowicryl (Chemische Werke Lowi, D-8264, Waldkraiborg, West Germany; see Chapter 9, pp. 113–121) is produced in a hydrophilic grade suitable for immunocytochemistry [9]. It consists of both hydroxypropyl and hydroxyethyl methacrylates cross-linked by an aliphatic glycol dimethacrylate to reduce sublimation in the electron microscope. The acrylic is cured by exposure to ultraviolet light using a benzoic alkyl ether as the activator.

The resin is very mobile even at low temperatures and is very suitable for techniques that require low temperature antigen preservation. It is, however, very similar to the standard 'glycol' formulae and has no special features that make it specially suited to electron microscopy.

2.2.2. Aromatic cross-linked acrylics

The polyhydroxy aromatic acrylic resin LR White (London Resin Company, P.O. Box 34, Basingstoke, Hants., U.K.) was formulated specifically to combine a high hydrophilic character with an electron beam stable cross-link. The resin is either heat-cured at 50°C or room temperature-cured using an aromatic tertiary amine accelerator. The cross-link density of the resin is critical for good immunocytochemical results and Newman and colleagues have demonstrated that the best results are obtained using a slow heat cure at 50°C [10] (see also Chapter 5, pp. 53–70).

The need for a low temperature curing, aromatically cross-linked acrylic resin was recognised some time ago, and in particular, one with a low solvent power so that it could be used to embed unfixed tissue for use in enzyme studies. A blue light curing system called LR Gold (London Resin Company, address above) has been developed using the quartz halogen lamp from a light microscope combined with a benzil/amine blue light activator. The system cures at −25°C in 20 h to give blocks of a suitable cross-link density and hydrophilicity to allow the successful use of avidin-biotin antibody conjugates.

2.2.3. General comments on acrylics

Acrylics cure by a radical chain reaction; oxgen can terminate this reaction, hence oxygen must be excluded from curing acrylic resins and gelatin capsules are probably the best moulds for this purpose. Acrylics have the advantage of tolerating water during polymerisation, hence dehydration of tissue need not be so stringent.

2.2.4. Antibody resin interaction

2.2.4.1. Non-specific adsorption? Non-specific adsorption of antibody onto the polymer is most likely to occur if its surface is either hydrophobic or it contains ionic groups. If the surface is highly hydrophilic it should swell and become hydrated, and under such conditions should exclude any other polymer from occupying the same space provided that it contains little or no ionic groups [11]. Hence, provided there are no strong chemical interactions between the resin and the antibody, there should be no non-specific adsorption of the antibody onto the embedding resin. With acrylics, provided that the monomers are free of ionic groups and high in hydroxy groups, the above criteria for non-specific adsorption should be fulfilled.

2.2.4.2. Penetration? Penetration of the resin by antibody is difficult for two reasons:

a) the antibody is a large molecule (molecular weight > 40,000) and
b) the antibody has rod-like components.

Work on the diffusion of IgG into swollen hydrophilic polymers reveals a lack of viscous interaction between molecules [12], but a large steric interaction. This implies that the cross-link density, and hence the mean free path between molecules, will be the major influence on the outcome of any diffusion process. Therefore, always use the lowest cross-link density possible to allow the greatest antibody/antigen interaction.

2.3. Points to consider when choosing a suitable resin system

a) What is the maximum temperature the antigen can tolerate? Higher temperatures lead to increased rapid diffusion processes, lower temperatures result in better tissue preservation. (See Table 1 for temperature ranges over which resins can be used).

TABLE 1

Temperature ranges over which embedding resins can be used.

Resin	*Temperature Range* (°C)	
	Minimum	*Maximum*
Epoxy	15	>200
Acrylics heat cured	40	100
Amine/peroxide cured	5	30
Ultraviolet light activated	−50	100
Blue light/α-diketone activated	−50	100

b) Is the final antibody conjugate large? Direct immunocytochemistry methods require less mean free space within the embedding resin than large conjugates, hence adapt cross-link density accordingly.
c) Is the antibody easily oxidised? If so, avoid epoxy resins which will need to be oxidised to increase their hydrophilicity.
d) Is the tissue highly coloured? Either avoid light curing systems or decolourise the tissue prior to embedding.

2.4. Some physico-chemical reasons why a resin may cause your immunocytochemical method to fail

2.4.1. Poor infiltration

a) Not enough time given at the temperature chosen for infiltration. (The diffusion coefficient varies as the log of $(\text{time})^{-1}$.)
b) Tissue sample too large. Time taken for infiltration increases as the square of the smallest dimension of the block, i.e. doubling the block size increases infiltration time 4-fold [13].
c) Diffusion coefficients of resin too low in the tissue chosen. Could be due to molecular size or chemical interaction with tissue.

2.4.2. Poor curing

a) Heat-cured systems. Epoxies: water or alcohol left in tissue. Acrylics: oxygen inhibition or acetone left in tissue.
b) Room temperature curing acrylics. Too much accelerator causing the resin to set before accelerator has had time to penetrate tissue. Oxygen inhibition.
c) Light-cured systems. Too thick a block. Tissue too highly coloured. Oxygen inhibition.

2.4.3. Immunocytochemical stain fails to work

a) Antigen lost due to: reaction with uncured resin (Fig. 3), oxidation by etching agent, too high a temperature during processing.
b) Antibody fails to penetrate resin. Cross-link density too high. Hydrophobicity too high causing the resin to remain only partially swollen in an aqueous environment.

2.4.4. Background staining unacceptably high

a) Resin too hydrophobic;
b) Resin too ionic;
c) Resin contaminated with reactive species.

3. References

1. Glauert, A.M., Rogers, G.E. and Glauert, R.H. (1956) A new embedding medium for electron microscopy. Nature (London) 178, 803.
2. Causton, B.E. (1980) The molecular structure of resins and its effect on the epoxy embedding resins. Proc. R.M.S. 15, 185–189.
3. Lee, H. and Neville, K. (1967) Handbook of Epoxy Resins. McGraw-Hill, Inc., New York, pp. 65.
4. Finck, H. (1960) Epoxy resins in electron microscopy. J. Biophys. Biochem. Cytol. 7, 27–30.
5. Causton, B.E., Ashurst, D.E., Butcher, R.G., Chapman, S.K., Thomson, D.J. and Webb, M.J.W. (1981) Resins: Toxicity and safe handling. Proc. R.M.S. 16, 265–268.
6. Spurr, A.R. (1969) A low-viscosity epoxy resin embedding medium for electron microscopy. J. Ultrastruct. Res. 26, 31–43.
7. Tanaka, Y. and Kakiuchi, H. (1961) Study of epoxy compounds. VI. J. Polymer Sci. (A) 2, 3405–3430.
8. Lee, H. and Neville, K. (1967) Handbook of Epoxy Resins. McGraw-Hill, Inc., New York, pp. 5.1–5.40.
9. Carlemalm, E., Garavito, R.M. and Villiger, W. (1982) Resin development for electron microscopy and an analysis of embedding at low temperature. J. Microsc. 126, 123–143.
10. Newman, G.R., Jasani, B. and Williams, E.D. (1983) A simple post-embedding system for rapid demonstration of tissue antigens under the electron microscope. Histochem. J. 15, 543–555.
11. Edmond, E. and Ogston, A.G. (1970) Phase separation in an aqueous quaternary system. Biochem. J. 117, 85–89.
12. Preston, B.N., Obrink, B. and Laurent, T.C. (1973) The rotational diffusion of albumin in solutions of connective tissue polysaccharides. Eur. J. Biochem. 33, 401–406.
13. Crank, J. and Park, G.S. (1968) Water in polymers. In: J.A. Barrie (Ed.) Diffusion in Polymers. Academic Press Inc., New York, pp. 259–313.

Immunolabelling for Electron Microscopy (Polak/Varndell, eds)

CHAPTER 4

Pre-embedding ultrastructural immunocytochemistry: Immunoenzyme techniques

John V. Priestley

Neuroanatomy-Neuropharmacology Group, University Departments of Pharmacology and Human Anatomy, South Parks Road, Oxford, U.K.

CONTENTS

1. Introduction

In pre-embedding ultrastructural immunocytochemistry, immunostaining is carried out on relatively thick unembedded tissue sections. Only after immunostaining are the tissue slices embedded in plastic (Fig. 1). The plastic-embedded tissue is then sectioned and prepared for electron microscopical examination using standard preparative procedures. In contrast, post-embedding immunocytochemistry involves the immunostaining of ultrathin plastic embedded, grid-mounted, sections.

The penetration of antibodies into thick, fixed, tissue slices is slight and this is a major disadvantage with pre-embedding staining. Many of the variations in pre-embedding protocol which will be discussed in this chapter relate to the problem of antibody penetration. However, the pre-embedding approach also has two important advantages. Firstly, antigens may be destroyed or obscured by the dehydration and plastic embedding processes required for post-embedding electron immunocytochemistry and so the pre-embedding procedure is useful in situations where antigen is present in low concentrations or where the antigen is particularly sensitive to embedding. Secondly, many antigens are very sensitive to osmium and so this highly effective membrane contrasting agent is normally avoided in

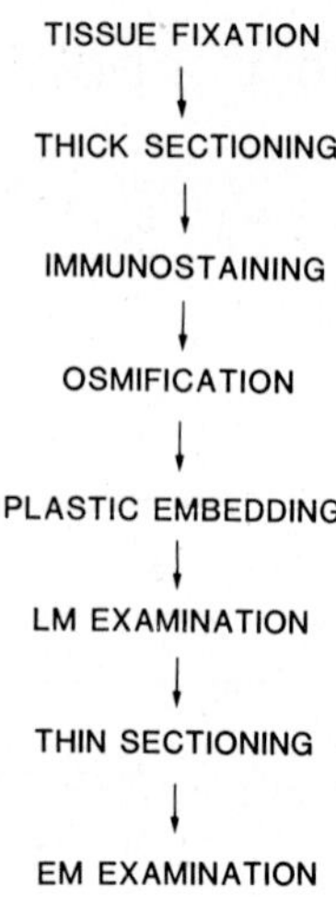

Fig. 1. Stages of pre-embedding immunocytochemistry.

post-embedding procedures. However, osmium can be added after pre-embedding immunostaining has been successfully completed but before dehydration and embedding (Fig. 1). Therefore, the pre-embedding procedure is normally the method of choice in situations where osmium is necessary in order to obtain adequate membrane fixation and contrasting.

Pre-embedding peroxidase-antiperoxidase (PAP) immunocytochemistry was introduced by Pickel and colleagues for the ultrastructural localisation of monoamine synthesising enzymes [1, 2] and it has remained the most widely used immunocytochemical procedure for the electron microscopical demonstration of central nervous system (CNS) transmitters and transmitter markers. Illustrative examples for this chapter will be taken from the same general field, namely the localisation of neurotransmitters and neuropeptides. Several recent reviews cover similar ground and for further examples or for more detailed information relating to particular antigens the reader is referred to these other publications [3–5]. The present review will concentrate on the use of enzyme-based immunolabels. Non-enzymic labels can be used in pre-embedding immunocytochemistry [6] but have not yet been widely represented in the literature. The various stages involved in pre-embedding electron immunocytochemistry are shown schematically in Figure 1. These stages will be discussed in turn in the main body of this text and a general protocol is presented in the Appendix to this chapter.

2. Variables in pre-embedding immunocytochemistry

2.1. Tissue fixation

Tissue fixation has two main functions, namely to retain the antigen in situ and to preserve tissue structure. Fixatives introduce intra- and inter-molecular cross-links which are likely to decrease antigenicity. So in choosing a fixative for electron immunocytochemistry a compromise has normally to be sought between preservation of good ultrastructure and retention of antigenicity. With pre-embedding staining there is the additional problem that fixative-induced cross-linking inhibits antibody penetration. This is particularly a problem with bifunctional fixatives such as glutaraldehyde. Higher concentrations of glutaraldehyde can be used for post-embedding staining than are normally used in pre-embedding immunocytochemistry and this suggests that the effects of fixative on antibody access may be a more significant problem than fixative effects on antigenicity per se. Various model systems have been used to assess the effects of fixative on antigenicity [7–9] but it is difficult to assess the effects of fixation on antibody penetration. Procedures for increasing antibody penetration into well-fixed tissue sections will be dealt with in section 2.2 below. Regarding the selection of fixative, most workers have used a mixture of glutaraldehyde and paraformaldehyde for pre-embedding immunocytochemistry. In their original pre-embedding protocol, Pickel and colleagues used 1% glutaraldehyde plus 1% paraformaldehyde in 0.1 M sodium cacodylate buffer (pH 7.2) to localise successfully the enzymes tyrosine hydroxylase, dopamine-β-hydroxylase, tryptophan hydroxylase [2] and the peptide substance P [10]. A similar fixative has been used for the localisation of neurophysin [11], vasopressin and oxytocin [12]. More recently, several research groups have used 4% paraformaldehyde plus 0.05–0.5% glutaraldehyde, the exact proportion of glutaraldehyde depending on the particular antigen and primary antibody. Transmitter and related antigens successfully localised using such a mixture include: neuropeptides (substance P [13], enkephalin [14], thyrotropin-releasing hormone (TRH) [15], somatostatin [16], cholecystokinin (CCK) [17], adrenocorticotropic hormone (ACTH) [18], vasoactive intestinal peptide (VIP) [19]), transmitters, 5-hydroxytryptamine (5-HT) [20], enzymes (tyrosine hydroxylase [21], glutamic acid decarboxylase (GAD) [22], choline acetyltransferase [23]) and cyclic nucleotides [24]. Figure 2 shows an example of the localisation of substance P immunoreactivity in the spinal trigeminal nucleus of the rat using 4% paraformaldehyde and 0.5% glutaraldehyde as fixative. When higher concentrations of glutaraldehyde (1% or 5%) were used to localise glutamic acid decarboxylase, Vaughn et al. [4] observed increased non-specific staining. It is important to note that the sections should be washed

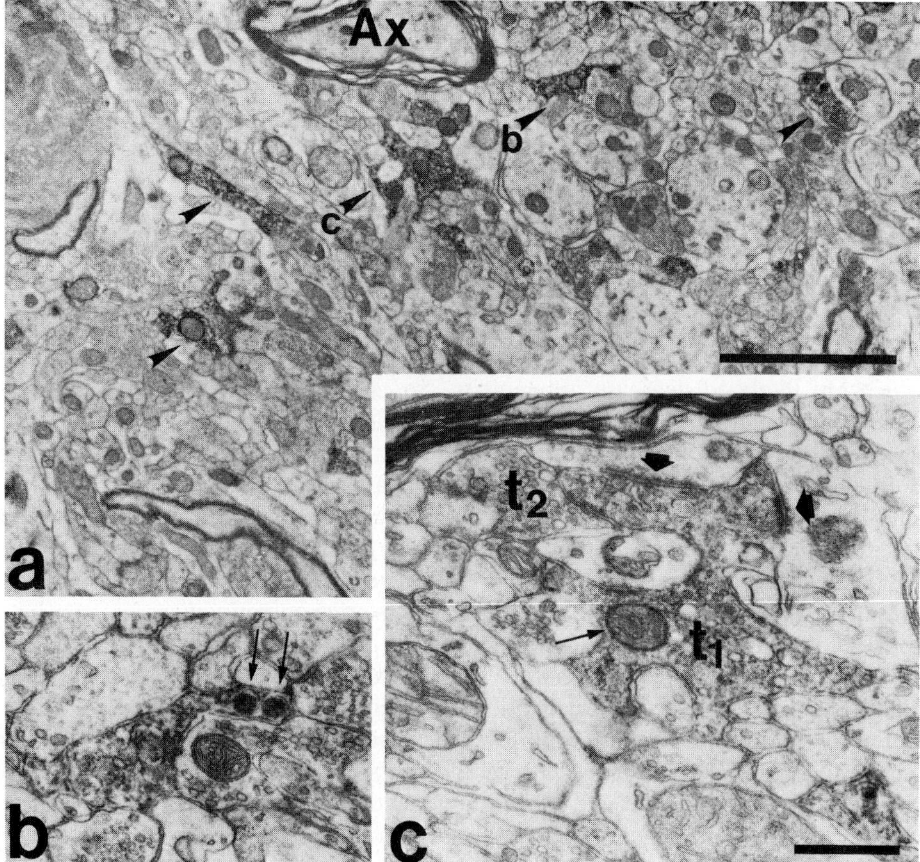

Fig. 2. Ultrastructural localisation of substance P in the spinal trigeminal nucleus. 4% paraformaldehyde and 0.5% glutaraldehyde fixed tissue. a. Low magnification micrograph showing a number of immunostained profiles (arrow heads) in an uncontrasted thin section. Two of the profiles (a and b) are shown at higher magnification in (b) and (c). Ax indicates a myelinated axon. The layers of myelin are separated. This artifact is difficult to avoid in immunostained material. Scale bar = 2.5 μm. b. Profile (b) shown in a contrasted semi-serial section. Immunostaining is seen over two dense cored vesicles and diffusely in the cytoplasm. c. Profile (c) in a contrasted semi-serial section. An immunostained terminal (t_1) adjoins an unstained terminal (t_2). The immunostaining is more difficult to see in the contrasted section than in the uncontrasted section (a). In the contrasted section (c) ultrastructural features can be clearly seen including membrane thickenings (arrow heads). Immunostaining occurs on a mitochondrial membrane (arrow). Scale bar = 0.5 μm.

thoroughly in order to remove unreacted fixative before commencing immunostaining [4].

A particular antigen may require different proportions of fixative in different tissues or even in different areas of the brain. For example, lower concentrations of glutaraldehyde were necessary to localise substance P in the substantia nigra [25] than were used to localise the peptide in the substantia gelatinosa of the spinal trigeminal nucleus [13]. Recently, several detailed studies have examined the optimal conditions for pre-embedding electron immunocytochemistry using aldehyde fixatives. Berod and colleagues [26] reported that optimal paraformaldehyde fixation is obtained by first perfusing at near neutral pH (pH 6.5) and then changing to a more basic pH (pH 11). This allows rapid and uniform penetration of fixative before full fixation takes place. These results have been confirmed by other workers [27]. Several studies have also confirmed the report by Weber and colleagues [28] that antigenicity in glutaraldehyde-fixed material can be partially restored by treatment of tissue with sodium borohydride [27, 29].

Other fixatives have been used with success to localise CNS markers, and these include acrolein [30], a periodate-lysine-paraformaldehyde fixative [31, 32] first introduced by McLean and Nakane for post-embedding staining [33] and a buffered picric acid-paraformaldehyde-glutaraldehyde fixative based on the solution described by Zamboni and de Martino [34] but modified to make it more suitable for electron microscopy [35]. The modified Zamboni fixative has been used by Somogyi and collaborators in various combined electron microscopical procedures [36] and has been successfully employed in immunocytochemical studies to localise substance P [35], enkephalin [35, 37], somatostatin [35], GAD [36], ACTH [38] and VIP [39]. Before embarking on an electron microscopical study, various fixatives should be screened by light microscopical immunohistochemistry in order to ascertain which are compatible with the particular antigen and primary antibody under investigation. Whichever fixative is used, fixation should be carried out in a way which allows rapid and uniform penetration of the tissue by the fixative while at the same time preserving optimal ultrastructure. For CNS studies this normally means carrying out perfusion fixation via the vascular system (for details see references [4, 5, 40]).

2.2. *Tissue sectioning and penetration of immunoreagents*

The Vibratome® (Oxford Instruments) is the most suitable instrument for cutting tissue slices prior to immunostaining because it will produce thicknesses ranging from 20 μm to several hundred microns with minimal damage to ultrastructure. Frozen tissue sectioning procedures such as those involving a cryostat or freezing microtome can be used [31, 32] but generally produce an unacceptable amount of ultrastructural damage [5].

It is widely recognised that penetration of immunoreagents into well fixed Vibratome® sections is slight, although few specific publications exist in which an attempt has been made to quantify the degree of penetration. Recently, in a study of neurophysin staining, Piekut and Casey [41] reported that immunoreagents penetrated only the superficial 8–9 μm of a 80 μm thick Vibratome® section of brain tissue fixed in 2% paraformaldehyde, 1% glutaraldehyde and 1% picric acid. In the above study the primary antibodies were diluted in buffer containing 0.04%–0.4% Triton X-100. Without the addition of Triton the penetration of the immunoreagents would undoubtedly have been much less. Detergents are commonly used to enhance penetration and can be applied at various stages in the staining. However, Triton treatment causes destruction of membranes and so many workers prefer to avoid detergents altogether and instead use other techniques for enhancing penetration. There is also some evidence that saponin causes much less ultrastructural damage than does Triton [42], however, as yet few electron immunocytochemical studies have been published employing saponin treatment. Ironically, it may not be possible to get adequate immunostaining of certain membrane-bound antigens without using detergents and in this situation a short preincubation in Triton [3] or saponin can be used.

A method which has been successfully used by several research groups to enhance penetration without recourse to detergents is to give small cryoprotected tissue blocks a rapid freeze-thaw [4, 5, 22, 35]. Vibratome® sections are then cut from the blocks. This procedure appears to produce microfractures in the tissue and thus facilitates antibody penetration but without seriously compromising membrane morphology. In our hands [5] the freeze-thaw approach still does not allow antibody penetration through the whole thickness of a 40 μm Vibratome® section, but the degree of penetration is much greater than into sections which have not been subjected to the freeze-thaw treatment. Another approach which has been used to enhance penetration is to dehydrate partially and then rehydrate sections in a graded alcohol series prior to immunostaining [27, 43], however, this approach produces rather indistinct membrane morphology [43].

2.3. Immunostaining

The PAP immunostaining procedure used for electron immunocytochemistry is not significantly different from that employed for light microscopical immunohistochemistry and so will not be dealt with in in this review. Full details of the PAP procedure can be found in the book by Sternberger [44]. However, a few general principles which differ from light microscopical immunohistochemistry should be noted. As indicated in section 2.2, detergents are usually omitted in electron immunocytochemistry or are only applied sparingly (for example, in a brief pre-incubation stage). In addition, long incubation times should be avoided because of the gradual ultrastructural deterioration which occurs in buffer solutions. We normally try to complete all the stages from perfusion to resin embedding within 2 days [5]. It is difficult to accomplish this in less time because steps such as infiltration in 30% sucrose (see section 2.2 above), osmication and dehydration all require several hours. This time schedule also includes incubation in the primary antibody for about 10 h and incubation in link antibody and PAP reagent for 1–1.5 h each (see Appendix). Penetration of immunoreagents is enhanced if sections are continuously agitated during incubation. Although a large number of chromogens are available for localisation of peroxidase in light microscopical immunohistochemistry [45–47], most are unsuitable for electron immunocytochemical work either because they are soluble in alcohols or because they are not osmiophilic or because they do not give a sufficiently discrete localisation at the ultrastructural level. The only chromogen to have been widely used for electron immunocytochemistry is 3,3′-diaminobenzidine. This reagent is thought to be carcinogenic and should be handled with extreme caution.

2.4. *Electron microscopical processing*

Following PAP immunostaining any standard osmication, dehydration and plastic embedding procedure can be used. However, several research groups have found it useful to adopt a two stage flat embedding procedure which allows direct correlation between light- and electron microscopical observations [5, 25, 48]. The procedure is described in some detail by Priestley and Cuello [5] and an application is illustrated in Figure 3. The correlative approach has two main advantages. Firstly, it allows large areas of tissue to be scanned in the light microscope and then suitable areas may be selected for electron microscopical analysis. This is particularly important if electron immunocytochemistry is being combined with some other experimental procedure (see section 4). It also means that full benefit can be taken of the pre-embedding procedure. Large areas of tissue can be processed and kept flat embedded until required. The second advantage of the two stage embedding approach is that it facilitates handling of the tissue in the ultramicrotome. The limited antibody penetration obtained with Vibratome® tissue slices (see sections 2.1 and 2.2 above) means that only a narrow depth of the tissue will have both suitable immunostaining and ultrastructure and so it is very important to know the distribution of immunostaining in the specimen before commencing trimming and before semithin or ultrathin sectioning. The surface of slices usually exhibits very poor ultrastructure and non-specific staining and may therefore have to be discarded. Depth of reaction product deposition in the tissue needs to be continually monitored during thin sectioning and to this end it is useful if the tissue block can be illuminated through the microtome chuck using a system such as that described by Peters for correlative Golgi electron microscopical studies [49]. This allows observation of the tissue block using a light microscope without having to remove the block from the chuck and therefore without having to align the block before recommencing sectioning. Another approach that we have found useful is to mount serial sections on to Formvar-coated single slot grids and to contrast a selection of these grids with lead citrate. In the uncontrasted grids the immunostaining can be seen clearly while in the contrasted series more details of the membrane can be observed [5, 13] (Fig. 4). *En bloc* contrasting with uranyl acetate can also be used if required.

3. Interpretation of results

As with any experimental procedure the value of immunocytochemistry depends as much on correct interpretation of the results as it does on achieving them. Crucial questions of interpretation in immunocytochemistry centre on the specificity of the staining and the nature of the tissue antigen which is being localised. These questions are common to both light and electron microscopical immunocytochemistry and are thoroughly dealt with in several recent reviews [50–54]. However, there are other problems of interpretation which are peculiar to pre-embedding immunocytochemistry and which any potential worker should be aware of. These

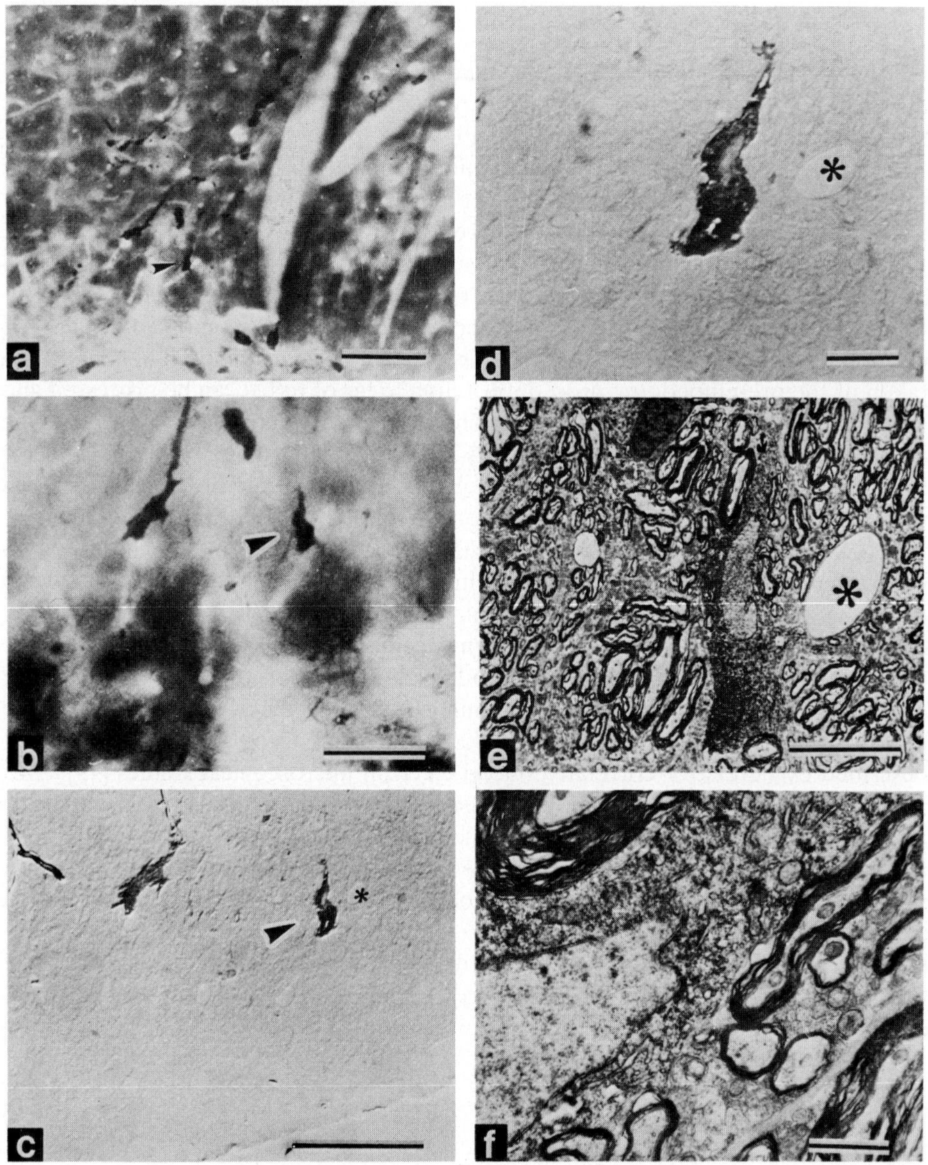

Fig. 3. Localisation of 5-hydroxytryptamine immunostained neurons in nucleus reticularis paragigantocellularis. Use of a two-stage flat embedding procedure allows a single cell (arrow) situated close to a blood vessel (asterisk) to be examined at both light (a–d) and electron (e, f) microscopical levels. (a, b) Immunostained cells in the 40 μm thick slices embedded in plastic on glass microscope slides. (c, d) The same cells as they appear in a 2 μm thick semi-thin section cut on the ultramicrotome following re-embedding of the 40 μm thick slices into electron microscope embedding capsules. Interference contrast illumination. (e, f) One of the cells photographed in the electron microscope. Diffuse reaction product fills the cytoplasm and there is heavy staining of membranes. Scale bars = 100 μm (a), 50 μm (b, c), 10 μm (d, e), 1 μm (f). Reproduced from Priestley and Cuello [5] with permission.

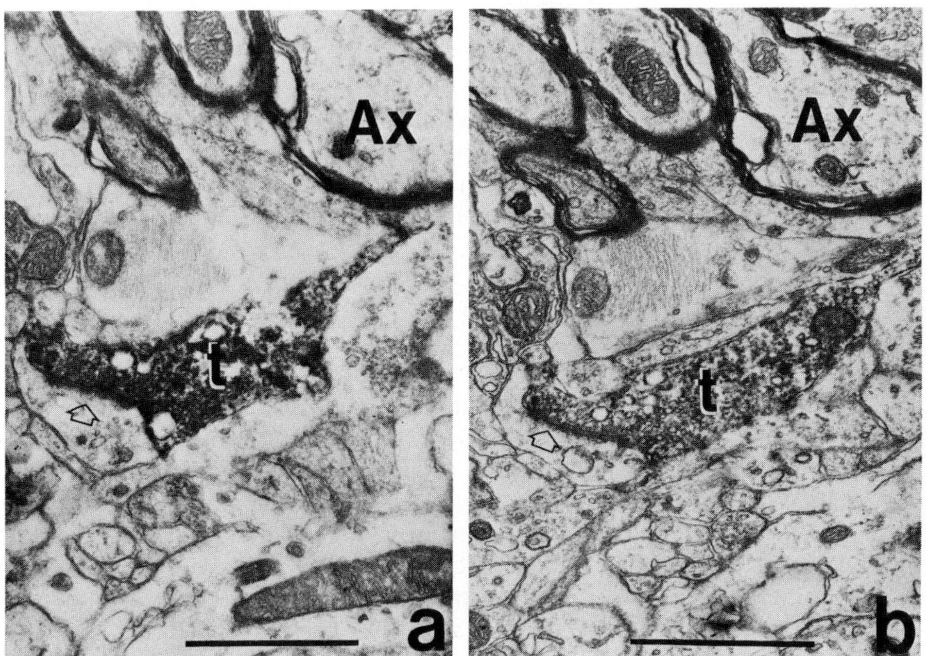

Fig. 4. A substance P immunostained terminal (t) observed in serial uncontrasted (a) and contrasted (b) sections. The terminal synapses (arrow head) with a small dendrite. Ax indicates myelinated axon. Scale bars = 1 μm. Reproduced from Priestley et al. [13] with permission.

problems are dealt with at some length in an earlier review [5] and so only one, but nevertheless central, problem will be discussed here. This problem is the significance of the subcellular localisation of the immunoreaction deposit. In some early pre-embedding studies on transmitter-related CNS antigens the ultrastructural localisation of the reaction deposit was assumed to be representative of the in vivo distribution of the antigens [1, 2] and this assumption is often implicit in the discussion sections of more recent papers. However, although the localisation of the oxidised diaminobenzidine (ODAB) reaction product may authentically reflect the distribution of the antigen this is not necessarily the case and cannot be assumed to be so. In pre-embedding immunocytochemistry diffusion and relocation of some antigens within the cell takes place during fixation [5] and subsequently diffusion and relocation of the ODAB reaction deposit may also occur [55–57]. Diffusion of antigen during fixation is a particular problem with cytoplasmic antigens. However, it is also a problem in situations in which the antigen is associated with a particular organelle but is not firmly bound to the organelle or does not have appropriate residues for interaction with the fixative. Reaction product relocation takes place because the ODAB polymer will precipitate on conveniently situated membranes. There are many published micrographs showing a localisation of reaction deposit which does not agree with what is known about the physiological localisation of the particular antigen. For example, the cytoplasmic enzyme glutamic acid decarboxylase (GAD) appears to be membrane bound [4, 22], the putative transmitters 5-hydroxytryptamine (5-HT) and enkephalin are observed in nuclei [5, 14, 58] and staining for neuropeptides is frequently seen on mitochondrial membranes [5] (see Fig. 2). Before firm conclusions are drawn on the basis of electron immunocytochemistry, results should be compared with those obtained using other techniques in order to determine whether the subcellular localisation of the immunostaining is likely to be artifactual.

4. Combined procedures

As long as care is exercised in interpretation (see section 3 above), pre-embedding electron immunocytochemistry can provide information about the ultrastructural distribution of an antigen and there are good illustrations of this in the literature. For example, Broadwell and colleagues used electron immunocytochemistry to examine the organelles associated with neurophysin synthesis and packaging [11]. In CNS studies pre-embedding immunocytochemistry has also been used as a tool for exploring the synaptic circuitry of transmitter-characterised neurons [13, 14, 37]. More recently, this latter type of study has been greatly facilitated by the development of techniques in which pre-embedding procedures are combined with other marking techniques. These combined techniques fall into two main classes. The first class involves the development of procedures for simultaneously staining two different antigens. Cuello, Milstein and colleagues have developed techniques for immunocytochemistry using radioactively labelled monoclonal antibodies [6, 59–61] and we have combined this approach (termed radioimmunocytochemistry) with PAP immunocytochemistry to allow double antigen pre-embedding staining at the electron microscopical level [6, 58, 62]. At present no other technique is available for double staining using the pre-embedding procedure. The other major class of combined technique involves using electron immunocytochemistry together with another marking procedure. For example, pre-embedding immunostaining has been combined with anterograde degeneration [31, 39], anterograde autoradiography [63], transmitter uptake autoradiography [64], Golgi impregnation [36] and retrograde peroxidase tracing [65, 66]. Figure 5 shows an example of this last approach used to identify immunostained peptide-containing terminals in synaptic contact with identified projection neurons. The combined techniques are quite complicated, however, they provide an additional level of information and can give solutions to problems which are inaccessible to electron immunocytochemistry when used alone.

5. Conclusion

This chapter has attempted to review some of the variables in pre-embedding immunocytochemistry and has highlighted both difficulties and advantages in the

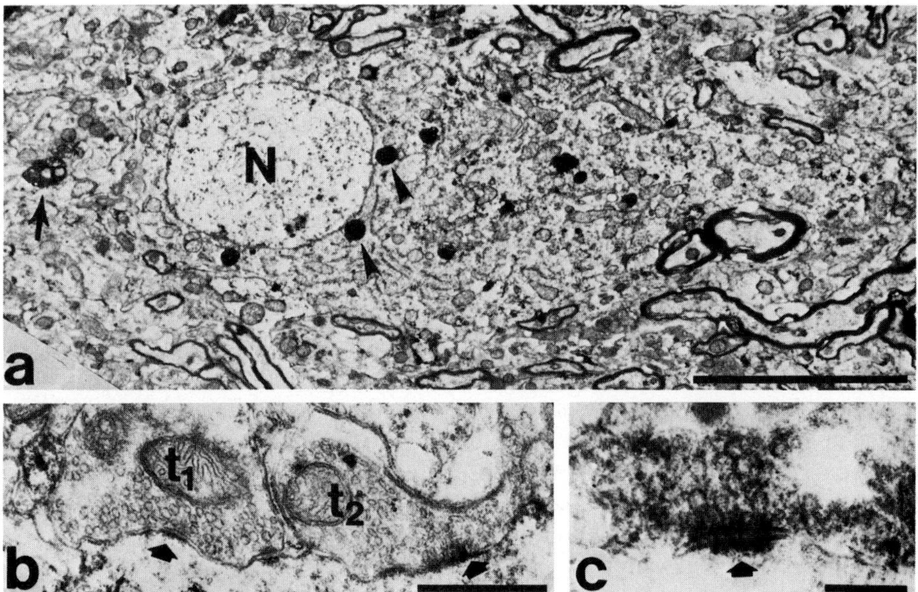

Fig. 5. Substance P immunostaining combined with horseradish peroxidase retrograde transport. (a) A retrogradely labelled trigemino-thalamic projection neuron (N) is shown in the same field as a substance P immunostained profile (arrow). The neuron is identified as retrogradely labelled by the presence of numerous dense bodies (arrow heads). Scale bar = 5 μm. (b) The retrogradely labelled neuron shown in (a) receives synapses (arrow heads) from two unstained axon terminals (t_1, t_2). Scale bar = 0.5 μm. (c) The neuron also receives a synapse (arrow head) from an immunostained terminal. Scale bar = 0.2 μm. Reproduced from Priestley and Cuello [66] with permission.

procedure. Although emphasis has been given to transmitter-related antigens, similar principles apply to any biological system and hopefully this review will prove useful to workers interested in areas other than the CNS. It is not easy to obtain both good ultrastructure and immunostaining, however, the technique is potentially very powerful and is well worth the time invested in it. For CNS transmitter studies no other technique can provide comparable membrane preservation and resolution at the time of writing. Further methodological improvements are anticipated in the coming years and pre-embedding immunocytochemistry seems likely to continue to be an important tool for ultrastructural immunocytochemistry.

6. Appendix

6.1. Protocol for pre-embedding immunocytochemistry

As indicated in the main body of this chapter, experimental details in pre-embedding immunocytochemistry can vary widely and are largely dependent on the particular antigen, primary antibody and biological system under study. However, we have found the following general protocol useful for localising a number of different antigens and it can be used as a general guide in planning an experiment. Antigens successfully localised by us using this procedure include substance P [13, 25, 58, 66], enkephalin [6, 37, 67], 5-hydroxytryptamine [58] and choline acetyltransferase (Connaughton and colleagues, unpublished).

Day 1. Perfuse through the ascending aorta [40] with 4% paraformaldehyde and 0.2% glutaraldehyde in 0.1 M phosphate buffer (pH 7.4). For a rat we first perfuse with 20 ml phosphate buffer followed by 200 ml of the fixative. Dissect out the tissue of interest and leave in fixative for a further 3 h. Cut tissue into blocks, transfer to phosphate buffer containing 30% sucrose and leave until fully infiltrated (6–8 h). Rapidly freeze blocks by plunging into liquid nitrogen, thaw blocks by returning them to 30% sucrose and then transfer them to Tris/phosphate buffer [47]. Cut 40 μm thick slices on a Vibratome®. Incubate slices in 10% normal serum (e.g. goat) for 30 min, wash in Tris/phosphate buffer, incubate in primary antibody (e.g. rabbit) overnight. All antisera are diluted in Tris/phosphate buffer containing 0.7% lambda carrageenan [47] and incubations are carried out at 4°C.

Day 2. Wash slices in Tris/phosphate, incubate in link antibody for 1 h (e.g. 1:10 goat anti-rabbit IgG), wash, incubate in PAP reagent for 1.5 h (e.g. 1:50 rabbit PAP), wash. Stain slices using 3,3′-diaminobenzidine (0.06%) in 0.05 M Tris-HCl buffer pH 7.6, room temperature for 10 min. Add hydrogen peroxide to a final volume of 0.01%. Terminate reaction (after 5–10 min) by replacing the diaminobenzidine-hydrogen peroxide solution with Tris/phosphate. Wash sections thoroughly, stain for 1.5 h with 1% osmium tetroxide, dehydrate and bring sections to resin. For flat embedding, Durcupan is recommended because the polymerised resin will easily lift off glass microscope slides (see below).

Day 3. Place slices in a drop of Durcupan on a warmed glass microscope slide. Cover with a plastic cover slip. Polymerise resin (56°C, 2 days).

The slide-mounted sections can be examined in a light microscope. Subsequently the slide is warmed on a hot plate and areas of interest cut from the resin embedded tissue. Such areas can be mounted in an electron microscope embedding capsule using fresh resin [5] or can simply be glued onto a polymerised blank capsule using a rapid setting epoxy adhesive. Thin sections are cut for electron microscopy using standard procedures (see section 2.4).

7. Acknowledgements

J. V. Priestley is a Beit Memorial Research Fellow. The work for this chapter was carried out in collaboration with Dr. A. C. Cuello and was supported by grants to ACC from the Medical Research Council (U.K.), the Wellcome Trust and the E. P. Abraham Cephalosporin Trust (Oxford). ACC is thanked for advice, encouragement and generous provision of facilities. Thanks to M. A. Connaughton for rapid and expert reading of the manuscript.

8. References

1. Pickel, V.M., Joh, T.H. and Reis, D.J. (1975) Ultrastructural localisation of tyrosine hydroxylase in noradrenergic neurons of brain. Proc. Natl. Acad. Sci. U.S.A. 72, 659–663.
2. Pickel, V.M., Joh, T.H. and Reis, D.J. (1976) Monoamine-synthesising enzymes in central dopaminergic, noradrenergic and serotonergic neurons. Immunocytochemical localisation by light and electron microscopy. J. Histochem. Cytochem. 24, 792–806.
3. Pickel, V.M. (1981) Immunocytochemical methods. In: L. Heimer and M.J. Robards (Eds.) Neuroanatomical Tract Tracing Methods. Plenum Press, New York, pp. 483–509.
4. Vaughn, J.E., Barber, R.P., Ribak, C.E. and Houser, C.R. (1981) Methods for the immunocytochemical localisation of proteins and peptides involved in neurotransmission. In: J.E. Johnson (Ed.) Current Trends in Morphological Techniques, Vol. III. CRC Press, Florida, pp. 33–70.
5. Priestley, J.V. and Cuello, A.C. (1983) Electron microscopic immunocytochemistry for CNS transmitters and transmitter markers. In: A.C. Cuello (Ed.) Immunohistochemistry. John Wiley, Chichester, pp. 273–321.
6. Cuello, A.C., Priestley, J.V. and Milstein, C. (1982) Immunocytochemistry with internally labelled monoclonal antibodies. Proc. Natl. Acad. Sci. USA 79, 665–669.
7. Kraehenbuhl, J.P. and Jamieson, J.D. (1974) Localisation of intracellular antigens by immunoelectron microscopy. Int. Rev. Exp. Pathol. 13, 1–53.
8. Larsson, L.I. (1981) A novel immunocytochemical model for specificity and sensitivity screening of antisera against multiple antigens. J. Histochem. Cytochem. 29, 408–410.
9. Schipper, J. and Tilders, F.J.H. (1983) A new technique for studying specificity of immunocytochemical procedures: specificity of serotonin immunostaining. J. Histochem. Cytochem. 31, 12–18.
10. Pickel, V.M., Reis, D.J. and Leeman, S.E. (1977) Ultrastructural localisation of substance P in neurons of rat spinal cord. Brain Res. 122, 534–540.
11. Broadwell, R.D., Oliver, C. and Brightman, M.W. (1979) Localization of neurophysin within organelles associated with protein synthesis and packaging in the hypothalamo-neurohypophysial system: An immunocytochemical study. Proc. Natl. Acad. Sci. USA 76, 5999–6003.
12. Buijs, R.M. and Swaab, D.F. (1979) Immuno-electron microscopical demonstration of vasopressin and oxytocin synapses in the limbic system of the rat. Cell Tiss. Res. 204, 355–365.
13. Priestley, J.V., Somogyi, P. and Cuello, A.C. (1982) Immunocytochemical localisation of substance P in the spinal trigeminal nucleus of the rat: a light and electron microscopic study. J. Comp. Neurol. 211, 31–49.
14. Pickel, V.M., Sumal, K.K., Beckley, S.C., Miller, R.J. and Reis, D.J. (1980) Immunocytochemical localisation of enkephalin in the neo-striatum of rat brain: a light and electron microscopic study. J. Comp. Neurol. 189, 721–740.
15. Johansson, O., Hökfelt, T., Jeffcoate, N., White, N. and Sternberger, L.A. (1980) Ultrastructural localisation of TRH-like immunoreactivity. Exp. Brain Res. 38, 1–10.
16. DiFiglia, M. and Aronin, N. (1982) Ultrastructural features of immunoreactive somatostatin neurons in the rat caudate nucleus. J. Neurosci. 2, 1267–1274.
17. Hendry, S.H.C., Jones, E.G. and Beinfeld, M.C. (1983) Cholecystokinin-immunoreactive neurons in rat and monkey cerebral cortex make symmetric synapses and have intimate associations with blood vessels. Proc. Natl. Acad. Sci. USA 80, 2400–2404.
18. Leranth, C., Williams, T.H., Hamori, J. and Chretien, M. (1981) Light and electron microscopic immunocytochemical localization of adrenocorticotrophin-like activity in rat cerebellar and red nuclei. Neuroscience 6, 481–487.
19. Pelletier, G., Leclerc, R., Puviani, R. and Polak, J.M. (1981) Electron immunocytochemistry of vasoactive intestinal peptide (VIP) in the rat brain. Brain Res. 210, 356–360.
20. Ruda, M.A., Coffield, J. and Steinbusch, H.W.M. (1982) Immunocytochemical analysis of serotonergic axons in laminae I and II of the lumbar spinal cord of the cat. J. Neurosci. 2, 1660–1671.
21. Pickel, V.M., Joh, T.H., Reis, D.J., Leeman, S.E. and Miller, R.J. (1979) Electron microscopic

localisation of substance P and enkephalin in axon terminals related to dendrites of catecholaminergic neurons. Brain Res. 160, 387–400.

22. Ribak, C.E., Vaughn, J.E. and Barber, R.P. (1981) Immunocytochemical localisation of GABAergic neurons at the electron microscopic level. Histochem. J. 13, 555–582.
23. Houser, C.R., Crawford, G.D., Barber, R.P., Salvaterra, P.M. and Vaughn, J.E. (1983) Organization and morphological characteristics of cholinergic neurons: an immunocytochemical study with a monoclonal antibody to choline acetyltransferase. Brain Res. 266, 97–119.
24. Chan-Palay, V. and Palay, S.L. (1979) Immunocytochemical localization of cyclic GMP: Light and electron microscope evidence for involvement of neuroglia. Proc. Natl. Acad. Sci. USA 76, 1485–1488.
25. Somogyi, P., Priestley, J.V., Cuello, A.C., Smith, A.D. and Bolam, J.P. (1982) Synaptic connections of substance P immunoreactive terminals in the substantia nigra of the rat: a correlated light and electron microscopic study. Cell Tissue Res. 223, 469–486.
26. Berod, A., Hartman, B.K. and Pujol, J.F. (1981) Importance of fixation in immunohistochemistry: Use of formaldehyde solutions at variable pH for the localization of tyrosine hydroxylase. J. Histochem. Cytochem. 29, 844–850.
27. Eldred, W.D., Zucker, C., Karten, H.J. and Yazulla, S. (1983) Comparison of fixation and penetration enhancement techniques for use in ultrastructural immunocytochemistry. J. Histochem. Cytochem. 31, 285–292.
28. Weber, K., Rathke, P.C. and Osborn M. (1978) Cytoplasmic microtubular images in glutaraldehyde-fixed tissue culture cells by electron microscopy and by immunofluorescence microscopy. Proc. Natl. Acad. Sci. USA 75, 1820–1824.
29. Willingham, M.C. (1983) An alternative fixation-processing method for preembedding ultrastructural immunocytochemistry of cytoplasmic antigens: The GBS (glutaraldehyde-borohydride-saponin) procedure. J. Histochem. Cytochem. 31, 791–798.
30. King, J.C., Lechan, R.M., Kugel, G. and Anthony, E.L.P. (1983) Acrolein: A fixative for immunocytochemical localization of peptides in the central nervous system. J. Histochem. Cytochem. 31, 62–68.
31. Hunt, S.P., Kelly, J.S. and Emson, P.C. (1980) The electron microscopic localisation of methionine-enkephalin within the superficial layers (I and II) of the spinal cord. Neuroscience 5, 1871–1890.
32. Ninkovic, M., Hunt, S.P. and Kelly, J.S. (1981) Effect of dorsal rhizotomy on the autoradiographic distribution of opiate and neurotensin receptors and neurotensin-like immunoreactivity within the rat spinal cord. Brain Res. 230, 111–119.
33. McLean, I.W. and Nakane, P.K. (1974) Periodate-lysine-paraformaldehyde fixative. A new fixative for immuno-electron microscopy. J. Histochem. Cytochem. 22, 1077–1083.
34. Zamboni, L. and de Martino, C. (1967) Buffered picric-acid formaldehyde: a new rapid fixative for electron-microscopy. J. Cell Biol. 35, 148A.
35. Somogyi, P. and Takagi, H. (1982) A note on the use of picric acid-paraformaldehyde-glutaraldehyde fixative for correlated light and electron microscopic immunocytochemistry. Neuroscience 7, 1779–1783.
36. Somogyi, P., Freund, T.F., Wu, J.-Y. and Smith, A.D. (1983) The section-Golgi impregnation procedure. 2. Immunocytochemical demonstration of glutamate decarboxylase in Golgi-impregnated neurons and in their afferent synaptic boutons in the visual cortex of the cat. Neuroscience 9, 475–490.
37. Somogyi, P., Priestley, J.V., Cuello, A.C., Smith, A.D. and Takagi, H. (1982) Synaptic connections of enkephalin-immunoreactive nerve terminals in the neostriatum: a correlated light and electron microscopic study. J. Neurocytol. 11, 779–897.
38. Kiss, J.Z. and Williams, T.H. (1983) ACTH-immunoreactive boutons form synaptic contacts in the hypothalamic arcuate nucleus of rat: evidence for local opiocortin connections. Brain Res. 263, 142–146.
39. Leranth, C.S. and Frotscher, M. (1983) Commissural afferents to the rat hippocampus terminate on VIP-like immunoreactive non-pyramidal neurons. An EM immunocytochemical degenerative study. Brain Res. 276, 357–361.

40. Friedrich, V.L. and Mugnaini, E. (1981) Electron microscopy: Preparation of neural tissues for electron microscopy. In: L. Heimer and M.J. Robards (Eds). Neuroanatomical Tract Tracing Methods. Plenum Press, New York, pp. 345–375.
41. Piekut, D.T. and Casey, S.M. (1983) Penetration of immunoreagents in vibratome-sectioned brain: a light and electron microscope study. J. Histochem. Cytochem. 31, 669–674.
42. Ohtsuki, I., Manzi, R.M., Palade, G.E. and Jamieson, J.D. (1978) Entry of macromolecular tracers into cells fixed with low concentrations of aldehydes. Biol. Cell 31, 119–126.
43. Light, A.R., Kavookjian, A.M. and Petrusz, P. (1983) The ultrastructure and synaptic connections of serotonin-immunoreactive terminals in spinal laminae I and II. Somatosensory Res. 1, 33–50.
44. Sternberger, L.A. (1979) Immunocytochemistry, 2nd Edn. John Wiley, New York.
45. Nakane, P.K. (1968) Simultaneous localisation of multiple tissue antigens using the peroxidase-labelled antibody method: a study on pituitary glands of the rat. J. Histochem. Cytochem. 16, 557–560.
46. Vandesande, F., Dierickx, K. and De Mey, J. (1977) The origin of the vasopressinergic and oxytocinergic fibres of the external regions of the median eminence of the rat hypothalamus. Cell Tissue Res 180, 443–452.
47. Sofroniew, M.V. and Schrell, U. (1982) Long-term storage and regular repeated use of diluted antisera in glass staining jars for increased sensitivity, reproducibility, and convenience of single- and two-color light microscopic immunocytochemistry. J. Histochem. Cytochem. 30, 504–511.
48. Chan-Palay, V. and Palay, S.L. (1977) Ultrastructural identification of substance P cells and their processes in rat sensory ganglia and their terminals in the spinal cord by immunocytochemistry. Proc. Natl. Acad. Sci. USA 74, 4050–4054.
49. Peters, A. (1981) The Golgi-electron microscope technique. In: J.E. Johnson (Ed.) Current Trends in Morphological Techniques, Vol I. CRC Press, Florida, pp. 187–212.
50. Swaab, D.F., Pool, C.W. and van Leeuwen, F.W. (1977) Can specificity ever be proved in immunocytochemical staining? J. Histochem. Cytochem. 25, 388–391.
51. Hutson, J.C., Childs, G.V. and Gardner, P.J. (1979) Considerations for establishing the validity of immunocytochemical studies. J. Histochem. Cytochem. 27, 1201–1202.
52. Pool, C.W., Buijs, R.M., Swaab, D.F., Boer, G.J. and van Leeuwen, F.W. (1983) On the way to a specific immunocytochemical localization. In: A.C. Cuello (Ed.) Immunohistochemistry. John Wiley, Chichester, pp. 1–46.
53. Milstein, C., Wright, B. and Cuello, A.C. (1983) The discrepancy between the cross-reactivity of a monoclonal antibody to serotonin and its immunohistochemical specificity. Mol. Immunol. 20, 113–123.
54. Sofroniew, M.V., Couture, R. and Cuello, A.C. (1983) Immunocytochemistry: preparation of antibodies and staining specificity. In: A. Björklund and T. Hökfelt (Eds.) Handbook of Chemical Neuroanatomy, Vol. 1: Methods in Chemical Neuroanatomy. Elsevier, Amsterdam, pp. 210–227.
55. Novikoff, A.B., Novikoff, P.M., Quintana, N. and Davis, C. (1972) Diffusion artifacts in 3,3′-diaminobenzidine cytochemistry. J. Histochem. Cytochem. 27, 1438–1444.
56. Novikoff, A.B. (1980) DAB cytochemistry: artifact problems in its current uses. J. Histochem. Cytochem. 28, 1036–1038.
57. Dourmashkin, R., Patterson, S., Shah, D. and Oxford, J.S. (1982) Evidence of diffusion artefacts in diaminobenzidine immunocytochemistry revealed during immune electron microscope studies of the early interactions between influenza virus and cells. J. Virol. Meth. 5, 27–34.
58. Priestley, J.V. and Cuello, A.C. (1983) Studies on the cellular and subcellular immunocytochemical localization of substance P and 5-HT in raphe neurons. In: V. Chan-Palay and S.L. Palay (Eds.) Coexistence of Neuroactive Substances. John Wiley, New York (In press).
59. Cuello, A.C., Milstein, C. and Priestley, J.V. (1980) Use of monoclonal antibodies in immunocytochemistry with special reference to the central nervous system. Brain Res. Bull. 5, 575–587.
60. Cuello, A.C., Milstein, C. and Galfre, G. (1983) Immunocytochemistry with monoclonal antibodies. In: A.C. Cuello (Ed.) Immunohistochemistry. John Wiley, Chichester, pp.215–255.
61. MacMillan, F.M., Sofroniew, M.V., Sidebottom, E. and Cuello, A.C. (1984) Immunocytoche-

mistry with monoclonal anti-immunoglobulin as a developing agent: application to immunoperoxidase staining and radioimmunocytochemistry. J. Histochem. Cytochem. 32, 76–82.

62. Priestley, J.V. and Cuello, A.C. (1982) Co-existence of neuroactive substances as revealed by immunohistochemistry with monoclonal antibodies. In: A.C. Cuello (Ed.) Co-Transmission. Macmillan Press, London, pp. 165–188.
63. Sumal, K.K., Blessing, W.W., Joh, T.H., Reis, D.J. and Pickel, V.M. (1983) Synaptic interaction of vagal afferents and catecholaminergic neurons in the rat nucleus tractus solitarius. Brain Res. 277, 31–40.
64. Basbaum, A.I., Glazer, E.J. and Lord, B.A.P. (1982) Simultaneous ultrastructural localization of tritiated serotonin and immunoreactive peptides. J. Histochem. Cytochem. 30, 780–784.
65. Ruda, M.A. (1982) Opiates and pain pathways: demonstration of enkephalin synapses on dorsal horn projection neurons. Science 215, 1523–1525.
66. Priestley, J.V. and Cuello, A.C. (1983) Substance P immunoreactive terminals in the spinal trigeminal nucleus synapse with lamina I neurons projecting to the thalamus. In: P. Skrabanek and D. Powell (Eds.) Substance P Dublin 1983. Boole Press, Dublin, pp. 251–252.
67. Priestley, J.V. (1981) Ultrastructural localisation of substance P and enkephalin in the substantia gelatinosa of the spinal trigeminal nucleus. Br. J. Pharmacol. 74, 893P.

Immunolabelling for Electron Microscopy (Polak/Varndell, eds)

CHAPTER 5

Post-embedding immunoenzyme techniques

Geoffrey R. Newman[1] and Bharat Jasani[2]

[1]*University Hospital of Wales and* [2]*Welsh National School of Medicine, Heath Park, Cardiff, U.K.*

CONTENTS

1. Introduction

Pre- and post-embedding systems for immunoelectron microscopy have historically evolved almost in parallel. Pre-embedding systems require that tissue is immunostained before it is embedded and sectioned for the electron microscope. Early attempts understandably evolved from methods applied at the light microscope level which were simply extended to enable the results to be seen by transmission electron microscopy [1–3]. Overall, they provided very poor standards of ultrastructure. Sternberger's PAP technique [4], although a great immunocytochemical innovation, added to the problems of pre-embedding immunoelectron microscopy because it was felt that the massive PAP complex (large enough to be visible with the electron microscope!) would be unable to penetrate fresh or even lightly fixed tissue. Interest was therefore stimulated in post-embedding methods in which ultrathin tissue sections, usually mounted on nickel or gold grids, were immunostained. However, in early studies tissue which was prepared routinely for electron microscopy, whilst offering good ultrastructure with very suitable cell and organelle profiles in ultrathin sections, demonstrated little or no antigenic activity. Modern methods, which include cryoultramicrotomy, strive to find the ideal compromise between preserving tissue ultrastructure and retaining its antigenicity.

2. Why post-embedding?

2.1. Penetration

While it is preferable to expose tissue antigens to antibodies before they can be denatured or even eluted by processing and embedding, it remains difficult to guarantee adequate antibody penetration into fresh or fixed tissue and therefore it may only be possible to immunostain some antigens using post-embedding methods. This is in spite of a number of approaches which have been employed to deal with pre-embedding penetration problems. Tissues may have to be dehydrated and rehydrated [5], frozen and thawed and/or treated with various chemicals such as dimethyl sulphoxide, digitonin or detergents in order to disrupt membranes and effect the entry of antibodies [4]. Some of these techniques are necessarily very damaging to ultrastructure although when using glutaraldehyde-fixed cultured mouse fibroblasts, treatment with borohydride and saponin appears to allow the penetration of antibodies and ferritin into them whilst preserving good ultrastructure [6]. It was not clear to what depth in solid tissue ferritin or other immunoreagents could be expected to penetrate but generally only 1–3 μm is considered acceptable for pre-embedding methods [7]. These problems do not exist for post-embedding systems because antigens are exposed when cell and organelle membranes are cut open during ultrathin sectioning; however, the impenetrability of many embedding agents still makes etching essential (see section 2.3 and Chapter 3, pp. 29–36.

2.2. *Simple, rapid, reproducible technology*

At its simplest, post-embedding technology only requires that fixed tissue is dehydrated and embedded. Some embedding plastics are water soluble so that the tissue may not even need dehydration but other methods may require more complicated processing. For example, tissue can be snap-frozen, vapour-fixed and embedded by freeze substitution [8]. Tissue thus prepared is said to retain better antigenic reactivity [8, 9]. Some hydrophilic plastics allow 'cold cure' methods using ultraviolet light sources for their polymerisation [10] and others require anaerobic conditions [11]. In the main, however, tissue handling and processing is very uncomplicated. This contrasts markedly with other systems where the process of immunostaining becomes intricately involved in the fixation and processing of the tissue before its eventual ultramicrotomy. Except in the case of cell culture and monolayers, tissue has to be sectioned or sliced to facilitate immunostaining so that it may be necessary to use free-floating Vibratome ® sections, which are not easy to prepare at a constant thickness, or to use frozen [1], polyethylene glycol [3], or even paraffin-wax embedded [2] tissue. Free-floating sections are delicate and difficult to manoeuvre between solutions. They may be wrinkled when re-embedded in plastic for electron microscopical sectioning although embedding sections between coverslips [12] usually results in acceptably flat preparations, however, it must be remembered that only the surface (1–3 μm depth) will be worth sectioning (see Chapter 4, pp. 37–52). Sections adhering to glass-slides will be flatter but have only one surface through which immunostaining can occur, and this limits the penetration of immunoreagents, making results unpredictable and superficial. These sections also have to be re-embedded and reorientated for ultramicrotomy. The much simpler technology of post-embedding gives it great reliability and easy reproducibility.

2.3. *Controls*

Controls are of enormous importance in immunocytochemistry, particularly where a new antigen whose characteristics are vague, is being sought. With tissue which is embedded but not immunostained almost any of the many and varied controls can be employed on sections from the *same* tissue block. For example, the primary antibody specificity can be checked by absorption with pure extracted or, preferably, synthetic antigen, and method specificity can be checked by omitting the primary antibody or substituting it with an inappropriate antibody. Dilution profiles, whereby the effects of ranges of dilution of primary or secondary immunoreagents are examined, can be completed with reasonable ease, and all on the same piece of tissue. A useful guide to the specificity of controls in post-embedding immunocytochemistry is given by Childs [13]. Similar controls for pre-embedding methods, even if possible, are more laborious and of course have to be performed on *different* pieces of tissue. They often have to be regulated under the light microscope.

2.4. *Variety of immunolabelling methods*

Almost all the numerous immunostaining techniques can be used in post-embedding systems. Morphologically distinctive markers have been used such as haemocyanin [14] and ferritin [15, 16], although the last of these has been associated with non-specific binding to some of the commoner embedding plastics which has made it unpopular for post-embedding use [4]. The marker of choice for post-embedding immunocytochemistry has been colloidal gold, where particles can be accurately made to size and firmly linked to specific proteins [17–20]. Both ferritin and colloidal gold particles are important as markers for cell surface antigens but because of their relatively large particle sizes (ferritin 12 nm, colloidal gold 3–20 nm) and inflexibility, their penetrability into tissue is restricted and they have therefore been of limited use in pre-embedding methods. Nevertheless, both have been successfully used to localise intracellular antigens (see section 2.1) [6, 21]. Immunoperoxidase methods therefore, as well as being popular for post-embedding immunocytochemistry, dominate the pre-embedding methods, where the accent on diffusion favours the use of relatively small peroxidase conjugates (for example, those manufactured from Fab fragments) [22].

2.5. *Multiple antigen demonstration*

The wide choice of immunomarkers available for use on ultrathin sections means that more than one antigen can be localised on the same section. Different antigens may be shown by different markers or in the case of colloidal gold by different sizes of particle ([23, 24] see also Chapter 12, pp. 143–154 and Chapter 13, pp. 155–177). Post-embedding also allows the possibility of the same marker being used to identify different antigens in serial sections which can then be compared for sites of positivity. (Figs. 1 and 2).

3. Problems with post-embedding

3.1. *Fixatives*

Almost all methods use fixation of some sort, so any losses here could be considered a 'common denominator'. However, post-fixation in osmium, commonly used in pre-embedding methods after immunostaining, is generally avoided in post-embedding systems because of its adverse effect on antigenicity. This leaves tissue, lipids in particular, unprotected during processing and embedding. Therefore, to minimise tissue extraction, post-embedding systems usually require a greater strength of fixation. This is achieved by using more reactive aldehydes, higher concentrations of fixative, or longer times of fixation. Again a compromise has to be struck so that the cross-linking of proteins is minimised, leaving them recognisable to antibodies, and to this end a variety of aldehyde-based methods [25–28] and one non-aldehyde method [29] have been suggested. In any event it has

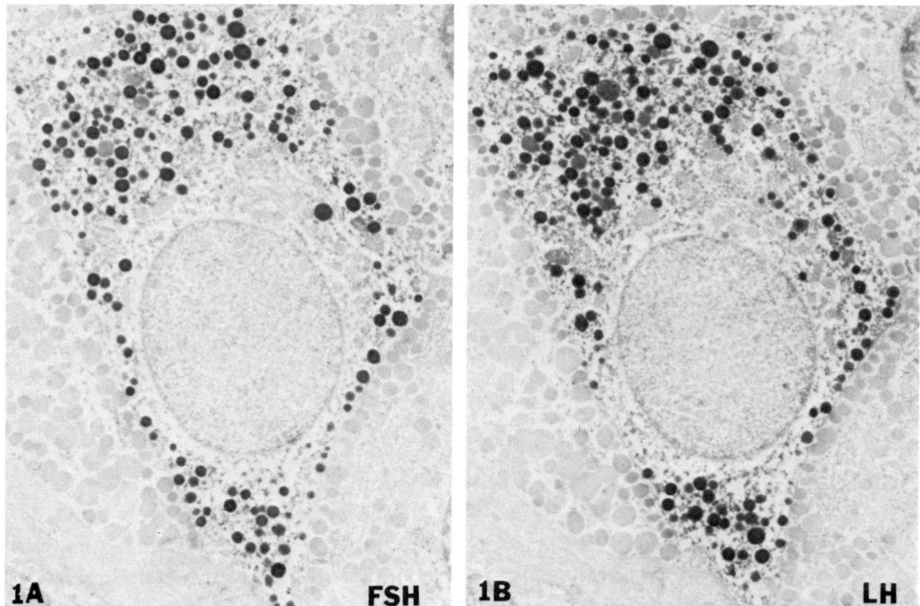

Fig. 1. BGPA immersion-fixed, 'LR White'-embedded human pituitary. Ultrathin serial sections have been immunostained with two different DNP-labelled antibodies, both at 1:800 final dilution (standard DHSS procedure, DAB/gold chloride; no counterstaining). (A) anti-FSH; magnification = ×4,500, (B) anti-LH; magnification = ×4,500. The same cell is positive with both antibodies.

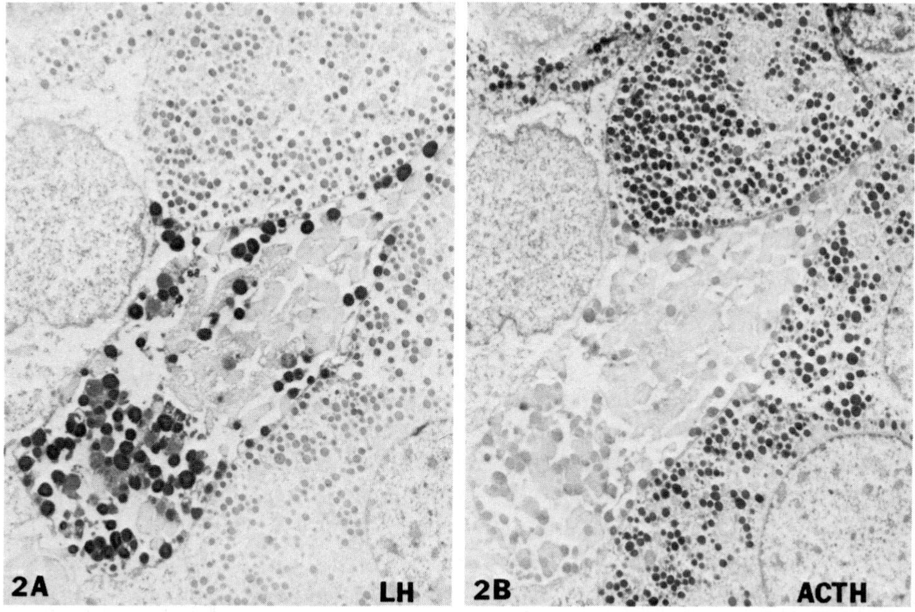

Fig. 2. BGPA immersion-fixed, 'LR White'-embedded human pituitary. Ultrathin serial sections have been immunostained with two different DNP-labelled antibodies, both at 1:800 final dilution (standard DHSS procedure, DAB/gold chloride; no counterstaining). (A) anti-LH; magnification = ×3,750, (B) anti-ACTH; magnification = ×3,750. Different cells are positive with the two antibodies used.

to be assumed that, because of the loss of and damage to tissue elements caused by the processing and embedding of unosmicated tissue, there will be an inevitable reduction in antigenicity.

3.2. Processing

In addition to the dehydration and clearing agents used in processing, epoxides and polyesters are themselves very lipophilic. Even when water-soluble forms are used there can be a gross disruption to membranes and cytosolic organelles. The net result of these agents and procedures is that the remaining ultrastructure can be quite poor, with the tissue retaining little contrast.

3.3. Plastics

Polyesters and epoxides are very tightly bonded and virtually impermeable to aqueous solutions at neutral pH. As a result, tissue antigens are isolated from their antibodies by a hydrophobic barrier. 'Etching' therefore becomes a prerequisite for antibody binding, and ultrathin sections are commonly exposed to 10% hydrogen peroxide for several minutes [4, 30]. Ironically, etching may also be deleterious to some antigens, for example myelin (Sternberger N., personal communication). Methacrylate embedding plastics have been used in attempts to overcome these problems, but they introduce problems of their own. Their erratic polymerisation and shrinkage is notorious and can give rise to severe polymerisation artefact. Unlike epoxides and polyesters, they are not stable in the electron beam, and they need support films and carbon coating. When used for light microscopy, they appear to allow the immunostaining of some antigens but not of others and etching with alcohol or benzene, or digestion with a protease enzyme have been reported to be necessary [4, 31].

4. Can post-embedding problems be overcome?

4.1. Plastics

In recent years a number of new resin embedding media has appeared, of which at least two have enabled distinct improvements to be made in post-embedding technology. The development of the 'Lowicryl' range (see Chapter 3, pp. 29–36) has made possible the embedding of tissue at temperatures of −30°C or below, where the denaturation and extraction of protein during dehydration is minimised [32]. The recently formulated acrylic resin 'LR White' (London Resin Company, Basingstoke, Hants., U.K.) has a number of advantageous characteristics and is simpler to use. It is not used at low temperatures, but like 'Lowicryl K4M', 'LR White' is hydrophilic, so that ultrathin sections permit full penetration of aqueous solutions. It allows the use of partially dehydrated tissue, and will fully infiltrate tissue from 70% ethyl alcohol and sometimes from much lower alcohol dilutions. Tissue thus prepared shows an improved antigenicity over its fully dehydrated counterpart [33, 34]. Its lipid solvency appears to be significantly lower than that of the epoxides and polyesters and therefore membranes and organelles remain for observation even when the tissue is not post-fixed in osmium. Importantly, the plastic itself does not have an avidity for antibodies nor does it prevent them from binding to tissue antigens. In experiments to date no 'etching' or protease digestion has been required to 'unmask' antigens even though some, such as immunoglobulins, do require protease digestion when processed as paraffin sections. Finally, it is stable in the electron beam and needs no supporting film on the grid, immunostaining can therefore take place through both sides of ultrathin sections.

4.2. Fixatives

Assuming that there is full penetrability of 'LR White' ultrathin sections by immunoreagents, a number of things follow. There must be a greater extent of antibody binding to tissue in 'LR White' than can occur in tissue embedded in hydrophobic plastics, because antigen throughout the thickness of 'LR White' sections will be available to antibody, not just that which is superficially exposed by etching. This increased immunoreactivity means that for the demonstration of antigen at many tissue sites, in particular those which have a high antigen content such as storage granules, the operator can afford a slight loss in antigenicity if the ultrastructure is better preserved as a result. Neutral buffered glutaraldehyde at 1–2% for 3–4 h will give far better ultrastructural preservation than can equivalent formaldehyde solutions. In the absence of post-fixation in osmium, the addition of a protein precipitant which does not impair antigenicity, such as picric acid, makes excellent standards of membrane and cytosolic preservation routine, whilst retaining good antigenic reactivity [33]. When possible, perfusion-fixation of buffered glutaraldehyde/picric acid (BGPA) will raise the standards of ultrastructure still further [34]. If, however, antigenicity is the priority then, of course, formaldehyde-based fixatives or formaldehyde/glutaraldehyde mixtures are strongly recommended.

4.3. Immunotechniques

4.3.1. Immunoperoxidase or colloidal gold?

To exploit fully the increased immunosensitivity of 'LR White' embedded tissue, it is necessary to use an immunomarker which is capable of diffusion into the plastic. The morphological and electron-dense markers such as ferritin or colloidal gold particles may reduce the penetrability of immunoreagents and then once again only superficial antigen will be localised. For this reason, immunoperoxidase methods are preferred, and a hapten sandwich technique has been adopted for use at the electron microscope level (see section 4.3.2). As in the vast majority of peroxidase methods, the oxidised polymer of 3,3′-diaminobenzidine (DAB) is precipitated at the site of the marker peroxidase, but osmium tetroxide need no longer be used to

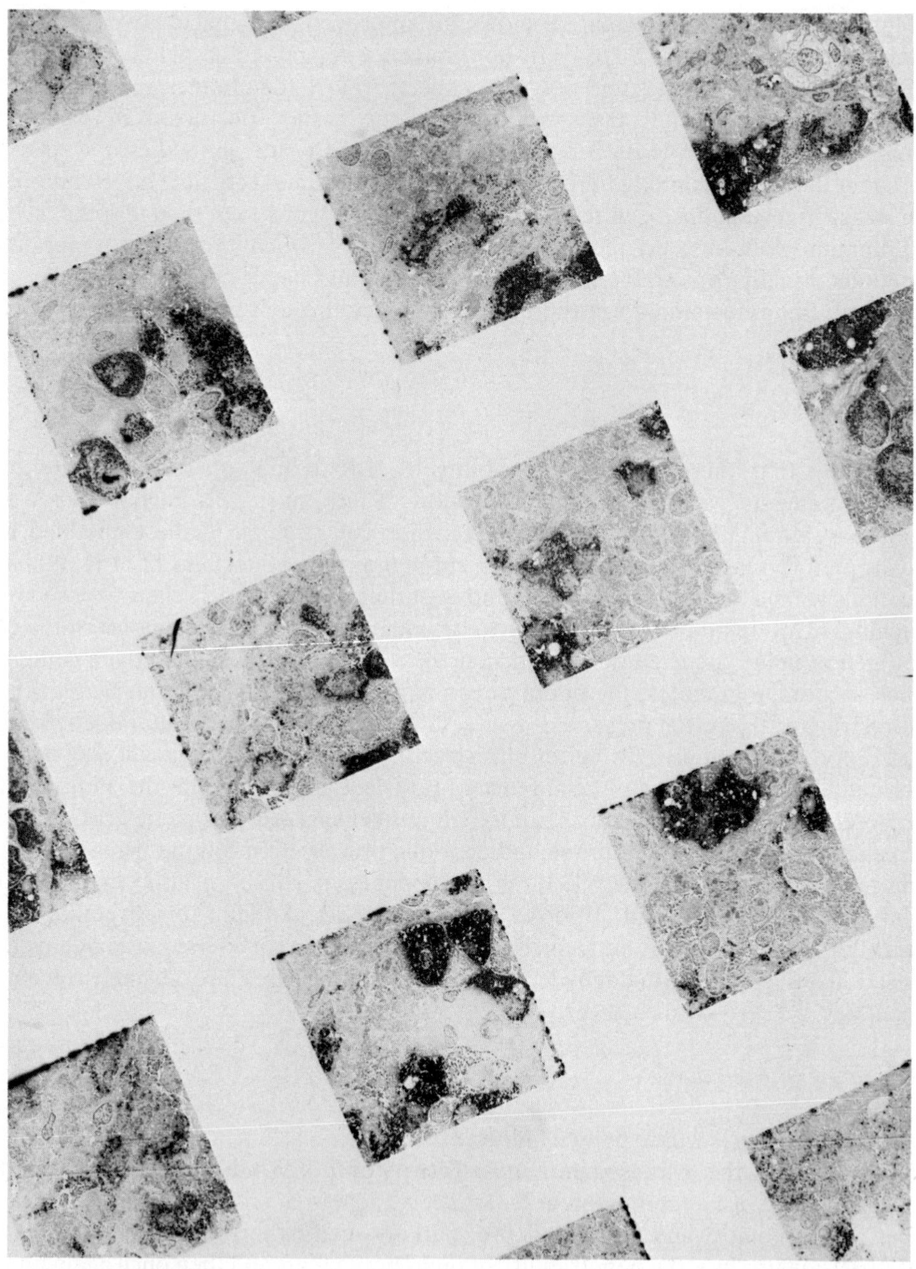

Fig. 3. BGPA immersion-fixed 'LR White'-embedded human pituitary. Ultrathin section immunostained using DNP-labelled ACTH antibody at 1:800 final dilution (standard DHSS procedure; DAB/gold chloride; no counterstaining). Magnification = ×400. Note easy identification of positive cells even at this very low magnification.

increase the electron density of the DAB for viewing under the electron microscope. Instead, 0.1% gold chloride is preferred, giving greater electron density than osmium and creating less background density. Gold chloride is also much more stable and safer to store and use than osmium tetroxide [35]. It thus becomes possible to fix, process, embed and immunoperoxidase-stain tissues, producing high standards of ultrastructure and antigenicity, without their ever having come into contact with osmium. Of course, colloidal gold techniques are perfectly usable with 'LR White' embedded tissue and, like peroxidase methods, can be applied directly to ultrathin sections without the need for etching. Indeed, the obvious difference between the amorphous electron-dense deposit produced by the immunoperoxidase method and the particulate colloidal gold left by immunogold methods makes possible the sequential localisation of two antigens in the same section (see section 2.5). A serious drawback with colloidal gold methods, however, is the difficulty of observing the tiny particles on counterstained ultrathin sections at low magnifications. The searching of sections, particularly where the localisation site of the antigen is unknown, becomes very tedious when it has to be conducted at high magnifications. No counterstaining is used after the peroxidase technique because the tissue usually has sufficient density without it and even at very low magnifications the sites of positivity are often quite obvious (Fig. 3).

4.3.2. Dinitrophenyl (DNP) Hapten Sandwich Staining (DHSS) procedure

A sensitive and versatile three-layer sandwich immunoperoxidase method has been developed and optimised for use under the electron microscope. It is capable of localising the specific tissue binding sites of a whole variety of DNP-labelled antibodies [36–38] and a number of DNP-labelled non-antibody ligands such as thyroid-stimulating hormone (TSH) (Fig. 4) [39] and the Fc component of immunoglobulin E (IgE-Fc) [40]. In a quantitative system under the light microscope, the antigen detection sensitivity of the DHSS procedure has been shown to be significantly greater than two already established three-layer sandwich immunoperoxidase methods, one employing peroxidase-antiperoxidase (PAP) complexes, and the other using avidin biotin (AB) based reagents [41].

The high sensitivity and versatility of the DHSS procedure derives from the use of three very efficient reagents:

a) an imidoester derivative of the DNP group capable of non-deleterious labelling of the primary ligand reagents [42];
b) a DNP specific IgM monoclonal bridge antibody which has the capacity to bind, with great specificity, as many as 10 DNP groups simultaneously, and
c) a stable low molecular weight (its minimum theoretical size is 54,000 daltons) DNP-peroxidase conjugate acting as the enzyme marker. A schematic representation of the three-layer interaction of these key reagents at the tissue section level is given in Figure 5.

The routine use of the DHSS procedure on thick, semithin and ultrathin sections has made obvious at least two distinct practical advantages of this method:

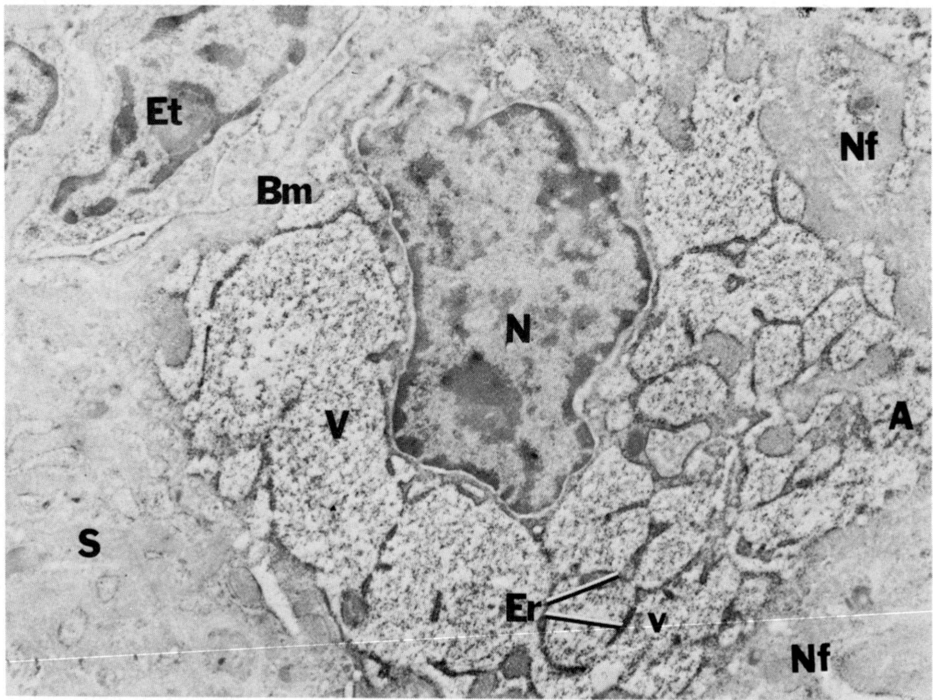

Fig. 4. BGPA fixed, 'LR White'-embedded, goitrous rat thyroid. The ultrathin section was exposed to 1:50,000 final dilution of DNP-labelled TSH. The TSH was localised by the DHSS procedure. (DAB/osmium; no counterstaining). Magnification = × 10,000. A = apex of positive follicular cell; Bm = basement membrane; Er = endoplasmic reticulum; Et = endothelium; N = nucleus of positive follicular cell; Nf = neighbouring follicular cells; S = stroma; V = vacuoles in endoplasmic reticulum. Note DAB reaction product deposited on profiles of rough endoplasmic reticulum (Er) and in the vacuolar contents of the endoplasmic reticulum (V). Adjacent stromal (S) and endothelial (Et) areas are negative.

a) the fact that any species of DNP-labelled primary antibody can be localised using the same bridge antibody and enzyme marker provides a very high level of quality control, particularly in terms of the method specificity. Furthermore, the key reagents have been found to be remarkably stable for extended periods (up to a year), especially in the presence of preservatives, i.e. thiomersal. They are stable even at room temperature, and in a relatively diluted state. Thus the long-term quality control of the technique in situations demanding repeated use of bulk amounts of the reagents (e.g. automated immunoperoxidase systems) is feasible;
b) the IgM bridge antibody, apart from having its unique specificity for the DNP group, has been found to bind to the hapten with a very high efficiency. Thus IgG antibody molecules (molecular weight 160,000) possessing as few as 3–4 DNP groups are as effectively localised by the IgM bridge system as those bearing a greater number of DNP groups. This is in striking contrast to the

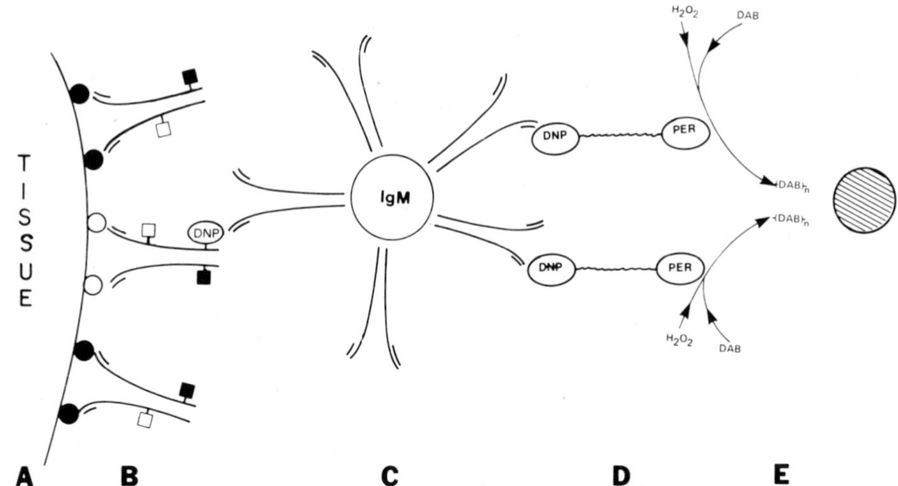

Fig. 5. Schematic representation of the interaction at the tissue section level of sequentially added unlabelled non-immune blocking antibody, DNP-labelled primary antibody, IgM monoclonal anti-DNP bridge antibody and DNP-labelled peroxidase conjugate (DNP-PER). (A) Tissue bearing antigen (○) and non-specific binding sites (●). (B) Blocking and DNP-labelled antibodies. (C) Monoclonal IgM anti-DNP bridge antibody. (D) DNP-labelled peroxidase conjugate. (E) Diaminobenzidine (DAB) reaction product.

avidin-based detection of biotin-labelled antibodies, where unless a special intermediate spacer group is incorporated [43], the biotin residues are not 'seen' efficiently by avidin except at very high substitution ratios [44].

The great specificity and detection efficiency of the IgM bridge antibody for DNP residues has permitted successful and economical routine use of DNP-labelled *whole* antisera, the basis of which has already been described [36].

The details of the DHSS procedure and reagent conditions currently used for staining ultrathin sections are given in Appendix, section 7.4.

5. Applicability of LR White/BGPA/DHSS post-embedding immunoenzyme system

Although very recent, this technology has already been applied to demonstrate a number of tissue antigens. For example, by using DNP-labelled antibodies, many of the thyroid, pituitary and pancreatic hormones such as calcitonin, TSH [33], ACTH [34, 35] (Figs 2 and 3), growth hormone, LH (Figs. 1 and 2), FSH (Fig. 1), prolactin, insulin [34], somatostatin and gastrin have been shown in endocrine granules contained by cells which have high standards of ultrastructural integrity. In immersion-fixed human tonsil [34] and lymph node, various immunoglobulins [34], alpha-1-antitrypsin, lysozyme and other antigens have been demonstrated. By contrast, binding sites for TSH have been localised to rough endoplasmic reticulum in thyroid follicular cells. The cells had been TSH stimulated by the administration of goitrogen, which was then withdrawn [39]. Ultrathin sections of ‘LR White’-embedded thyroid were exposed to DNP-labelled purified TSH and the DNP was localised by the usual hapten sandwich procedure (Fig. 4).

The simplicity of ‘LR White’ embedding, taken in conjunction with improved fixation and highly sensitive immunoenzyme immunocytochemistry, provides a system that has a very broad spectrum of applicability (for details, see Appendix). It also takes post-embedding immunocytochemistry out of the realms of the purely esoteric and makes it a routine laboratory method which can be extremely valuable, for example, in diagnostic pathology.

6. The future of electron microscope post-embedding immunoenzyme techniques

The significant advances which have been made recently in post-embedding technology will undoubtedly continue. The production of new plastics is no longer solely the province of large industrial companies whose products are modified for less profitable biological applications. Today a welcome co-operation between polymer chemists and biologists means that the properties of new embedding resins are matched to the problems in biology and this will ensure further developments in this area. As the scope for experimentation is widened by these new embedding media, so novel tissue preparation methods will emerge, and many of the delicate tissue antigens that are regarded as beyond the scope of contemporary post-embedding systems will become demonstrable. The demonstration of such antigens will also be attributable to the ingenuity of immunocytochemists, who have steadily increased the sensitivity, reliability, specificity, versatility and ease of use of modern immunocytochemical methods.

7. Appendix

7.1. Recommended fixative and schedule

Buffered glutaraldehyde/picric acid (BGPA)	
Vacuum distilled monomeric 50% glutaraldehyde* (stored at −20°C)	2 ml
Picric acid (sat. aq.)	15 ml
0.1 M Sorensen's phosphate buffer pH 7.3	83 ml
	100 ml

This produces a 1% glutaraldehyde 0.2% picric acid solution. It should be freshly made and can be used under perfusion conditions, when the phosphate buffer

* Obtained from EM Chemicals Ltd., Aldreth Farm, Ely, Cambs.

molarity may need to be decreased to 0.05 M if cell shrinkage is observed. Fixation is at room temperature for 3–4 h. For immersion-fixation, tissue blocks should not exceed 1–2 mm^3.

7.2. Dehydration

Tissue can be taken directly into 70% ethyl alcohol where two one hour washes will remove active picrates and also free aldehydes. Where delicate tissue is being handled, shrinkage caused by dehydration and embedding can be lessened by first washing it in 50% ethyl alcohol for 15 min before taking it to 70% ethyl alcohol as usual. Where aldehydes alone are used as fixatives, exposure to alcohol can be reduced even further by using a buffer wash before a brief dehydration in 50% and 70% ethyl alcohol. A buffer rinse is not advised if picric acid is included in the fixative.

7.3. Embedding

Directly from 70% ethyl alcohol, the tissue is placed into its first change of 'LR White' resin (see text). Again, if delicate, the tissue may be given an intermediate step in 1:2 70% ethyl alcohol:'LR White' resin (care should be taken not to carry over much 70% ethyl alcohol when changing the tissue). Another change of 'LR White' resin should be made and the tissue left on a roller mixer overnight. The following morning after a third change for one hour the tissue may be embedded. Anaerobic conditions are recommended for polymerisation, but in fact if gelatin capsules are used and they are well filled and tightly capped, this is unnecessary. To minimise protein denaturation, the polymerisation temperature should be kept low. 'LR White' will polymerise quite sufficiently if kept in an accurately set oven at 50°C for 24 h.

It is possible to embed tissue from even lower alcohol dilutions (e.g. from 30% ethyl alcohol) but this cannot be guaranteed, and is therefore open to experimentation. The greatest gain in antigenicity is seen between tissue taken from 70% ethyl alcohol and that which has been fully dehydrated, but when tissue is embedded from alcohol dilutions which are lower than 70% improvements can be made even on this.

7.4. Dinitrophenyl Hapten Sandwich Staining (DHSS)

7.4.1. DNP-labelling reagent

This is an imidoester derivative of the DNP group [45] named 3- (2,4-dinitrophenylamino) propionimidate hydrochloride (DNP-N-IE). It is readily synthesisable according to the method of Hewlins and Jasani [42].

7.4.2. DNP-labelling of the primary antibody reagent

Antiserum (125 μl) or fractionated or monoclonal antibody preparation (125 μl of 10 mg protein/ml) solution is equilibrated to pH 9.0 using triethanolamine-HCl buffer (425 μl; pH 9.0; 0.3 M) and cooled to 0–2°C in an ice-water bath. Freshly made DNP-N-IE solution (80 μl aliquot of 1 mg per 100 μl of distilled H_2O at 0–2°C) is then added to initiate the DNP-labelling reaction. This is allowed to proceed for 2½ h for fractionated (or monoclonal antibody) or 4 h for whole antiserum. The product thereof is separated from the unreacted reagents by passing the reaction mixture down a Sephadex G-25 column (10 × 0.8 cm; equilibrated with PBS, pH 7.2; 0.01 M at room temperature). The first of the two discrete yellow bands eluted in the column is collected to represent the DNP-labelled antibody reagent (usually present at 1:15 dilution of the original). This is storable for long periods (e.g. over a year) at 4°C with a small amount of preservative added (thiomersal 1×10^{-5} M) or in small aliquots at −70°C without any preservative.

7.4.3. DNP-specific IgM monoclonal bridge antibody

Dilutions of 1:5 of the supernatant and 1:500 of the ascites fluid are considered optimal for these when using them in the DHSS procedure.

7.4.4. DNP-peroxidase conjugate

This is a DNP-labelled covalent conjugate of horseradish peroxidase (HRP; Boehringer) with cytochrome (CYT-C; bovine heart, Sigma). The latter is attached to the enzyme using the modified periodate method described by Boorsma et al. [46] and provides a lysyl residue-rich backbone for linking DNP groups away from the active sites of the peroxidase molecule (cf. Sternberger and Petrali [47]). Preliminary experiments helped to establish that under these conditions the conjugation of 8 mg of HRP with 4 mg CYT-C resulted in the highest yield of the labelled marker (DNP-CYTC-HRP) possessing substantial staining activity when incorporated into the DHSS procedure (see Fig. 6). Hence currently this conjugate is adapted to represent the optimal enzyme marker for the DHSS procedure. The DNP-labelling of the conjugate is performed along the same lines as that for the fractionated or monoclonal antibody preparation (see above).

7.4.5. Staining of ultrathin sections with the DHSS procedure

The current protocol is based essentially on a combination of previously published methods [34, 35] and involves the following steps and reagent conditions:

1) Neat blocking serum (same donor species as primary antiserum) – 15 min
2) DNP-labelled primary antiserum diluted in 0.6% BSA in PBS – 15 min
3) IgM monoclonal DNP specific antibody diluted appropriately in 0.2% rabbit IgG in PBS – 15 min
4) DNP-CYTC-HRP conjugate diluted 1:800 in 0.6% BSA + 0.5N NaCl in PBS – 15 min
5) 0.05% Diaminobenzidine in Graham and Karnovsky's buffer [48] with hydrogen peroxide (0.01% v/v) – 3 min
6) 0.1% Gold chloride – 1 min

Reagents in steps (1) to (4) are centrifuged at 60,000 rpm in a microfuge. They are used in 40 μl droplets on dental wax in which the grids holding the ultrathin sections are totally immersed. The grids are rinsed in PBS (200 μl droplets) between each step up to step (5) and in distilled water before and after step (6). They are then air-dried and without any further treatment viewed and photographed at 80 KV and with a 50 μm objective lens aperture under the electron microscope.

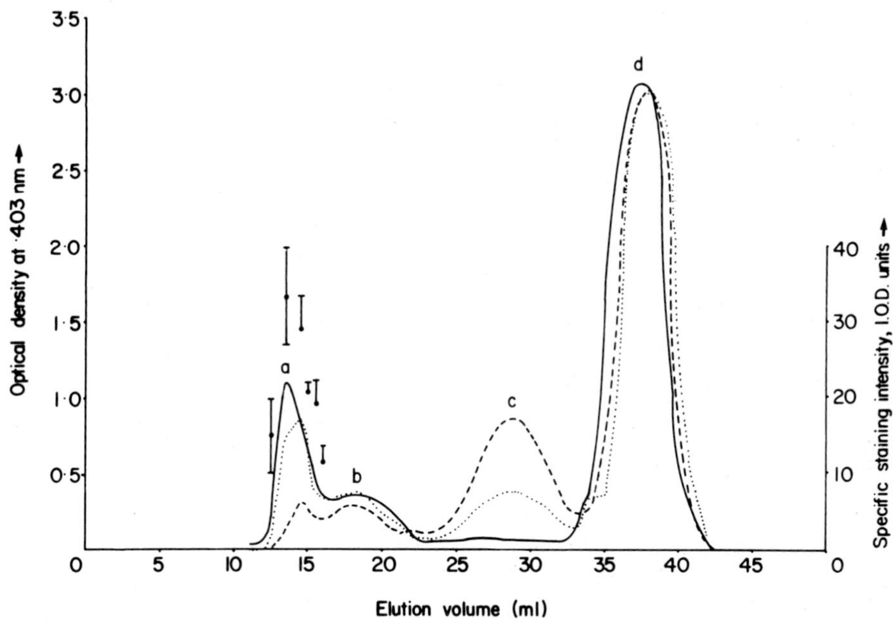

Fig. 6. Ultrogel ACA-44 column (30 × 0.8 cm; PBS) elution profiles of DNP-CYTC-HRP conjugate preparations obtained using HRP (8 mg) and 4 mg (———), 8 mg (.....) and 16 mg (----) of CYTC, respectively. In each case, the four peaks represent (a) DNP-CYTC-HRP conjugate; (b) unconjugated DNP-HRP; (c) DNP-CYTC; and (d) unreacted DNP-N-IE. The bars around peak (a) represent the staining activity of the fractions of the optimal conjugate (derived from 8 mg HRP and 4 mg CYTC) when they were used in the DHSS procedure.

The DHSS procedure shows a bridge antibody directed prozone phenomenon similar to that described by Bigbee et al. [49] for the PAP technique. Thus at higher concentrations of DNP-labelled antiserum, there is a paradoxical diminution in the staining. This is due to a progressive exhaustion of the bridge antibody valency (available at a fixed IgM concentration) by the larger numbers of DNP groups presented as a result of the increasing level of primary antibody binding to the tissue section. It is therefore essential to examine a range of primary antibody dilutions (usually 1:150 to 1:15,000) in any given study in order to exclude poor levels of staining arising from the prozone phenomenon.

8. References

1. Chang, J.P. and Yokoyama, M. (1970) A modified section freeze-substitution technique. J. Histochem. Cytochem. 18, 683–684.
2. Kobernick, S.D. and Thomas, M. (1970) Selecting and preparing tissues directly from paraffin blocks for electron microscopy: a new method. J. Cell. Biol. 47, 108a.
3. Mazurkiewicz, J.E. and Nakane, P.K. (1972) Light and electron microscopic localisation of antigens in tissues embedded in polyethylene glycol with a peroxidase labelled antibody method. J. Histochem. Cytochem. 20, 969–974.
4. Sternberger, L.A. (1979) Immunocytochemistry, 2nd Edn. John Wiley & Sons Inc., New York.
5. Van Noorden, S. and Polak, J.M. (1983) Immunocytochemistry today: techniques and practice. In: J.M. Polak and S. Van Noorden (Eds.) Immunocytochemistry: Practical Applications in Pathology and Biology. Wright, P.S.G., Bristol–London–Boston, pp. 11–42.
6. Willingham, M.C. (1983) An alternative fixation-processing method for pre-embedding ultrastructural immunocytochemistry of cytoplasmic antigens: The GBS (glutaraldehyde-borohydride-saponin) procedure. J. Histochem. Cytochem. 31, 791–798.
7. Pickel., V.M., Joh, T.H. and Reis, D.J. (1975) Ultrastructural localisation of tyrosine hydroxylase in noradrenergic neurons of brain. Proc. Natl. Acad. Sci. U.S.A. 72, 659–663.
8. Dudek, R.W., Childs, G.V. and Boyne, A.F. (1982) Quick-freezing and freeze-drying in preparation for high quality morphology and immunocytochemistry at the ultrastructural level: Application to pancreatic Beta cell. J. Histochem. Cytochem. 30, 129–138.
9. Dudek, R.W., Varndell, I.M. and Polak, J.M. (1984) Quick freeze-fixation and freeze drying for electron immunocytochemistry. In: J.M. Polak and I.M. Varndell (Eds) Immunolabelling for Electron Microscopy. Elsevier Science Publ., Amsterdam, pp. 235–248.
10. Roth, J., Bendayan, M., Carlemalm, E., Villiger, W. and Garavito, M. (1981) Enhancement of structural preservation and immunocytochemical staining in low temperature embedded pancreatic tissue. J. Histochem. Cytochem. 29, 663–671.
11. Wynford-Thomas, D., Stringer, B. and Newman, G.R. (1981) Hydroxyethylmethacrylate embedding: an improved technique. Med. Lab. Sci. 38, 121–122.
12. Pickel, V.M., Sumal, K.K., Beckley, S.C., Miller, R.J. and Reis, D.J. (1980) Immunocytochemical localization of enkephalin in the neostriatum of rat brain: A light and electron microscopic study. J. Comp. Neurol. 189, 721–740.
13. Childs, G.V. (1983) The use of multiple methods to validate immunocytochemical stains. J. Histochem. Cytochem. 31, 168–176.
14. Gonda, M.A., Gregg, M., Elser, J.E. and Hsu, K.C. (1980) Sensitivity and specificity of viron and cell surface labelling using the unlabelled antibody-haemocyanin bridge method. J. Histochem. Cytochem. 28, 710–713.
15. Rifkind, R.A., Hsu, K.C., Morgan, C., Seegal, B.C., Knox, A.W. and Rose, H.M. (1960) Use of ferritin-conjugated antibody to localise antigen by electron microscopy. Nature (London) 187, 1094–1095.
16. Singer, S.J. and Schick, A.F. (1961) The properties of specific stains for electron microscopy prepared by the conjugation of antibody molecules with ferritin. J. Biophys. Biochem. Cytol. 9, 519–537.
17. Romano, E.L., Stolinski, C. and Hughes-Jones, N.C. (1974) An antiglobulin reagent labelled with colloidal gold for use in electron microscopy. Immunochemistry 11, 521–522.
18. Roth, J., Bendayan, M. and Orci, L. (1978) Ultrastructural localisation of intracellular antigens by the use of protein A-gold complex. J. Histochem. Cytochem. 26, 1074–1081.
19. Larsson, L.I. (1979) Simultaneous ultrastructural demonstration of multiple peptides in endocrine cells by a novel immunocytochemical method. Nature (London) 282, 743–746.
20. De Mey, J. (1983) Colloidal gold probes in immunocytochemistry. In: J.M. Polak and S. Van Noorden (Eds.) Immunocytochemistry: Practical Applications in Pathology and Biology. Wright, P.S.G., Bristol–London–Boston, pp. 82–112.
21. De Mey, J., Moeremans, M., Geveens, G., Nuydens, R. and De Brabander, M. (1981) High

resolution light and electron microscopic localisation of tubulin with the IGS (immuno gold staining) method. Cell Biol. Int. Rep. 5, 889–899.

22. Avrameas, S. and Ternyck, T. (1971) Peroxidase labelled antibody and Fab conjugates with enhanced intracellular penetration. Immunochemistry 8, 1175–1179.
23. Bendayan, M. (1982) Double immunocytochemical labeling applying the protein A-gold technique. J. Histochem. Cytochem. 30, 81–85.
24. Tapia, F.J., Varndell, I.M., Probert, L., De Mey, J. and Polak, J.M. (1983) Double immunogold staining method for the simultaneous ultrastructural localisation of regulatory peptides. J. Histochem. Cytochem. 31, 977–981.
25. Ito, S. and Karnovsky, M.J. (1968) Formaldehyde/glutaraldehyde fixatives containing trinitro compounds. J. Cell. Biol. 39, 418a.
26. McLean, I.W. and Nakane, P.K. (1974) Periodate-lysine-paraformaldehyde fixative. A new fixative for immunoelectron microscopy. J. Histochem. Cytochem. 22, 1077–1083.
27. Stefanini, M., De Martino, C. and Zamboni, I. (1967) Fixation of ejaculated spermatozoa for electron microscopy. Nature (London) 216, 173–174.
28. Somogyi, P. and Takagi, H. (1982) A note on the use of picric acid-paraformaldehyde-glutaraldehyde fixative for correlated light and electron microscopic immunocytochemistry. Neuroscience 7, 1779–1783.
29. Hassel, J. and Hand, A.R. (1974) Tissue fixation with diimidoesters as an alternative to aldehydes. I. Comparison of crosslinking and ultrastructure obtained with dimethylsuberimidate and glutaraldehyde. J. Histochem. Cytochem. 22, 223–239.
30. Moriarty, G.C. and Halmi, N. (1972) Electron microscopic localisation of the adrenocorticotropin-producing cell with use of unlabelled antibody and the peroxidase-anti-peroxidase complex. J. Histochem. Cytochem. 20, 590–603.
31. Mozdzen, Jr., J.J. and Keren, D.F. (1982) Detection of immunoglobulin A by immunofluorescence in glycol methacrylate embedded human colon. J. Histochem. Cytochem. 30, 532–535.
32. Carlemalm, E., Garavito, R.M. and Villiger, W. (1982) Resin development for electron microscopy and an analysis of embedding at low temperature. J. Microsc. 126, 123–143.
33. Newman, G.R., Jasani, B. and Williams, E.D. (1982) The preservation of ultrastructure and antigenicity. J. Microsc. 127, RP5-RP6.
34. Newman, G.R., Jasani, B. and Williams, E.D. (1983) A simple postembedding system for the rapid demonstration of tissue antigens under the electron microscope. Histochem. J. 15, 543–555.
35. Newman, G.R., Jasani, B. and Williams, E.D. (1983) Metal compound intensification of the electron density of diaminobenzidine. J. Histochem. Cytochem. 31, 1430–1434.
36. Jasani, B. and Williams, E.D. (1980) DNP as a hapten in immunolocalisation studies. J. Med. Microbiol. 13, xv.
37. Jasani, B., Wynford-Thomas, D. and Williams, E.D. (1981) Use of monoclonal anti-hapten antibodies for immunolocalisation of tissue antigens. J. Clin. Pathol. 34, 1000–1002.
38. Jasani, B., Thomas, N.D., Newman, G.R. and Williams, E.D. (1983) DNP-hapten sandwich staining (DHSS) procedure: design, sensitivity, versatility and applications. Immunol. Commun. 12, 50.
39. Jasani, B., Newman, G.R. and Williams, E.D. (1981) The localisation of TSH receptors in thyroid follicular cells. Ann. Endocrinol. 42, 50A.
40. Jasani, B., Newman, G.R., Stanworth, D.R. and Williams, E.D. (1982) Immunohistochemical localisation of tissue receptors using the dinitrophenyl (DNP) hapten sandwich procedure. J. Pathol. 138, 50.
41. Jasani, B., Millar, D.A. and Williams, E.D. (1983) A quantitative comparison of the sensitivity of three immunoperoxidase procedures. Histochem. J. (Suppl.), March, 1983.
42. Hewlins, M.J.E. and Jasani, B. (1984) Non-deleterious dinitrophenyl (DNP) hapten labelling of antibody protein. I. Preparation and properties of some short-chain DNP imidoesters. J. Immunol. Methods (Submitted for publication).
43. Hutchison, N.J., Langer-Safer, P.R., Ward, D.C. and Hamkalo, B.A. (1982) In situ hybridization at the electron microscope level: Hybrid detection by autoradiography and colloidal gold. J. Cell. Biol. 95, 609–618.

44. Bayer, E.A., Wilcheck, M. and Skutelsky, E. (1976) Affinity cytochemistry: The localisation of lectin and antibody receptors on erythrocytes via the avidin-biotin complex. FEBS Lett. 68, 240–244.
45. Schramm. H.J. (1967) Use of imidoesters for the chemical modification of protein. I. Synthesis of coloured nitriles, dinitriles and bifunctional imidoesters. Hoppe Seylers Z. Physiol. Chem. 348, 289–292.
46. Boorsma, D.M. and Streefkerk, J.G. (1979) Periodate or glutaraldehyde for preparing peroxidase conjugates? J. Immunol. Methods 30, 245–255.
47. Sternberger, L.A. and Petrali, J.P. (1977) The unlabelled antibody enzyme method. Attempted use of peroxidase-conjugated antigen as the third layer in the technique. J. Histochem. Cytochem. 25, 1036–1042.
48. Graham, R.C. and Karnovsky, M.J. (1966) The early stages of absorption of injected horseradish peroxidase in the proximal tissues of mouse kidney with ultrastructural cytochemistry by a new technique. J. Histochem. Cytochem. 14, 291–302.
49. Bigbee, J.W., Kosek, J.C. and Eng, L.F. (1977) Effects of primary antiserum dilution on staining of 'antigen rich' tissues with the peroxidase antiperoxidase technique. J. Histochem. Cytochem. 25, 443–447.

Immunolabelling for Electron Microscopy (Polak/Varndell, eds)

CHAPTER 6

Immuno-cryoultramicrotomy

K.T. Tokuyasu

Department of Biology, University of California at San Diego, La Jolla, CA 92093, U.S.A.

CONTENTS

1. INTRODUCTION

Freeze-sectioning at room temperature and in moist air creates many more opportunities to frustrate the operator of a microtome than conventional ultrathin sectioning. However, difficulties involved in the technique are now deemed to be peripheral and the frustration to be a matter for habituation. In fact, after overcoming the initial difficulties, many investigators in our laboratory now prefer cryoultramicrotomy over the conventional ultramicrotomy technique. We frequently complete the entire process from specimen preparation to the examination of immunostained ultrathin frozen sections within one day. Thus, cryoultramicrotomy is no longer viewed as a technique beset with intrinsic frustrations and resorted to in exceptional circumstances, but rather as the method of preference.

Practical aspects of the technique have been well documented in several recent articles [1–4]. The aim of this chapter is therefore more to discuss the applicability of cryoultramicrotomy to different problems than to describe experimental details.

1.1. Pre- and post-embedding immunocytochemistry

In pre-embedding immunostaining of intracellular antigens, antibodies must first pass through the plasma membrane and then organellar membranes, to locate the target antigens. Although novel approaches such as fusing antibody-containing liposomes with the cells of interest are available, they are not widely applicable. In general, therefore, the membranes need to be damaged either chemically (for example, by treating with detergents or chemical solvents) or physically (for example, by freeze-thawing or osmotic shock) to make them permeable to antibodies. Also, cells generally contain a high concentration of protein and if all of the protein molecules are rigidly cross-linked, macromolecules, such as antibodies, will not be able to penetrate into the cell interior. Fixation, therefore, must be appropriately light to enable a large portion of the cytosolic protein to be extracted. This is necessarily accompanied by difficulties in preserving or even retaining many sub-cellular structures. The ultimate problem with the pre-embedding immunostaining method is that the interior of dense structures such as secretory granules or dense plaques of intercellular junctions is not accessible to antibodies, unless they have been damaged or sectioned before the infusion of antibodies.

Many of the difficulties inherent with the pre-embedding method may be overcome by the post-embedding technique. In this case, however, several problems arise from the embedding itself. The first is that the matrix of the embedding material becomes an obstruction to the entry of antibodies into the sections. We reported that even the presence of a 5% polyacrylamide gel caused a significant reduction in the sensitivity of antigen detection [5]. Secondly, dehydration of specimens by chemical solvents which is required for embedding can be harmful to the antigenicity of many proteins. In addition, when membranes are exposed to chemical solvents, not only their structural integrity but also the conformation of integral proteins residing within them will be damaged. Membrane structures are well preserved by the application of osmium tetroxide, however, this also severely reduces the antigenicity of most proteins.

For a number of subjects, the difficulties mentioned above may not be too serious and the pre- and post-embedding techniques can be quite successfully applied. However, there will be many other subjects which cannot be easily studied using these methods. The need for immuno-cryoultramicrotomy then arises.

1.2. Concept of cryoultramicrotomy

In cryoultramicrotomy, ice provides the rigidity that is required for cutting tissues or isolated cells into ultrathin sections. Upon returning to room temperature, the sections regain their original hydrated state.

Actually, to improve the freezing characteristics and the plasticity of frozen pieces, fixed specimens may be infused with a solution containing hydrophilic substances [6] before freezing. Low molecular weight substances such as sugars or polyethylene glycol are effective modifiers of the physical characteristics of the cell interior, whereas macromolecular substances such as proteins (e.g. gelatin or bovine serum albumin) or polysaccharides (e.g. dextran) may be used to adjust the physical nature of the intercellular spaces. When sections are thawed and exposed to water, these hydrophilic substances dissolve. If proteins are included in the infusion solution, they can be cross-linked with an aldehyde fixative to form an insoluble network, which may aid the preservation of intercellular relationships. Agarose can also be used to preserve these relationships [7].

Tissue blocks or isolated cells can be embedded before freezing to improve the cutting quality of the frozen pieces or the preservation of cell structures in the thawed sections, or both. The embedding in this case does not need to attain a completely dehydrated and highly rigid state, which is essential in conventional ultramicrotomy, and the embedding material can also be highly hydrophilic. Thus, the concept of embedding in cryoultramicrotomy becomes much broader than in conventional microtomy. For instance, we often embed fragile specimens such as embryonic tissues in a 5% polyacrylamide gel prior to freezing with the purpose of preserving intercellular and interorganellar relationships in thawed sections.

If ultrathin frozen sections are air-dried, cellular structures will be largely destroyed by surface tension. This fragility of the sections is often considered to be a basic weakness of the concept of cryoultramicrotomy. However, protection against surface tension can be easily provided by submerging (or embedding) sections in a thin layer of some non-volatile material. In fact, only if such protection is provided does the freedom of choosing the staining method become much greater in ultrathin frozen sections than in plastic-embedded sections. Frozen sections can be stained positively by choosing appropriate hydrophilic [8, 9] or hydrophobic [8, 10] embedding materials, or negatively by embedding the sections in a thin layer of a heavy metal compound with [11, 12] or without [6, 13] being mixed with organic compounds. The delineation of ultrastructure in frozen sections so embedded is often equal or superior to that in conventional sections. For instance, it was demonstrated that the A-line of striated muscle was more finely resolved in negatively stained ultrathin frozen sections than in positively stained conventional sections [13]. Conversely, positively stained ultrathin frozen sections are often equal to conventional sections in the definition of cell structures [10, 14–16].

It should be noted that both in pre-freezing embedding of tissue pieces or post-immunostaining embedding of ultrathin frozen sections harsh treatments such as osmication, dehydration in chemical solvents or embedding in hydrophobic materials are not required, although if desirable for some reason, these treatments can be performed [8, 10]. Thus, cryoultramicrotomy is not a peculiar addition to the conventional techniques but offers advantages that open many new possibilities in both the observation of cellular structure and the localisation of specific macromolecules in tissues and cells by immunocytochemistry and other affinity-binding methods.

2. Technical aspects of immuno-cryoultramicrotomy

2.1. *Infusion of sucrose and freezing*

Glycerol is known to be an excellent cryoprotectant but when infused into specimens it does not appreciably improve the plasticity of the frozen tissue. Sugars, in particular, disaccharides such as sucrose or maltose, serve both purposes well. If sugars cannot be used for some reason, polyethylene glycol of low molecular weight (300–500) is a good substitute. We commonly infuse 0.6–2.3 M sucrose into appropriately fixed specimens of 0.5–1 mm in height and width. When the fixation is very light or the specimen is highly hydrated, a high sucrose concentration may provide, upon freezing, the plasticity necessary for sectioning. Many laboratories which employ fixation with formaldehyde or a low concentration of glutaraldehyde use 2.3 M sucrose as the routine infusion solution. In this case, freezing of the specimens may simply be carried out by rapidly immersing it into liquid nitrogen and sectioning at −90 to −110°C. On the other hand, in some specimens, infusion with 2.3 M sucrose may result in a texture too soft for ultrathin sectioning. If it is impractical to lower the sectioning temperature below −110°C, the sucrose concentration needs to be lowered, sometimes even to 0.6 M. Freezing is then carried out by immersing the tissue into liquid Freon-12 (CCl_2F_2) or Freon-22 ($CHClF_2$) chilled over liquid nitrogen to minimise the freezing damage.

2.2. *Sectioning and recovery of sections*

To obtain the best sections, slow sectioning is essential just as in conventional microtomy. A variability in section thickness may occur when soft specimens are slowly sectioned, again as in conventional microtomy, but it is the softness of the specimens rather than the speed of sectioning that is to be blamed. Fast sectioning can provide a section in each cutting stroke, even when the specimens are soft, but it tends to create an irreversible compression in the thin tissue slice. Only when portions of sections are to be studied, can one choose a reasonably fast sectioning speed. On the other hand, slow sectioning can cause wrinkling of sections, particularly when specimens are soft. However, when sections are recovered to room temperature and expanded on a sucrose droplet (see below) they will often regain a flat form. The build up of static electricity and mechanical strains created within the sections particularly during fast sectioning may result in curling or jumping of sections. Slow sectioning often avoids these problems. However, it is interesting to note that in our experience diamond knives apparently generate more static electricity than glass knives and so are less suitable for cryosectioning.

The quality of glass knives is very important for successful cryosectioning. It is now customary to produce 45° knives by diagonally scoring square glass pieces and breaking them along the scores. To obtain good knives, the break should occur very close to the diagonal points (within 0.1–0.2 mm of the points) [17]. To ensure this, the shape of the glass pieces should be very accurately square and the scoring line should be drawn very close to the geometrical diagonal line.

The conventional concept of section recovery, that is floating sections on an aqueous solution and attaching them directly or indirectly onto grids, is difficult to apply to cryoultramicrotomy, since no medium has been hitherto found to remain liquid below −50°C and have surface tension sufficient to allow floating of sections. Our method to recover sections is to suspend a 2.3 M sucrose droplet on a wire loop of 1.5–2 mm diameter, insert it quickly into the cryochamber, attach sections on the knife edge on the sucrose droplet before it is completely solidified, bring it out to room temperature, and thaw the sucrose droplet to expand the sections on it. To recover fragile sections, one may use a mixture of 2 M sucrose and 0.75% gelatin which has a lower surface tension than 2.3 M sucrose. The sections are then transferred to the supporting film on the grid by touching the sucrose droplet to the film.

Grids with sections are subsequently placed face down on a wet plate of 0.3% agarose and 1% gelatin to remove the sucrose by diffusion. Phosphate-buffered saline with 0.01 M glycine (PBS-glycine) is gently added to the plate to float the grids. They are carefully removed from the plate, placed on a puddle of 2% gelatin (in PBS-glycine) on a Parafilm® sheet and left for 10 min. The purpose of this step is to 'condition' the sections to reduce the background level of immunostaining. They are washed on a few droplets of PBS-glycine by successively transferring from one to the next. Throughout these steps, a wire loop of 3–4 mm diameter can be conveniently used to transfer the grids.

2.3. *Immunostaining*

Each grid picked up with forceps is placed on a 10 μl droplet of the primary antibody on a Parafilm® sheet. The reason for using forceps instead of a loop is to minimise the volume of buffer carried with the grid thus avoiding excessive dilution of the antibody. A pair of anti-capillary forceps is preferable to the ordinary type for the same reason. We usually immunostain for 10 min, but it may be extended to a longer period. Each grid is then washed on several droplets of PBS by transferring it, with a loop, between the solutions. Staining with the secondary antibody conjugated with ferritin or adsorbed to gold particles or with a protein A-gold complex is similarly carried out. After a thorough washing with PBS, sections on the grids may be fixed with 2% glutaraldehyde for 5 min to secure the antibodies to the sections. They are then thoroughly washed on several droplets of distilled water.

2.4. *Staining*

Our standard method of positive staining, which is called the 'adsorption staining method' [1, 9], is carried out as follows. Sections on grids are first stained for 10 min with 2% neutral uranyl acetate, which is made by mixing equal volumes of aqueous solutions of 4% uranyl acetate and 0.3 M potassium oxalate and adjusting the pH of the mixture to 7–8 by adding a small volume of 10% ammonium hydroxide. This step serves the purpose of stabilising the membranes. They are washed quickly on several droplets of water and then floated on an embedding mixture of 1.8%

carbowax (polyethylene glycol, molecular weight 1,540), 0.2% methylcellulose (400 cps) and 0.01–0.1% acidic uranyl acetate for 10 min. Each grid is then picked up with a loop of 2.5–3 mm inner diameter and after removing the excess embedding mixture with a piece of filter paper, it is air-dried. The removal of the excess should be such that when the grid is dried, the film formed on it shows a gold or gold-blue colour. A stock solution of 2–8% methylcellulose is made by first suspending the powder in hot water (60°C) and then cooling the suspension to 4°C. Recently, we have successfully used polyvinyl pyrrolidone (molecular weight 10,000 or 40,000) or dextran (molecular weight 10,000) in place of carbowax in the above embedding medium (unpublished data).

A modification of this method [12] is to stain, after neutral uranyl acetate staining and washing, with 2% acidic uranyl acetate for 3–5 min. The grid is then washed on three puddles of 1.5% methylcellulose. This should be done quickly, spending a total of only 20 sec, to retain an appropriate amount of uranyl acetate on the sections. The grid is subsequently picked up with a loop and after removal of the excess, dried in the same manner as in the positive staining method described above. This method produces mostly a negative staining effect due to a greater amount of uranyl acetate left on the grid and is suitable for counterstaining after gold-labelling. We also use a mixture of 0.2% sodium phosphotungstate (pH 7) and 0.8% dextran (molecular weight 10,000) for negative staining. In this case, staining is carried out just as in the standard negative staining procedure: grids floating on water after immunostaining are picked up with forceps, placed on a puddle of the mixture, left for 10–30 sec, picked up individually with forceps, and after removing the excess mixture, dried in air.

The chance of sectioning membranes obliquely is greater than that of cutting exactly across them. Cross-sections through lightly fixed membranes may also tend to collapse and become obliquely oriented against the supporting film. Negative staining is much more effective to delineate oblique profiles of membranes than positive staining. On the other hand, the effect of negative staining tends to be all or nothing, whereas positive staining can provide a wide range of densities and is suitable to differentiate multiple structures. Choice of the staining method or the degree of the staining intensity is also related to the density of the immuno-markers. If ferritin particles, Imposil particles [18] or gold particles smaller than 5 nm are used, the background density of negative staining may be too dense to recognise them clearly.

3. Problems of fixation

3.1. *Formaldehyde*

Chemical fixation is important in any approach to observe cellular structures but it bears an additional importance in cryoultramicrotomy in that it is largely responsible for the maintenance of thawed ultrathin frozen sections in an intact state. Glutaraldehyde and formaldehyde are the two most commonly used fixatives in biology. Glutaraldehyde has been more frequently used in ultrastructural studies, mainly because the preservation of ultrastructure is better and more stable after glutaraldehyde than after formaldehyde fixation. However, formaldehyde has many advantages over glutaraldehyde fixation in immunocytochemistry. Here, the characteristics of formaldehyde fixation will be discussed first.

We have observed that human erythrocytes fixed with formaldehyde (freshly prepared from paraformaldehyde) became as rigid as those fixed with glutaraldehyde but when they were exposed to an isotonic buffer, they began to lyse within 1–2 h [11]. This occurred even when the fixation was extended to 24 h in 4% formaldehyde. It was rather surprising then that this apparent reversibility of formaldehyde fixation had not been clearly documented in text books of electron microscopy. It is, however, to be recalled that before the introduction of glutaraldehyde [19], biological specimens were commonly preserved by keeping them in a solution of formalin, a mixture of aqueous formaldehyde and methanol. The difficulties in defining the concept of bound formaldehyde had in fact been discussed before commercial electron microscopes became available [20]. It is therefore advisable to store all formaldehyde-fixed preparations in 0.4–1% formaldehyde at 4°C to prevent the reversal of cross-linking.

This problem of reversibility of formaldehyde fixation is complicated by the variation of its magnitude in different proteins or cells. For instance, a gelatin block formed by fixation of a concentrated solution of the proteins with formaldehyde is much more stable than a human haemoglobin block formed in a similar manner, when they are exposed to an isotonic buffer. A significant difference is also found between human and chicken erythrocytes. Human erythrocytes fixed with formaldehyde need to be constantly exposed to a low concentration of formaldehyde throughout, and after, freeze-sectioning in order to maintain their structural integrity. Chicken erythrocytes similarly fixed with formaldehyde are, on the other hand, much more stable and do not require such critical treatment for the preservation of their structure. This variation also reflects on the method of immunolabelling. To immunostain sections of formaldehyde-fixed human erythrocytes in an intact state, the exposure of the sections to a formaldehyde-free solution of the antibody has to be minimal and the antibody solution should contain a high concentration of glycine or ammonium chloride to reduce the background staining. On the other hand, immunostaining of sections of formaldehyde-fixed chicken erythrocytes can be carried out much more easily in the manner described in the previous section.

Formaldehyde fixation is commonly used in immunofluorescence microscopy without such precautions as described above, but the same casualness will not always be appropriate in immunoelectron microscopy. First of all, in qualitative terms, the minimum degree of cross-linking necessary to preserve structures will be inversely related to the section thickness; and the sections used in immunoelectron microscopy are in excess of one order of magnitude thinner than those used in immunolight microscopy. Secondly, the necessary degree of structural preservation is directly related to the resolution of observation; and the resolving power of a modern electron microscope is three orders of magnitude greater than that of most light microscopes. One example to illustrate these points involves formaldehyde, which is known to inadequately preserve the structure of microtubules for electron microscopical observation [21] but which is quite commonly used in immunofluorescence microscopy to study the distribution of microtubules.

3.2. *Glutaraldehyde*

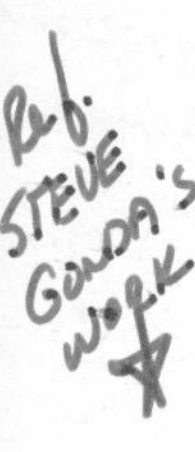

Human erythrocytes fixed in 0.1% glutaraldehyde for 10 min do not lyse when exposed to distilled water immediately after fixation but begin to lyse several hours after being suspended in an isotonic buffer (unpublished observations). Thus, glutaraldehyde fixation too is reversible but its rate is much slower than that of formaldehyde and for practical purposes can be considered to be irreversible. Using human erythrocytes as the specimen, we also observed that an initial exhaustive fixation with formaldehyde did not prevent a subsequent glutaraldehyde fixation from establishing an irreversible cross-linking [11]. Thus, as previously proposed [22], a mixture of a high concentration of formaldehyde and a relatively low concentration of glutaraldehyde can be more effective than the same low concentration of glutaraldehyde alone for the preservation of ultrastructure. However, our observation also indicates that a pre-treatment of specimens with formaldehyde may not be very effective in controlling the degree of the subsequent glutaraldehyde fixation.

3.3. *Ethylacetimidate*

As described above, the initial binding of formaldehyde with reactive sites does not prevent glutaraldehyde from subsequently reacting with the same sites. However, if imidates are initially bound to amino groups, they are not replaced by glutaraldehyde. We took advantage of this irreversibility of imidate-binding by using a monofunctional imidate, ethylacetimidate (EAI), to block a significant fraction of amino groups and thus restrict the degree of the subsequent glutaraldehyde fixation. This EAI treatment was effective in maintaining the antigen accessibility [11] and preserving the antigenicity of a protein, 95% of which was abolished by a one hour treatment with 0.1% glutaraldehyde [23]. It is noteworthy that a mere 5-min treatment with a mixture of 20 mM EAI and 8% formaldehyde was sufficient to protect the antigenicity of this protein against a subsequent, extensive fixation with a mixture of 4% glutaraldehyde and 4% formaldehyde [16, 24].

3.4. *Choice of fixation*

In choosing the optimal fixation conditions, one has to consider several factors: the target structure has to be intact, the antigen should be immobilised within it, the antigenicity should not be critically damaged and the antigen should be accessible to the antibody. Generally speaking, a strong fixation is better for the preservation of structures and immobilisation of antigens, but a weak one is better for the retention of antigenicity and the accessibility of antigens. It can also be said that in general, glutaraldehyde is more suited for structural preservation than formaldehyde but the opposite is true for antigenicity retention and antigen accessibility. Thus, if the antigen is plentiful and the antigenicity is resistant to glutaraldehyde, one can afford to employ a strong glutaraldehyde fixation but if the antigen is scarce and/or the antigenicity is sensitive to glutaraldehyde, then, the least possible degree of fixation simply to preserve the structure of interest (not necessarily all structures) needs to be chosen. This degree will be variable in different subjects. For instance, in the previously mentioned example of formaldehyde-fixed human erythrocytes, if the integral proteins of the plasma membrane are the subject of investigation, dissociation of haemoglobin from the ultrathin frozen sections may be desirable and one may use a formaldehyde solution as fixative. If, on the other hand, the subject requires that the frozen sections remain intact, then one may consider including a low concentration of glutaraldehyde in the fixative, as we did in our previous studies [23, 25, 26].

As a general rule, one may initially try a mixture of 0.1–1% glutaraldehyde and 2–4% formaldehyde and study the quality of morphological preservation and the degree of immunolabelling. Then, if necessary, 2–8% formaldehyde may be employed to observe whether or not the target structure remains relatively intact without a continual contact with formaldehyde. If the subject critically requires the satisfaction of all factors of fixation mentioned above, a two- [23] or three-stage [24] fixation involving the EAI treatment may need to be carried out.

The possibility that the antigenicity of the protein under investigation is sensitive to glutaraldehyde can be immunochemically studied [23] but, more conveniently, this can be determined by examining semithin frozen sections of the tissues or cells in question by immunofluorescence microscopy. If sections from formaldehyde-fixed preparations are specifically immunostained but those from glutaraldehyde-fixed tissue are not, then it is likely that the protein is rendered inactive by glutaraldehyde. In this case, a low concentration of glutaraldehyde should be used to ensure that the absence of immunostaining in the glutaraldehyde-fixed preparation is not due solely to a reduced accessibility (e.g. see reference [11]). If, on the other hand, the glutaraldehyde-fixed preparation is immunostained and the formaldehyde-fixed one is not, then the likelihood is that formaldehyde is incapable of fixing the antigen. Since glutaraldehyde emits fluorescence under ultraviolet illumination, frozen sections cut from glutaraldehyde-fixed tissue need to be treated with sodium borohydride ($NaBH_4$) before immunostaining to eliminate this effect [21, 27]. The same treatment is said to improve antigen accessibility and the retention of antigenicity, but we have not yet systematically studied this contention.

3.5. *Other methods of fixation*

Bifunctional imidates were successfully employed as fixatives in a few studies [28, 29] but the preservation of ultrastructure was not as satisfactory as that achieved by aldehyde fixation. Nevertheless, the combination of imidates with aldehyde fixatives may prove to be useful, as our study of EAI described above suggests. Carbodiimides were also used as fixatives [30, 31] and claimed to be as effective as glutaraldehyde in the preservation of ultrastructures and to be better than glutaraldehyde for the retention of activities of various enzymes [30]. In the periodic acid-lysine-paraformaldehyde (PLP) method [32], carbohydrates are oxidised to produce aldehyde groups, which are then cross-linked to each other or amino groups of proteins via lysine and paraformaldehyde. It is said that an appropriate mixture of periodate, lysine and paraformaldehyde can preserve antigenicity as well as paraformaldehyde and ultrastructure as well as glutaraldehyde. With the general precaution mentioned above on the reversibility of formaldehyde fixation in mind, PLP may be employed to study the distribution of glutaraldehyde-sensitive proteins.

Finally, the possibility of 'physical fixation' may need to be explored. If cellular structures could be trapped (or embedded) in a loose network of a physically strong but chemically inert material, then they would remain intact, even when they were cut into ultrathin frozen sections. Infusion of resin monomers into cells would involve various processes such as rapid freezing of unfixed specimens and substitution of ice with a glycerol-water mixture or chemical solvents such as acetone. Polymerisation of the monomer would also need to be carried out at low temperature. The conformation of proteins would be necessarily affected by these treatments, but the overall effects of the physical trapping could be milder than chemical fixation which is accomplished by binding chemical compounds to the reactive groups of proteins. The effect of chemical solvents on proteins is essentially the denaturation or conversion of proteins from a soluble to an insoluble state. It is likely that in some proteins, a certain degree of conformational change by chemical solvents at a very low temperature is preferable to the binding of chemical compounds to reactive sites in so far as recognition of the proteins by specific antibodies is concerned. In fact, it has been reported that at the light microscopical level, specimens fixed with cold acetone are often much more immunoreactive than those fixed with formaldehyde. Ultrathin frozen sections of specimens treated with such a denaturation fixation alone, however, will not remain intact when they are thawed. The necessity of physical fixation may then be realised. We embedded formaldehyde-fixed tissue pieces in a 5% polyacrylamide gel at −20°C and confirmed that the embedding was effective in preserving ultrastructure and did not appreciably damage the antigenicity of a protein which was quite sensitive to glutaraldehyde [5]. Polyacrylamide may thus be a possible candidate for 'physical fixation'.

There are cases in which chemical cross-linking cannot be tolerated. For instance, binding of any chemical compound to the reactive sites of some proteins may inhibit the reaction of those proteins with specific antibodies to them. Physical fixation and freeze-sectioning could be used to overcome this difficulty. The broad potential of the concept of cryoultramicrotomy should be fully explored.

4. Acknowledgements

The author would like to thank Dr. Pamela A. Maher for critical reading of the manuscript. The work from the Department of Biology at the University of

California at San Diego described in this chapter was supported by grants GM-15971 and HL-30282 from the U.S. National Institutes of Health.

5. References

1. Tokuyasu, K.T. (1980) Immunochemistry on ultrathin frozen sections. Histochem. J. 12, 381–403.
2. Singer, S.J., Tokuyasu, K.T., Dutton, A.H. and Chen, W.-T. (1982) High-resolution immunoelectron microscopy of cell and tissue ultrastructure. In: J. Griffiths (Ed.) Electron Microscopy in Biology, Vol. 2. John Wiley and Sons, New York, pp. 55–106.
3. Slot, J.W. and Geuze, H.J. (1983) The use of protein A-colloidal gold (PAG) complexes as immunolabels in ultrathin frozen sections. In: A.C. Cuello (Ed.) Immunohistochemistry IBRO, pp. 323–346.
4. Griffiths, G., Simons, K., Warren, G. and Tokuyasu, K.T. (1983) Immunoelectron microscopy using thin, frozen sections: application to studies of the intracellular transport of Semliki Forest virus spike glycoproteins. In: S. Fleischer and B. Fleischer (Eds.) Methods in Enzymology, Vol. 96. Academic Press, New York, pp. 466–485.
5. Tokuyasu, K.T. (1983) Present state of immunocryoultramicrotomy. J. Histochem. Cytochem. 31, 164–167.
6. Tokuyasu, K.T. (1973) A technique for ultracryotomy of cell suspensions and tissues. J. Cell Biol. 57, 551–565.
7. Bergmann, J.E., Tokuyasu, K.T. and Singer, S.J. (1981) Passage of an integral membrane protein, the vesicular stomatitis virus glycoprotein, through the Golgi apparatus en route to the plasma membrane. Proc. Natl. Acad. Sci. USA. 78, 1746–1750.
8. Tokuyasu, K.T. (1978) A study of positive staining of ultrathin frozen sections. J. Ultrastruct. Res. 63, 287–307.
9. Tokuyasu, K.T. (1980) Adsorption staining method for ultrathin frozen sections. In: G.W. Bailey (Ed.) Proceedings of 38th Meeting of Electron Microscopy Society of America, pp. 760–763.
10. Tokuyasu, K.T. (1983) Visualization of longitudinally-oriented intermediate filaments in frozen sections of chicken cardiac muscle by a new staining method. J. Cell Biol. 97, 562–565.
11. Tokuyasu, K.T. and Singer, S.J. (1976) Improved procedures for immunoferritin labeling of ultrathin frozen sections. J. Cell Biol. 71, 894–906.
12. Griffiths, G., Brands, R., Burke, B., Louvard, D. and Warren, G. (1982) Viral membrane proteins acquire galactose in *trans* Golgi cisternae during intracellular transport. J. Cell Biol. 95, 781–792.
13. Sjöström, M. and Squire, J.M. (1977) Fine structure of the A-band in cryosections: the structure of the A-band of human skeletal muscle fibers from ultrathin cryosections negatively stained. J. Molec. Biol. 109, 49–68.
14. Geuze, H.J., Slot, J. W. and Tokuyasu, K.T. (1979) Immunocytochemical localization of amylase and chymotrypsinogen in the exocrine pancreatic cell with special attention to the Golgi complex. J. Cell Biol. 82, 697–707.
15. Tokuyasu, K.T., Dutton, A.H. and Singer, S.J. (1983) Immunoelectron microscopic studies of desmin (skeletin) localization and intermediate filament organization in chicken skeletal muscle. J. Cell Biol. 96, 1727–1735.
16. Tokuyasu, K.T., Dutton, A.H. and Singer, S.J. (1983) Immunoelectron microscopic studies of desmin (skeletin) localization and intermediate filament organization in chicken cardiac muscle. J. Cell Biol. 96, 1736–1742.
17. Tokuyasu, K. and Okamura, S. (1959) A new method for making glass knives for thin sectioning. J. Biophys. Biochem. Cytol. 6, 305–308.
18. Dutton, A.H., Tokuyasu, K.T. and Singer, S.J. (1979) Iron-dextran antibody conjugates: general method for simultaneous staining of two components in high resolution immunoelectron microscopy. Proc. Natl. Acad. Sci. USA. 76, 3392–3396.
19. Sabatini, D.D., Bensch, K. and Barrnett, R.J. (1963) Cytochemistry and electron microscopy:

the preservation of cellular ultrastructure and enzyme activity by aldehyde fixation. J. Cell Biol. 17, 19–58.

20. French, D. and Edsall, J.T. (1945) The reactions of formaldehyde with amino acids and proteins. In: M.L. Anson and J.T. Edsall (Eds.) Advances in Protein Chemistry. Academic Press, New York, pp. 277–335.
21. Osborn, M., Webster, R.E. and Weber, K. (1978) Individual microtubules viewed by immunofluorescence and electron microscopy in the same PtK2 cell. J. Cell Biol. 77, R27–R34.
22. Karnovsky, M.J. (1965) A formaldehyde-glutaraldehyde fixative of high osmolarity for use in electron microscopy. J. Cell Biol. 27, 137a (Abstr.).
23. Geiger, B., Dutton, A.H., Tokuyasu, K.T. and Singer, S.J. (1981) Immunoelectron microscope studies of membrane-filament interactions: distributions of α-actinin, tropomyosin, and vinculin in intestinal epithelial brush border and chicken gizzard smooth muscle cells. J. Cell Biol. 91, 614–628.
24. Tokuyasu, K.T., Geiger, B., Dutton, A.H. and Singer, S.J. (1981) Ultrastructure of chicken cardiac muscle as studied by double immunolabelling in electron microscopy. Proc. Natl. Acad. Sci. USA 78, 7619–7623.
25. Tokuyasu, K.T., Schekman, R. and Singer, S.J. (1979) Domains of receptor mobility and endocytosis in the membranes of neonatal human erythrocytes and reticulocytes are deficient in spectrin. J. Cell Biol. 80, 481–485.
26. Zweig, S.E., Tokuyasu, K.T. and Singer, S.J. (1981) Membrane-associated changes during erythropoiesis: on the mechanism of maturation of reticulocytes to erythrocytes. J. Supramol. Struct. Cell Biochem. 17, 163–181.
27. Osborn, M. and Weber, K. (1982) Immunofluorescence and immunocytochemical procedures with affinity-purified antibodies: tubulin-containing structures. In: L. Wilson (Ed.) Methods in Cell Biology, Vol. 24. Academic Press, New York, pp. 97–132.
28. McLean, J.D. and Singer, S.J. (1970) A general method for the specific staining of intracellular antigens with ferritin-antibody conjugates. Proc. Natl. Acad. Sci. USA 65, 122–128.
29. Hassel, J. and Hand, A.R. (1974) Tissue fixation with diimidoesters as an alternative to aldehydes. I. Comparison of cross-linking and ultrastructure obtained with dimethylsuberimidate and glutaraldehyde. J. Histochem. Cytochem. 22, 223–239.
30. Yamamoto, N. and Yasuda, K. (1977) Use of a water soluble carbodiimide as a fixing reagent. Acta Histochem. Cytochem. 10, 14–37.
31. Willingham, M.C. and Yamada, S.S. (1979) Development of a new primary fixative for electron microscopic immunocytochemical localization of intracellular antigens in cultured cells. J. Histochem. Cytochem. 27, 947–960.
32. McLean, I.W. and Nakane, P.K. (1974) Periodate-lysine-paraformaldehyde fixative: a new fixative for immunoelectron microscopy. J. Histochem. Cytochem. 22, 1077–1083.

Immunolabelling for Electron Microscopy (Polak/Varndell, eds)

CHAPTER 7

Immunoenzyme techniques at the electron microscopical level

Georges Pelletier[1] and Gérard Morel[2]

[1]*MRC Group in Molecular Endocrinology, Le Centre Hospitalier de l'Université Laval, Quebec G1V 4G2, Canada and* [2]*Laboratoire d'Histologie-Embryologie, CNRS-ERA981, Faculté de Médecine, Lyon-Sud, Oullins, France*

CONTENTS

1. INTRODUCTION

The first serious attempts to apply immunoenzymatic techniques to electron microscopy were probably realised in the late sixties by Nakane and Pierce [1, 2] who used horseradish peroxidase (HRP) or acid phosphatase covalently bound to antibodies. Using pre-embedding immunostaining, these authors succeeded in identifying several of the cell types in the anterior pituitary. However, the ultrastructural preservation was not adequate for an extensive study of the organelles contained within these secretory cells.

In 1970, a very sensitive immunoperoxidase technique was developed by Sternberger et al. [3, 4] who used a peroxidase-antiperoxidase (PAP) complex as the labelling agent. This discovery prompted several research groups to use the PAP complex in different ways to detect antigens at the ultrastructural level. More recently, other immunoperoxidase techniques, such as those involving the use of protein A-peroxidase conjugates or avidin-biotin-peroxidase complexes (see Chapter 2, pp. 17–26) have also been successfully employed for immunoelectron microscopy. In this chapter we have tried to summarise what is known about the advantages and disadvantages of the immunoenzymatic techniques which are still the most commonly used procedures to detect antigens at the electron microscopical level. We have also discussed the combination of some techniques which can be applied to detect more than one substance in the same ultrathin section.

2. Post-embedding immunostaining technique

Since the early seventies, the PAP technique described by Sternberger [3, 4] has been largely used for staining antigens present in ultrathin sections of embedded tissue (post-embedding immunostaining). In early studies, Hardy et al. [5] and Moriarty and Halmi [6] found that better tissue preservation could be achieved by using epoxy resins (Epon or Araldite) for embedding, rather than glycol methacrylate even though the plastic section had to be 'etched' (with hydrogen peroxide solutions) to make antigenic determinants accessible to the applied antibodies.

Based on the results of early studies it was considered that osmium tetroxide denatured many antigenic determinants and, consequently, most post-embedding immunostaining was performed on aldehyde-fixed tissue without post-osmication (for a review, see Sternberger [4]). However, in 1978 Dacheux and Dubois [7] demonstrated that some pituitary hormones could be immunostained after primary fixation in osmium tetroxide and embedding in Epon. Subsequently, we have extended this work and have been able to demonstrate that not only all the currently known pituitary hormones but also several neuropeptides could be readily immunostained in tissue fixed in glutaraldehyde, post-fixed in osmium tetroxide and embedded in Araldite in the conventional way [8]. It now seems clear that in some cases conventionally fixed and embedded tissue can be used for immunoelectron microscopy, provided that a sensitive technique is used (Figs. 1, 2A). The two most sensitive immunoperoxidase techniques to date are generally accepted to be the PAP and the Avidin-Biotin Complex (ABC) techniques. These have been described in considerable detail [4, 9] and in other chapters in this volume. In our hands, both techniques applied for the localisation of pituitary hormones or neuropeptides seem to be equally sensitive, although, as reported by Childs and Unabia [9], the ABC procedure is relatively more sensitive when short-term incubations (1–3 h) of the primary antiserum are used.

The major advantages of the post-embedding immunoperoxidase staining techniques are:

a) the direct accessibility of antibodies to all cells and organelles which are exposed in a 70–100 nm section;
b) the potential to immunostain serial sections with different antibodies in order to study the co-existence of substances in the same cells or organelles [11–15];
c) the high degree of sensitivity which permits the low concentrations of primary antisera, normally at a dilution ranging from 1:1000 to 1:100,000 and occasionally in excess of this [10];

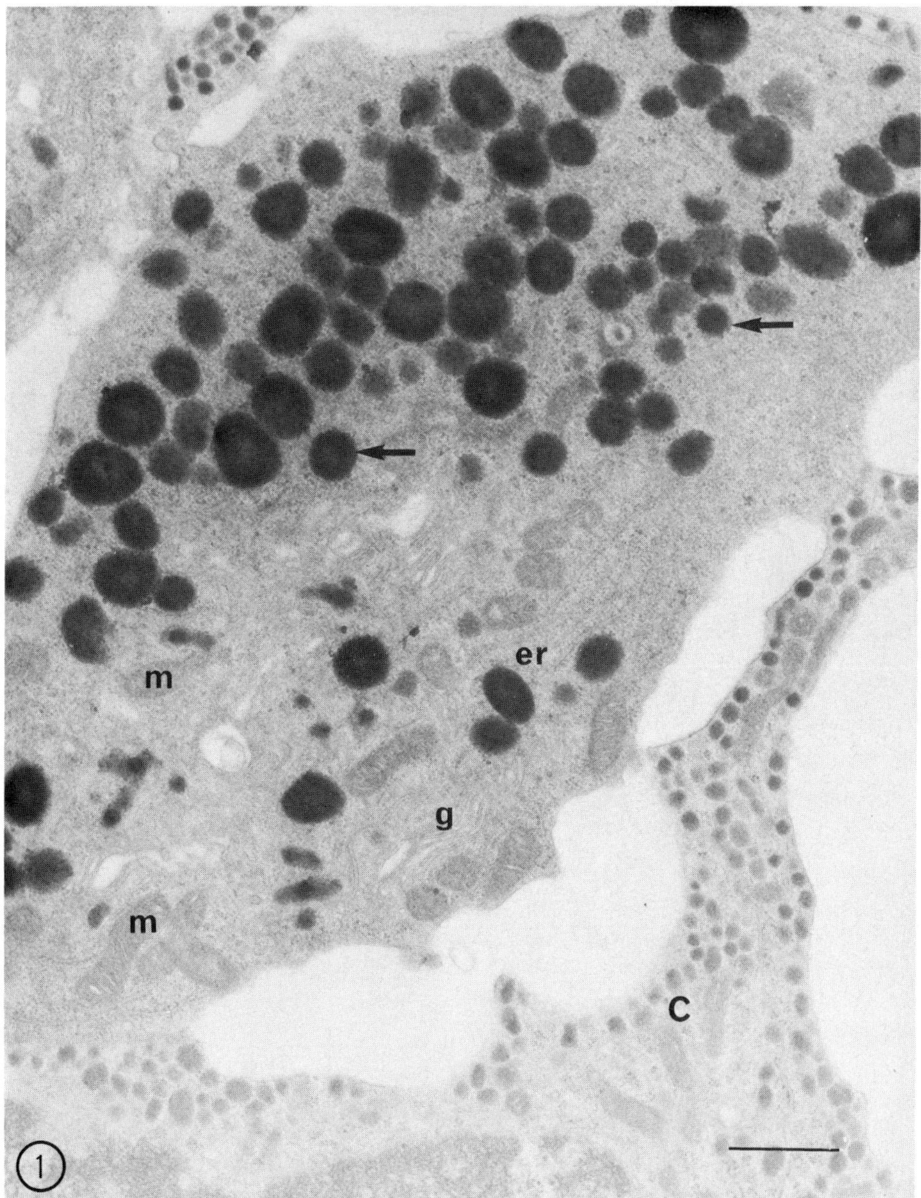

Fig. 1. Post-embedding staining for prolactin in rat pituitary tissue. Accumulation of molecules of PAP can be seen overlying the secretory granules (arrows). Note the excellent preservation of the rough endoplasmic reticulum (er), the Golgi apparatus (g) and the mitochondria (m). An adjacent corticotroph (C) is unlabelled. Glutaraldehyde-osmium fixation. Scale bar represents 1 μm.

d) the excellent tissue preservation, especially when conventional fixation regimes can be used.

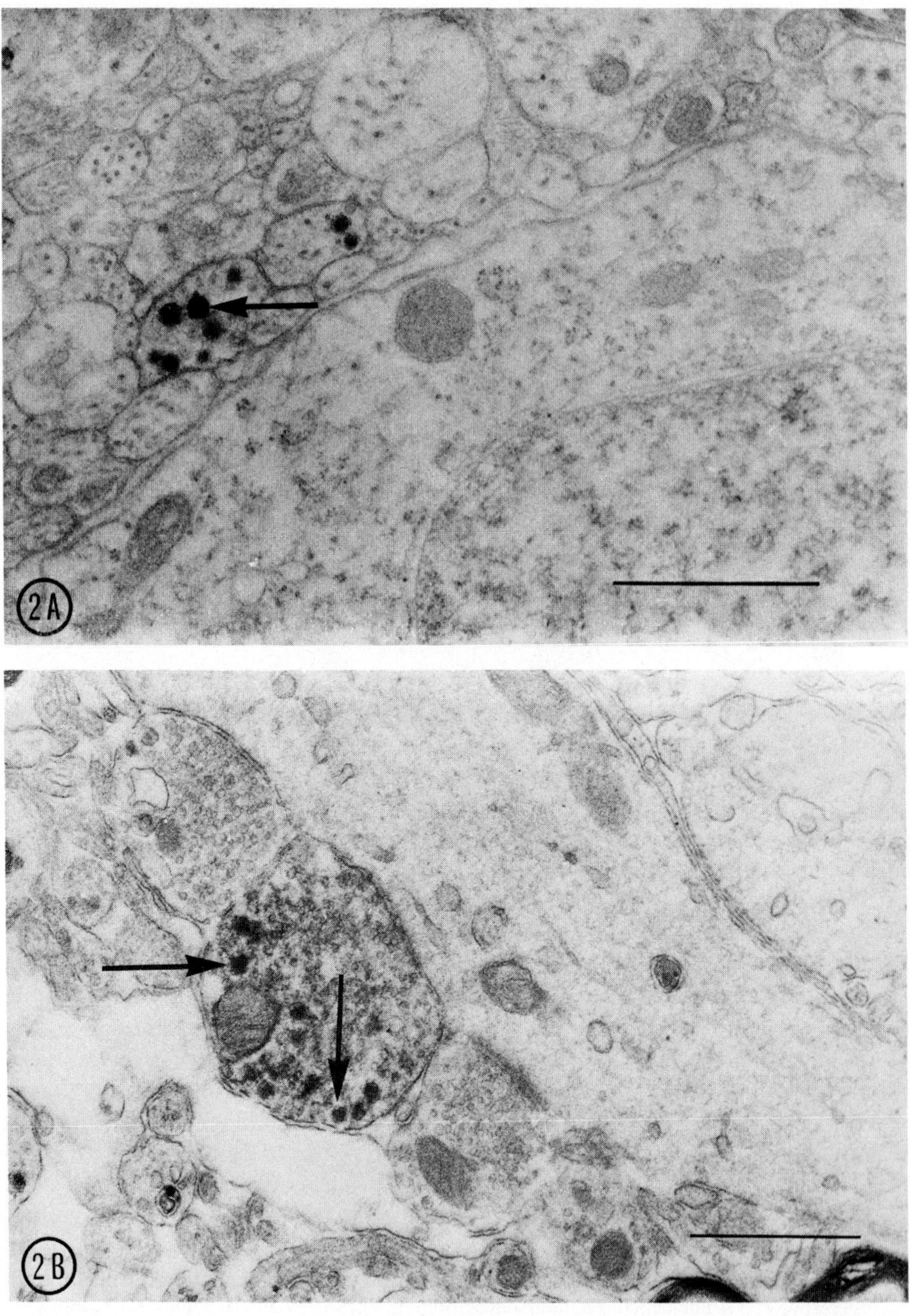

Fig. 2. Comparison between the pre- and post-embedding staining for α-MSH in the rat hypothalamus. With the post-embedding staining (A), the reaction product is restricted to dense core vesicles (arrows) in endings, whereas with the pre-embedding staining (B), the reaction product is not only found in dense core vesicles (arrows), but is also more diffusely located throughout the axoplasm. Glutaraldehyde-osmium fixation. Scale bars represent 1 μm.

The main disadvantage of these techniques is related to the fact that they cannot be applied to all antigens, for example, those that apparently do not survive dehydration and/or embedding. This is particularly the case for several enzymes, e.g. tyrosine hydroxylase (TH) [16], glutamic acid decarboxylase (GAD) [17] and neuron-specific enolase (NSE) [18]. Also, when paraformaldehyde alone is used as the primary fixative in order to maximise antigenicity, the ultrastructural preservation is often not suitable for adequate analysis of the tissue.

3. Pre-embedding immunostaining technique

For those antigens which do not survive conventional tissue processing procedures, one alternative is to perform immunostaining prior to embedding (pre-embedding immunostaining technique), the other one being to use ultrathin frozen sections (see section 4 below). Following initial attempts by Nakane's group [1, 2], Pickel et al [16] have demonstrated that strong staining and adequate tissue preservation could be obtained when an immunoperoxidase procedure was performed after fixation in aldehydes but prior to post-fixation with osmium tetroxide and plastic embedding. In this technique [4, 16] the tissue is generally fixed in a mixture of 4% paraformaldehyde and 0.1–0.3% glutaraldehyde. Vibratome® sections of the tissue, 20–40 μm thick, are generally pre-treated with 0.2% Triton X-100 to facilitate the penetration of antibodies.

The immunostaining technique involves the use of PAP complex and has been fully described by Pickel et al [16] and Sternberger [4]. We have also used the avidin-biotin-peroxidase complex and protein A-peroxidase techniques in place of the PAP reagent and we have observed no significant difference in staining intensity. Once immunostaining has been performed, the reactive areas are dissected out, post-fixed in osmium tetroxide and flat-embedded in Epon or Araldite.

Since only the outermost 3–4 μm of the sections are suitable for immunoelectron microscopy, ultrathin sectioning must be performed with caution. When pre-treatment with Triton is omitted in order to improve the preservation, only the outer 1–2 μm are optimally immunostained and thus suitable for ultrathin sectioning. It seems that antibodies are incapable of penetrating intact hydrophobic cell membranes. This is emphasised by the fact that De Mey and colleagues [19] used Triton to permeabilise the membranes of cells grown in monoculture prior to pre-embedding immunostaining in their early studies of cytoskeletal proteins.

The pre-embedding technique has the advantage of being suitable for most antigens since staining is performed prior to embedding, whilst providing good ultrastructural preservation due to the use of post-fixation in osmium (Fig. 2B). This is the technique of choice for several brain enzymes, such as tyrosine hydroxylase, glutamic acid decarboxylase and neuron-specific enolase [16–18] which do not survive osmium fixation and/or embedding. This technique also provides the possibility to observe and even photograph reactive elements at the light microscopical level and then to make a selection of the interesting areas prior to embedding. Moreover, the investigator knows in advance that there is immunoreactive material present in the ultrathin sections that he is planning to observe. This is not always the case with the post-embedding technique where one can fail to find any positive staining, especially when ultrathin sections are cut from areas which contain a low concentration of reactive elements.

The major inconvenience of the pre-embedding technique is that it is impossible to perform alternate staining of serial sections or to use an adjacent ultrathin section as an immunostaining control. Occasionally, one can experience technical difficulties when attempting to obtain ultrathin sections which combine convincing immunostaining with excellent ultrastructural preservation, since Triton destroys to some extent the ultrastructure. Unfortunately, the penetration of antibodies into tissue sections is very poor in the absence of the detergent.

4. Immunostaining of ultrathin frozen sections

Recently, another approach for immunostaining at the electron microscopical level has been developed. It involves the use of ultrathin frozen sections and combines some of the advantages of both the pre-embedding and post-embedding techniques, as non-dehydrated and non-embedded ultrathin sections can be immunostained. Since contact of sections with aqueous solutions is required for immunostaining, fixation is required prior to cryoultramicrotomy to enable thawing of sections without loss of ultrastructural integrity due to the effects of surface tension (see Chapter 6, pp. 71–82).

In our laboratories the immunostaining of frozen ultrathin sections is performed in the following way. After fixation in 2.5% glutaraldehyde and post-fixation in 1% osmium tetroxide, the tissues are cryoprotected by immersion in 0.4 M sucrose for 30 min. They are then placed in droplets of saturated sucrose on specimen supports (brass pins) and exposed to vapour phase nitrogen which forms a cold gradient above liquid nitrogen. The drop in temperature is monitored constantly. After 5–10 min, when the temperature has fallen to 269 K, the blocks are plunged into the liquid nitrogen. This method ensures good tissue preservation even deep down in the block (Morel and Dubois, in preparation). The tissues can then be stored in liquid nitrogen until required. Ultrathin sections are cut at 143 K with an Ultratome III (LKB, Stockholm, Sweden) fitted with a cryokit as described by Tokuyasu [20]. Sections are mounted on collodion-coated nickel grids and are then available for immunocytochemistry. The immunostaining technique is based on the method described by Moriarty and Halmi [6] and modified by Li et al. [21]. The ultrathin sections are incubated consecutively for 10 min periods with:

a) the primary antiserum raised in rabbit (dilution from $1:10^3$ to $1:10^8$);
b) sheep or goat antiserum to rabbit immunoglobulins (dilution $1:10^4$);
c) peroxidase-antiperoxidase complex (antiperoxidase raised in rabbit (dilution $1:10^3$);
d) 4-chloro-1-naphthol solution in Tris HCl-buffered saline containing 0.02% hydrogen peroxide;

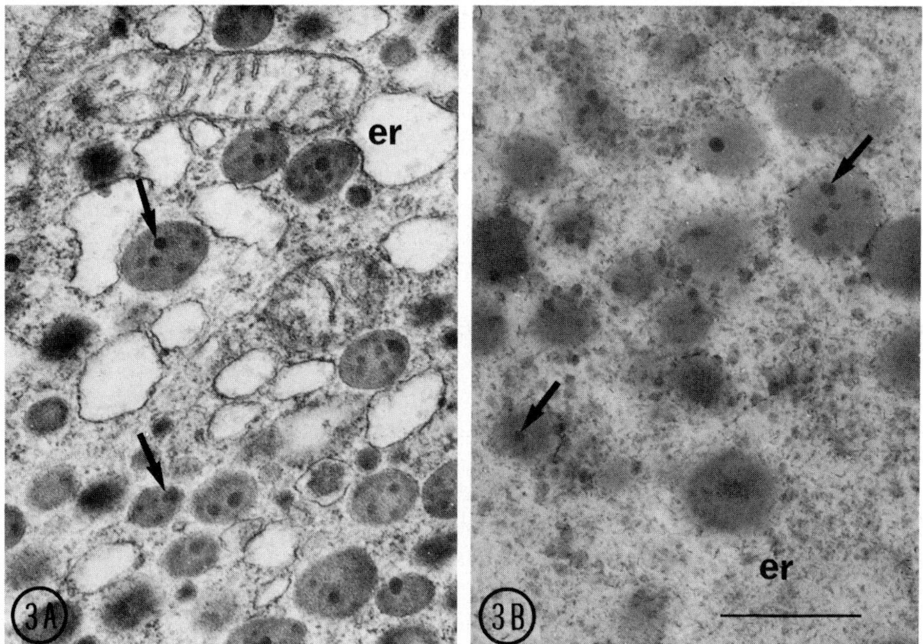

Fig. 3. Comparison between staining for β-luteinising hormone in rat gonadotrophs obtained with sections of plastic-embedded tissue and ultrathin frozen sections (cryoultramicrotomy). To immunostain plastic sections (A), the antiserum has been used at a dilution of 1:500. The staining is observed only in secretory granules (arrows) whereas the endoplasmic reticulum (er) remains unstained. Immunostaining of ultrathin frozen sections (B) has been performed with the antiserum diluted at $1{:}5 \times 10^5$. Reaction product can be detected in secretory granules (arrows) and endoplasmic reticulum (er). Glutaraldehyde-osmium fixation. Scale bars represent 0.5 μm.

e) 2% phosphate-buffered osmium tetroxide, and
f) uranyl acetate (2% aqueous solution).

After steps a, b and c, the grids are washed with 0.05 M Tris HCl buffer, pH 7.6 and after steps d, e and f, the grids are washed with distilled water. All washing steps are performed for 10 min.

This technique, which is technically more complex than the standard post- and pre-embedding techniques, combines the advantages of excellent antigenic preservation, very high sensitivity and direct accessibility of antibodies to organelles (Fig. 3). With this technique, serial ultrathin sectioning of non-embedded tissue is possible. In fact, the use of frozen sections has led to the use of much higher dilutions of the antisera (up to 10^3 times more than with the usual techniques; [22]) and has allowed the detection of endogenous brain peptides such as gonadotrophin-releasing hormone (GnRH; [23]), vasoactive intestinal peptide (VIP; [24]), substance P [25], somatostatin [26], corticotropin-releasing factor (CRF; [27]) and growth hormone-releasing factor (GRF; [28]) within their target pituitary cells. Except for GnRH [4], these localisations have not been observed previously with the other, less sensitive, immunostaining techniques. Moreover, steroids, which are soluble in dehydrating solutions (such as alcohol and acetone) have also been localised using frozen sections [29, 30]. Such localisation of steroids has not been possible to date with conventional immunoelectron microscopical techniques. The major disadvantage of this technique concerns the ultrastructural preservation of the tissue which is not as good as that obtained with the pre-embedding technique or with the post-embedding technique when post-fixation with osmium tetroxide is employed.

5. Immunoperoxidase techniques and double labelling

Although alternate staining of consecutive ultrathin sections can be generally used to study co-localisation of antigens at the ultrastructural level [11–15], double labelling of the same section is sometimes required. For example, double labelling is essential when antigens are associated with organelles too small to be serially sectioned or when close relationships between cells or their organelles must be studied in great detail. Immunoperoxidase techniques can then be combined with other labelling procedures.

Pre-embedding immunostaining can be performed in the usual way as the first step, and as a second step a post-embedding technique, involving colloidal gold adsorbed to protein A or immunoglobulins [30] can be used. Such a technique is particularly useful when the antigen stained in the first step does not survive post-fixation and embedding procedures. With this approach we have been able to localise dopamine-β-hydroxylase using the pre-embedding immunoperoxidase technique and GnRH by a post-embedding method involving protein A-gold [30]. When the antigens survive routine embedding procedures, appropriate controls must be used to rule out the possibility that the colloidal gold label could cross-react with the first set of immunoreagents.

Another procedure which can be reliably applied to the localisation of more than one substance is the combination of autoradiography and immunoperoxidase staining. We have recently developed a technique for the simultaneous localisation of biogenic amines and peptides in single ultrathin sections [31]. This technique has been used to study the relationship between endings containing dopamine and those containing vasopressin in the rat posterior pituitary (Fig. 4). The method involves the injection of a tritiated biogenic amine followed by fixation and embedding according to standard procedures [8]. Ultrathin sections are then placed onto collodion-coated slides [32] and immunostained with the post-embedding PAP procedure. Finally, autoradiography of sections previously immunostained are performed by coating the slides with diluted photographic emulsion, as described by Kopriwa [32].

Recently, a novel combination of autoradiography and immunocytochemistry has been successfully performed using the pre-embedding staining procedure [33, 34]. In this case, however, only the first sections coming from the surface of the blocks can be used for autoradiography, even after Triton permeabilisation.

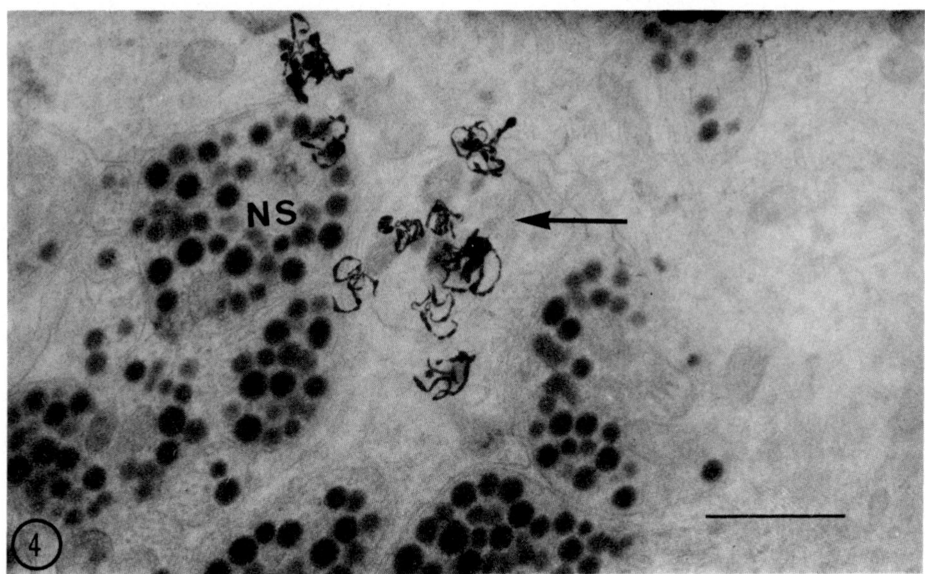

Fig. 4. Simultaneous localisation of dopamine and vasopressin in the same ultrathin section of the rat posterior pituitary. Dopamine has been localised with autoradiography after injection of [^{3}H] dopamine, whereas vasopressin was detected by immunocytochemistry. The autoradiographic reaction (arrows) can be easily distinguished from immunostaining occurring in neurosecretory (NS) granules. Glutaraldehyde-osmium fixation. Scale bar represents 1 μm.

6. Acknowledgements

Gérard Morel is the recipient of a fellowship from the Fondation pour la Recherche Médicale.

The authors wish to thank F. Hemming and P. Mesguich for the kind gift of the photographs used in Figure 3.

7. References

1. Nakane, P.K. and Pierce, B. (1968) Enzyme-labeled antibodies for the light and electron microscopic localization of tissue antigens. J. Cell Biol. 33, 307–318.
2. Nakane, P.K. (1970) Classification of anterior pituitary cell types with immunoenzyme histochemistry. J. Histochem. Cytochem. 18, 9–19.
3. Sternberger, L.A., Hardy, Jr., P.H., Cuculis, J.J. and Meyer, G.H. (1970) The unlabeled antibody-enzyme method of immunohistochemistry. Preparation and properties of soluble antigen-antibody complex (horseradish peroxidase anti-horseradish peroxidase) and its use in identification of spirochetes. J. Histochem. Cytochem. 18, 315–325.
4. Sternberger, L.A. (1979) Immunocytochemistry, 2nd Ed. John Wiley and Sons, New York.
5. Hardy, P.H., Meyer, H.G., Cuculis, J.J., Petrali, J.P. and Sternberger, L.A. (1970) Post-embedding staining for electron microscopy by the unlabeled antibody peroxidase method. J. Histochem. Cytochem. 18, 684–692.

6. Moriarty, G.C. and Halmi, N.S. (1972) Adrenocorticotropin production by the intermediate lobe of the rat pituitary. An electron microscopic immunohistochemical study. Z. Zellforsch Mikrosk. Anat. 132, 1–14.
7. Dacheux, F. and Dubois, M.P. (1976) Ultrastructural localization of prolactin, growth hormone and luteinizing hormone by immunocytochemical techniques in the bovine pituitary. Cell Tissue Res. 174, 313–322.
8. Pelletier, G., Puviani, R., Rosler, O. and Descarries, L. (1981) Immunocytochemical detection of peptides in osmicated and plastic-embedded tissue. J. Histochem. Cytochem. 29, 759–764.
9. Childs, G.V. and Unabia, G. (1982) Application of a rapid avidin-biotin-peroxidase complex (ABC) technique to the localization of pituitary hormones at the electron microscopic level. J. Histochem. Cytochem. 30, 1320–1329.
10. Moriarty, G.C., Moriarty, C.M. and Sternberger, L.A. (1973) Ultrastructural immunocytochemistry with unlabeled antibodies and the peroxidase-antiperoxidase complex. A technique more sensitive than radioimmunoassay. J. Histochem. Cytochem. 21, 825–836.
11. Vaudry, H., Pelletier, G., Guy, J., Leclerc, R. and Jegou, S. (1980) Immunohistochemical localization of γ-endorphin in the rat pituitary gland and hypothalamus. Endocrinology 106, 1512–1520.
12. Pelletier, G. (1980) Immunohistochemical localization of a fragment (16 K) of the common precursor for ACTH and β-LPH in the rat hypothalamus. Neurosci. Lett. 16, 85–90.
13. Guy, J., Leclerc, R. and Pelletier, G. (1980) Localization of a 16,000 dalton fragment of the common precursor of adrenocorticotropin and β-lipotropin in the rat and human pituitary gland. J. Cell Biol. 86, 825–830.
14. Pelletier, G., Steinbusch, H.W.M. and Verhofstad, A.A.J. (1981) Immunoreactive substance P and serotonin present in the same dense-core vesicles. Nature (London) 293, 71–72.
15. Leboulenger, F., Leroux, P., Tonon, M.C., Coy, D.H., Vaudry, H. and Pelletier, G. (1983) Co-existence of vasoactive intestinal peptide and enkephalins in the adrenal chromaffin granules of the frog. Neurosci. Lett. 37, 221–225.
16. Pickel, V.M., Joh, T.H. and Reis, D.J. (1975) Ultrastructural localization of tyrosine hydroxylase in noradrenergic neurons of brain. Proc. Natl. Acad. Sci. USA 72, 658–663.
17. Tappaz, M.L., Wassef, M., Oertel, W.H., Prout, L. and Pujol, J.F. (1981) Glutamic acid decarboxylase (GAD) immunocytochemistry in the medial basal hypothalamus: morphological evidence for GABAergic neuroendocrine regulations. Neurosci. Lett. Suppl. 7, S 255.
18. Langley, O.K., Ghandour, M.S., Vincendon, G. and Gambos, G. (1980) An ultrastructural immunocytochemical study of nerve-specific protein in rat cerebellum. J. Neurocytol. 9, 783–798.
19. De Mey, J., Hoekeke, J., De Brabander, M., Geuens, G. and Joniau, M. (1976) Immunoperoxidase visualization of microtubules and microtubular proteins. Nature (London) 264, 273–274.
20. Tokuyasu, K.T. (1973) A technique for ultracryotomy of cell suspensions and tissues. J. Cell Biol. 57, 551–565.
21. Li, S.Y., Dubois, M.P. and Dubois, P.M. (1977) Somatotrophs in the human fetal anterior pituitary. An electron microscopic immunocytochemical study. Cell Tissue Res. 181, 545–552.
22. Hemming, F.J., Mesguich, P., Morel, G. and Dubois, P.M. (1983) Cryoultramicrotomy versus plastic embedding: comparative immunocytochemistry of rat anterior pituitary cells. J. Microsc. 131, 25–33.
23. Morel, G. and Dubois, P.M. (1982) Immunocytochemical evidence for gonadoliberin in rat anterior pituitary gland. Neuroendocrinology 34, 197–206.
24. Morel, G., Besson, J., Rosselin, G. and Dubois, P.M. (1982) Ultrastructural evidence for endogenous vasoactive intestinal peptide-like immunoreactivity in the pituitary gland. Neuroendocrinology 34, 85–89.
25. Morel, G., Chayvialle, J.A., Kerdelhue, B. and Dubois, P.M. (1982) Ultrastructural evidence for endogenous substance P-like immunoreactivity in the rat pituitary gland. Neuroendocrinology 35, 86–92.
26. Morel, G., Mesguich, P., Dubois, M.P. and Dubois, P.M. (1983) Ultrastructural evidence for endogenous somatostatin-like immunoreactivity in pituitary gland. Neuroendocrinology 36, 291–299.

27. Morel, G., Hemming, F., Tonon, M.C., Vaudry, H., Dubois, M.P., Coy, D.H. and Dubois, P.M. (1982) Ultrastructural evidence for corticotropin-releasing factor (CRF)-like immunoreactivity in the rat pituitary gland. Biol. Cell. 44, 89–92.
28. Morel, G., Mesguich, P., Dubois, M.P. and Dubois, P.M. (1984) Ultrastructural evidence for endogenous growth hormone-releasing factor (GRF)-like immunoreactivity in the monkey pituitary gland. Neuroendocrinology 38, 123–133.
29. Morel, G., Forest, M.G. and Dubois, P.M. (1980) Etude des variations de la capture de la testostérone par les cellules gonadotrophes de l'antéhypophyse du rat. C.R. Acad. Sci. (Paris) 290, 1579–1582.
30. Morel, G., Forest, M.G. and Dubois, P.M. (1984) Ultrastructural evidence for endogenous testosterone immunoreactivity in rat pituitary gland. Cell. Tissue Res. 235, 159–169.
31. Pelletier, G. (1983) Identification of endings containing dopamine and vasopressin in the rat posterior pituitary by a combination of radioautography and immunocytochemistry at the ultrastructural level. J. Histochem. Cytochem. 31, 562–564.
32. Kopriwa, B.W. (1967) A semi-automatic instrument for the radioautographic technique. J. Histochem. Cytochem. 14, 923–928.
33. Bosler, O., Bloch, B., Bugnon, C. and Calas, H. (1982) Bases morphologiques des interactions monoamines-peptides dans l'hypothalamus neuroendocrine du rat. Données radioautographiques et immunocytochimiques. Coll. INSERM, 110, 17–36.
34. Nakai, Y., Shiada, S., Ochiai, H., Kudo, J. and Hashimoto, A. (1983) Ultrastructural relationship between monoamine- and TRH-containing axons in the rat median eminence as revealed by combined autoradiography and immunocytochemistry in the same tissue section. Cell Tissue Res. 230, 1–14.

Immunolabelling for Electron Microscopy (Polak/Varndell, eds)

CHAPTER 8

The streptavidin-biotin bridge technique: Application in light and electron microscope immunocytochemistry

Claude Bonnard[1], David S. Papermaster[2] and Jean-Pierre Kraehenbuhl[1]

[1]*Swiss Institute for Experimental Cancer Research, and the Institute for Biochemistry, University of Lausanne, CH-1066 Epalinges, Switzerland and* [2]*Department of Pathology, Yale Medical School, New Haven, CT 06510, U.S.A.*

CONTENTS

1. Introduction

The interaction of the egg white protein avidin with the coenzyme biotin known for more than 40 years, has simplified the problem of linking tracers to biological probes. The biotin-avidin system has been widely used in immunocytochemistry, and more recently in molecular biology. Avidin-ferritin conjugates have been exploited in determining the localisation of biotinyl lipids [1], proteins [2], nucleic acids [3], and opsin antibodies [4]. Avidin binds biotin with one of the strongest non-covalent bonds yet discovered (K_A $10^{15}\,M^{-1}$) [5], allowing rapid reactions with diluted biotinyl- or avidin conjugates. Despite this very high affinity, problems of non-specific binding have been reported which were attributed to the stickiness of the protein caused by its carbohydrate moiety [6]. In addition, the pI of avidin is 10 which may favour ionic interactions with anionic cell surfaces and hence partially explain non-specific binding of avidin conjugates [7].

Streptavidin, released from cultures of *Streptomyces avidini* has properties similar to avidin, i.e. both proteins are tetramers with subunit molecular weights of about 15,000, are rich in tryptophan, bind biotin with extremely high affinity and are stable to treatment with urea, guanidine-HCl and heat. Streptavidin is non-glycosylated and has a neutral pI [8]. These latter properties make streptavidin a superior reagent for the detection of biotinylated ligands. Conjugation of biotin to biological macromolecules is usually a simple procedure and numerous derivatives have been described [9]. Unfortunately, the conjugation of biotin to ligands using simple and conventional procedures can drastically reduce the affinity of the biotinylated ligand for avidin or streptavidin.

In this report we describe experimental procedures allowing:

a) optimal biotinylation of antibodies, and
b) the preparation of fluorochrome- or gold-streptavidin conjugates.

We illustrate the use of the streptavidin-biotin bridge technique in light and electron microscope immunocytochemistry by localising (Na^+, K^+)ATPase subunits in amphibian cells and tissues.

2. Technical procedures

2.1. *Material*

Streptavidin was purchased from BRL (Bethesda Research Laboratories, GmBH, Neu Isenburg, West Germany). Bovine serum albumin fraction V (BSA) was obtained from Armour Pharmaceutical Co. (Chicago, Illinois, U.S.A.). [^{14}C]biotin and Na^{125}I (carrier free) were obtained from Amersham International plc (Buckinghamshire, U.K.). Fluorescein isothiocyanate (FITC) was from Miles Laboratories (Kankakee, Illinois, U.S.A.) and lissamine rhodamine B-sulphonyl chloride from Polysciences, Inc. (Washington, PA, U.S.A.). *N*-biotinyl-ω-aminocaproyl-*N*-hydroxysuccinimide was purchased from Enzo Biochem., Inc. (New York, N.Y. U.S.A.).

2.2. *Methods*

2.2.1. *Synthesis of [^{14}C]biotinyl-N-hydroxysuccinimide*

[^{14}C]biotinyl-*N*-hydroxysuccinimide ester (B-NHS) was synthesised as previously described [2]. Briefly, 1 g of biotin and 250 μCi [^{14}C]biotin were dissolved into 60 ml of dry dimethylformamide. After addition of 0.51 g of *N*-hydroxysuccinimide (recrystallised from ethanol) and 0.93 g of dicyclohexylcarbodiimide, the mixture was stirred for 16 h at 50°C under nitrogen. After cooling to room temperature the reaction mixture was filtered and the filtrate evaporated under reduced pressure. After recrystallisation from isopropanol, [^{14}C]B-NHS was dissolved in dry dimethylformamide and stored at −20°C.

2.2.2. *Coupling of [^{14}C]B-NHS to proteins*

The presence of free amines in the buffers must be avoided, and a fresh preparation of antibodies passed on a Sephadex G-25 column equilibrated with 10 mM phosphate buffer, pH 7.4, containing 0.15 M NaCl (PBS) is required for optimal coupling. Approximately 100 μl [^{14}C]biotin (2.5 μCi; 10 mg/ml) in dimethylformamide was incubated for 2 h at room temperature with antibodies dissolved in PBS (1–2 mg/ml). Unreacted B-NHS was separated from biotinylated antibodies by gel filtration on Sephadex G-25 equilibrated with PBS. The number of biotin molecules bound per molecule of protein was estimated by scintillation counting. A biotin to protein molar ratio of 200 in the reaction mixture resulted in the binding of between 5 and 150 biotin molecules per molecule of protein. Increasing the protein concentration in the reaction mixture at constant molar ratio resulted in an increase in biotinylation.

2.2.3. *Coupling of N-biotinyl-ω-aminocaproyl-N-hydroxysuccinimide (B-ω-AC-NHS)*

The coupling was performed according to the supplier's instructions with modifications [10]. The reagent, dissolved at 1 mg/ml in dimethylformamide, was added at a molar excess of 50 to the antibody dissolved in PBS and incubated for 2 h at room temperature. The unbound reagent was separated from the antibody by gel filtration on a Sephadex G-25 column.

2.2.4. Coupling of fluorochromes to streptavidin

The coupling of both FITC or Rb200SC (lissamine-rhodamine B sulphonyl chloride) was performed according to Brandtzaeg [11] with the following modifications.

2.2.4.1. FITC-streptavidin. 200 μl of a 1 mg/ml FITC solution were added to 1 mg of streptavidin dissolved in 1 ml of 0.1 M carbonate pH 9.5 and incubated for 1 h at 37°C. Free fluorochrome was separated from FITC-streptavidin by Sephadex G-25 gel filtration.

2.2.4.2. Lissamine-streptavidin. 129 μl of a solution of 1.2 mg/ml Rb200SC were added to 1 mg of streptavidin dissolved in 1 ml of PBS. The pH was raised to 9.5 by addition of 0.1 M NaOH and the mixture was stirred for 1 h at room temperature. The conjugate was separated from unbound lissamine by Sephadex G-25 gel filtration.

2.2.5. Preparation of gold-protein complexes

2.2.5.1. 2 nm gold particles. Colloidal gold solutions were prepared by reduction of tetrachloroauric acid ($HAuCl_4 \cdot H_2O$) with sodium borohydride ($NaBH_4$). 150 μl of a 4% $HAuCl_4$ solution and 200 μl of 0.2 M K_2CO_3 were added to 40 ml double distilled water that had been pre-cooled to 4°C. Under rapid stirring, 400 μl aliquots (3–5) of freshly prepared $NaBH_4$ (0.5 mg/ml) were rapidly added until no further colour change from bluish-purple to reddish-orange was observed. The gold solution was then stirred for an additional 5 min.

2.2.5.2. 10 nm gold particles. Larger colloidal gold particles (10 nm) were prepared according to the procedure of Frens [12]. Three millilitres of a 1% sodium citrate solution was added to 100 ml of a 0.01% $HAuCl_4$ solution, boiled for 30 min and cooled to 4°C.

2.2.5.3. Gold-streptavidin complexes. The procedure was modified from Geoghegan and Ackerman [13]. To 20 ml of gold solution (2–5 or 10 nm gold particles), 200 μl of 1.0 M $NaHCO_3$ and 0.5 ml streptavidin (1 mg/ml) in 1 mM phosphate buffer, pH 7.4 were added and the solution stirred for 10 min at room temperature. Then, 200 μl of 2% polyethylene glycol 6,000 (Carbowax) was added. The 2–5 nm gold solution was subsequently centrifuged for 30 min at 10,000 rpm (JA21 Beckman rotor). The supernatant was loaded on top of a 37.5% sucrose cushion (in 0.1 M phosphate buffer, pH 7.4 containing 0.02% polyethylene glycol) and centrifuged in an SW40 Beckman rotor for 30 min at 40,000 rpm. The resulting pellet was resuspended in 0.1 M phosphate buffer pH 7.4 containing 0.02% polyethylene glycol and 0.05% sodium azide and stored at 4°C until use. All the glassware was siliconised. The preparation of 10 nm gold-streptavidin complexes was similar, except that the first centrifugation was performed at 3,000 rpm (JA21 Beckman rotor) for 30 min, and the second centrifugation at 11,000 rpm for 30 min, using the same rotor.

2.2.5.4. Biotinyl-BSA-gold complexes. Biotin was coupled to bovine serum albumin using the biotinyl-ω-aminocaproyl-*N*-hydroxysuccinimide ester linker arm as described above. The complexes were prepared as with streptavidin except that the concentration of biotinyl-BSA was 5 mg/ml.

2.3. *Microscopy*

2.3.1. *Light microscope immunocytochemistry*

Toad (*Bufo marinus*) cells and tissues were fixed in 3% formaldehyde and 0.5% glutaraldehyde buffered with 0.1 M phosphate buffer pH 7.4 for 4 h and processed for immunocytochemistry [14]. The sections were subsequently incubated with immunological reagents.

2.3.1.1. Single labelling. The sections were incubated for 30 min with affinity purified rabbit IgG (Na^+, K^+)ATPase α-subunit antiserum [15] at 0.1 mg/ml in PBS containing 0.2% gelatin, followed by three washes with PBS containing 0.2% gelatin. After a second incubation for 30 min with biotinylated sheep IgG anti-rabbit $F(ab')_2$ (0.1 mg/ml), the sections were washed three times with PBS-gelatin, and further incubated for 15 min with streptavidin-FITC, or streptavidin-lissamine (0.1 mg/ml).

2.3.1.2. Double labelling. The sections were incubated for 30 min with biotinylated mouse monoclonal IgG_1 anti-β-subunit (glycoprotein subunit from (Na^+, K^+)ATPase) at 0.1 mg/ml. After three washes with PBS-gelatin, the sections were further incubated for 15 min with streptavidin-lissamine (0.1 mg/ml), washed in PBS-gelatin, and incubated for 5 min with PBS-gelatin containing 1 mg/ml biotin. This step is required to quench any free biotin binding sites on the lissamine-streptavidin complex. The sections were then reacted for 30 min with biotinyl-rabbit IgG anti-(Na^+, K^+)ATPase α-subunit (0.1 mg/ml). After three washes with PBS-gelatin, the sections were reacted for 15 min with streptavidin-FITC (0.1 mg/ml), and washed three times with PBS-gelatin. The slides were then mounted in Citifluor mountant (Goodwin and Davidson, Dept. of Chemistry, The City University, London), and observed in a Zeiss photomicroscope II equipped with a fluorescence attachment.

2.3.2. *Electron microscope immunocytochemistry*

The tissues were fixed as described for light microscopy and embedded either in bovine serum albumin [16], Lowicryl [17] (see Chapter 3, pp. 29–36) or in Epon. Thin sections recovered on copper grids were incubated with immunological reagent as follows:

–15 min in PBS-1% bovine serum albumin (BSA); 30 min with biotinylated-antibodies; three washes in PBS-1% BSA; 30 min with streptavidin-colloidal gold; three washes in PBS-1% BSA. The sections were stained with uranyl acetate and lead citrate. Toad blood cells, fixed as described, were also incubated with immunological reagents prior to embedding. The cells were reacted as follows:
—a 30 min incubation with biotinylated antibodies, followed by extensive washes in PBS-0.5% BSA. After a 15-min incubation with streptavidin, the cells were washed in PBS-0.5% BSA, and finally incubated for 15 min with biotinyl-BSA-gold.

3. Results

3.1. *Properties of streptavidin-biotinylated macromolecule interactions*

In order to link streptavidin efficiently to biological probes, biotin bound to the macromolecular carrier should be fully accessible to streptavidin's biotin binding site with minimal steric hindrance. Hence the biotinylated linkers require the following properties. Firstly the spacer between biotin and the macromolecule should be at least 1 nm long, since the binding site of biotin in avidin, and presumably in streptavidin, resides within a deep (1 nm) depression [5], and secondly the polarity of the spacer should be compatible with a final reaction in an aqueous phase, i.e. the biotinylated linker should be soluble in a water-miscible solvent. We have developed an in vitro assay which allows one to assess the role of linker arms of various lengths on the interaction between streptavidin and biotinylated macromolecules. Red blood cells were derivatised with biotinylated linker arms and then tested for the capacity to bind radioiodinated streptavidin.

3.1.1. *Role of the length of biotinylated linkers on the kinetics of binding*

Red blood cells were derivatised with biotinyl-*N*-hydroxysuccinimide (B-NHS) or with biotinyl-ω-aminocaproyl-*N*-hydroxysuccinimide (B-ω-AC-NHS). 10^9 cells were incubated for 2 h at room temperature with 2×10^{-6} mole of B-NHS or B-ω-AC-NHS dissolved in 50 μl of dimethylformamide. The cells were then fixed with 3% formaldehyde-0.5% glutaraldehyde and unreacted aldehyde groups quenched with BSA. In a typical assay, aliquots of biotinylated cells (10^5) were incubated with ^{125}I-streptavidin (10^5 cpm) which corresponds to the concentration where the binding sites were half saturated. At different time points, four 100 μl aliquots of the cell suspension were centrifuged through a 12% polyethylene glycol cushion in order to separate bound from free streptavidin. The pellets were counted and the results expressed as percentage-bound. As shown in Fig. 1, the length of the spacer arm drastically increased the rate of the reaction. A 7-atom spacer was long enough to allow a significant increase in the rate. However, an additional elongation of the spacer, obtained with tetra-β-alanine (16 atoms) did not further increase the rate of the reaction (data not shown).

3.1.2. *pH dependency of the biotinyl-macromolecule-streptavidin interaction*

In order to determine the pH for optimal interaction between biotinylated macromolecules and streptavidin, biotinylated toad red blood cells were incubated with ^{125}I-streptavidin at three different pHs: 4.0; 7.4; and 11.0. When biotin was coupled via hydroxysuccinimide ester to red blood cells, the highest rate of binding was observed at pH 4.0 (Fig. 2). When a 7-atom spacer was introduced between the biotin and the red blood cell membrane, the maximum binding and highest rate were observed at pH 7.4 (data not shown).

Based on these observations, we propose that a 7-atom spacer should be used to link biotin to macromolecules including antibodies, and that the reaction should be performed with streptavidin at pH 7.4. Unfortunately, [^{14}C]biotin-ω-aminocaproyl-

N-hydroxysuccinimide ester has not been synthesised so far. Therefore it is not possible to monitor the degree of biotinylation with this linker arm, a parameter which should be determined in order to optimise the system.

3.2. Light microscope immunocytochemistry

Light microscope immunocytochemistry still remains an attractive and valuable method to correlate the function of a macromolecule with its cellular or tissue

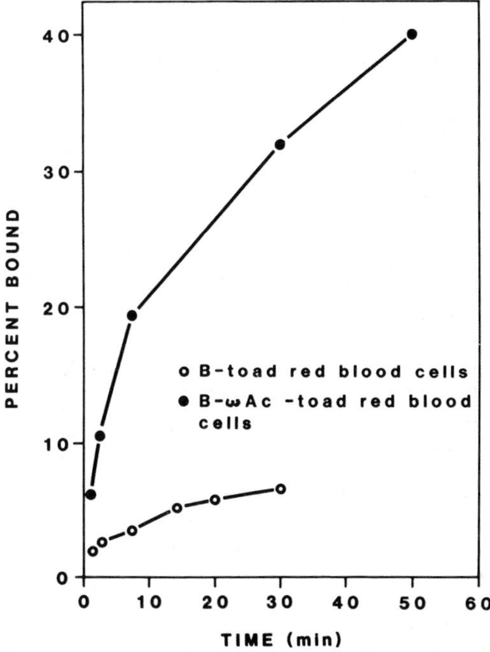

Fig. 1. Influence of a linker arm on the binding of ^{125}I-streptavidin to biotinylated toad erythrocytes. Red blood cells (10^5) biotinylated with biotinyl-N-hydroxysuccinimide ester (○—○), or biotinyl-ω-aminocaproyl-N-hydroxysuccinimide, were incubated at room temperature and pH 6.5 with ^{125}I-streptavidin (10^5 cpm). At the indicated time, the cells were processed as described in section 3 and the data expressed as percentage ^{125}I-streptavidin bound to cells.

localisation. In many cases, the high resolution provided by 0.5 μm thick frozen sections [18] is sufficient to analyse complex biological processes without the need for the electron microscope. In addition, since large areas of tissue can be screened under the light microscope, topological information related to tissue or organ organisation can be obtained. We have been motivated to evaluate the capacity of the biotin-streptavidin bridge technique to detect antigens moderately or poorly represented in cells or tissues. We have examined the localisation of the (Na^+, K^+)ATPase subunits in various transporting epithelia from *Bufo marinus*, a columbian toad. In all animal species examined so far, the oligomeric enzyme

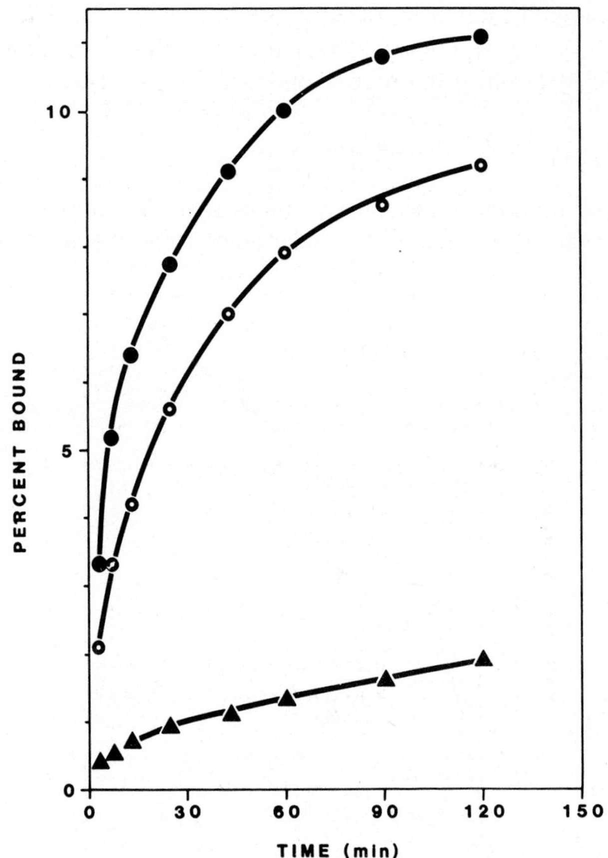

Fig. 2. Influence of pH on the binding of ^{125}I-streptavidin to biotinylated toad erythrocytes. Red blood cells (10^5) biotinylated with biotinyl-N-hydroxysuccinimide ester were incubated at room temperature with ^{125}I-streptavidin (10^5 cpm) at pH 4.0 (●—●), 7.4 (○—○), and 11.0 (▲—▲). At the indicated time, the cells were processed as described in section 3. The data are expressed as percentage ^{125}I-streptavidin bound to cells.

consists of at least two subunits, a 100,000 dalton catalytic α-subunit, and a smaller (40,000–60,000 dalton) glycoprotein β-subunit [19, 20]. Using rabbit polyclonal anti-α- or β-subunit antibodies [15], and monoclonal β-subunit antibodies, we have compared the labelling intensity between a direct procedure in which the detecting antibodies were biotinylated and visualised with fluorochrome-labelled streptavidin, and an indirect procedure in which a second step antibody directed against the detecting antibodies was biotinylated. With the polyclonal antibodies we found that the fluorescence intensity was increased in the indirect procedure, whereas with the monoclonal antibodies no significant change was noticed. In a typical experiment, illustrated in Figure 3, the α-subunit antibody labelled the basolateral membrane of a distal tubule. All cells were stained. In contrast, only some cells were labelled

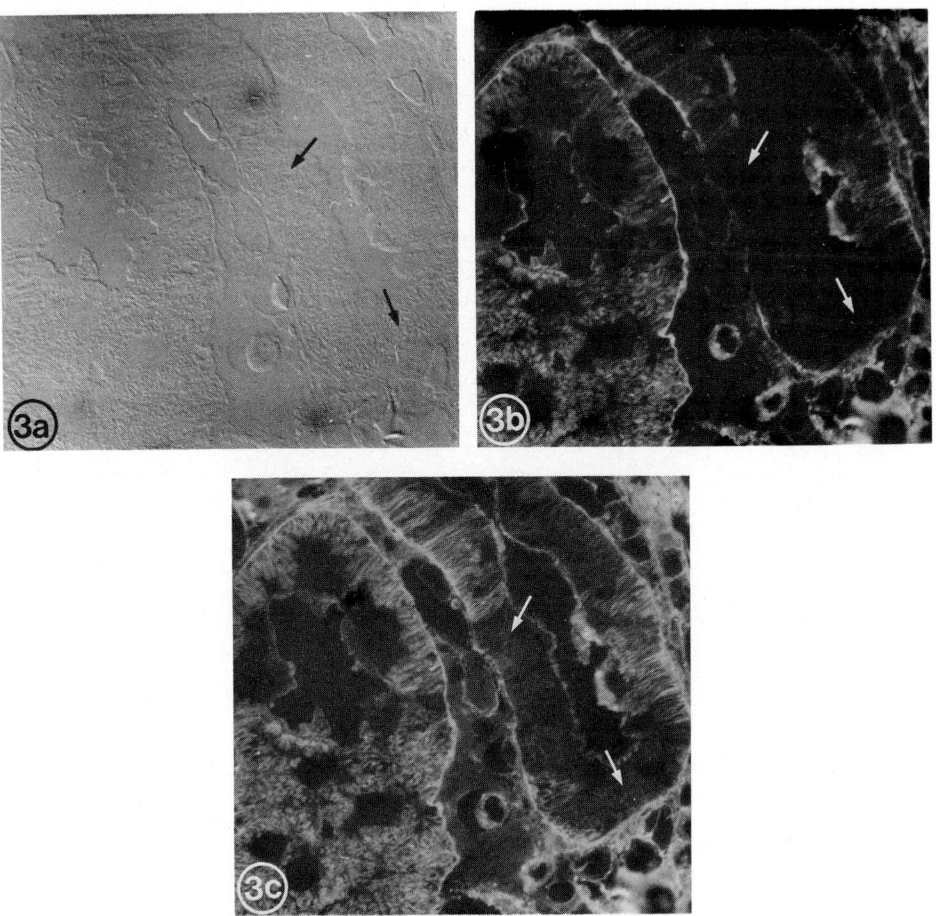

Fig. 3. Immunocytochemical localisation of the α- and β-subunit of (Na^+, K^+)ATPase in toad kidney using a double labelling procedure. Nomarski (a) and immunofluorescence (b and c) micrographs of 0.5 μm thick frozen sections processed as described in section 2. The sections were first incubated with biotinylated mouse monoclonal IgG_1 β-subunit antibodies, then with streptavidin-lissamine, followed by affinity-purified rabbit IgG anti-α-subunit, biotinylated sheep IgG anti rabbit IgG, and finally streptavidin-FITC. (b) Red fluorescence indicating the presence of β-subunit and (c) green fluorescence for the detection of α-subunit. Note that some cells from distal tubules are labelled with the α-subunit antibody, but not with the β-subunit antibody (arrows). The interstitial labelling is due to the presence of (Na^+, K^+)ATPase α-subunit epitopes. This labelling disappears when the α-subunit antibody preparation is depleted in antibodies recognising α-subunit ectoplasmic epitopes.

with the monoclonal anti-β-subunit. The biological significance of these findings will be discussed in a later publication. The detecting antibodies were absorbed with excess purified corresponding antigen in order to serve as controls. In all cases the controls were negative provided that the double labelling procedure was initiated with streptavidin-lissamine. When the enzyme was first visualised with streptavidin-FITC, there was a significant non-specific binding of streptavidin to

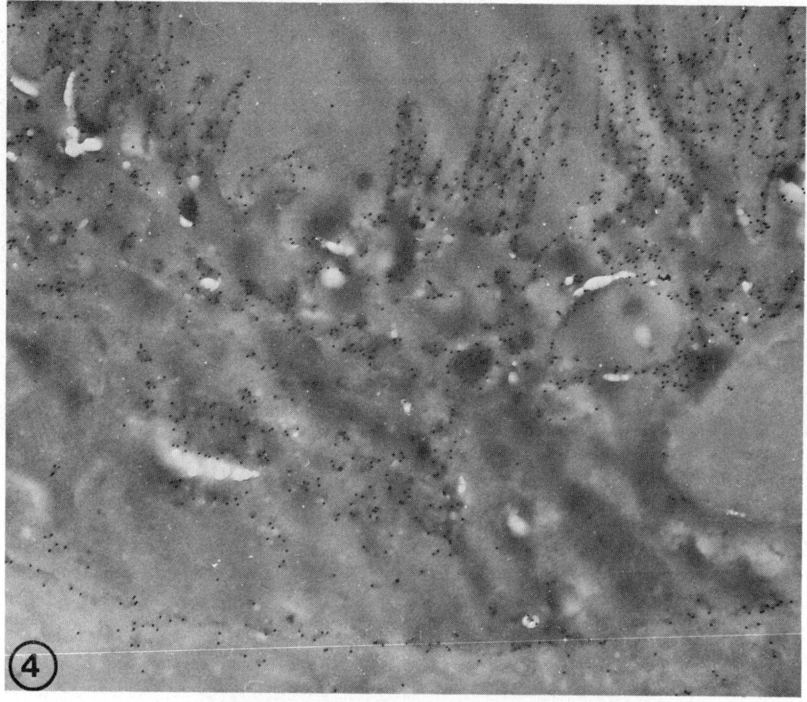

Fig. 4. Electron micrograph of toad kidney embedded in glutaraldehyde cross-linked BSA. The thin sections were labelled with biotinylated affinity-purified rabbit IgG anti-α-subunit and streptavidin-10 nm gold complexes as described in section 2. In cells from the proximal tubule gold particles decorate the microvilli in the apical membrane, the basal membrane, apical vesicles, and some regions reminiscent of Golgi complexes. The nuclei, the mitochondria and the cytosol remain unlabelled. Section stained with lead citrate and uranyl acetate. Magnification = ×16,200.

lissamine, even when the biotinylated detecting antibody was omitted. These results indicate that a unique sequence is required with the streptavidin-biotin bridge technique when double labelling is performed. Recently, the streptavidin-biotin bridge technique has allowed the subcellular detection of the mucosal antibody receptor in the rabbit mammary gland using antibodies directed against various domains of the same receptor molecule [21]. Also, and in combination with light microscope autoradiography, the technique has allowed the correlation of DNA synthesis activity of mammary cells in primary cultures with their to synthesise and store milk proteins [22, 23].

3.3. *Electron microscope immunocytochemistry*

The resolution limitations imposed by light microscopy has prompted the search for high resolution techniques which allow in situ detection at the subcellular or molecular level. Pre-embedding labelling procedures are inappropriate for in-

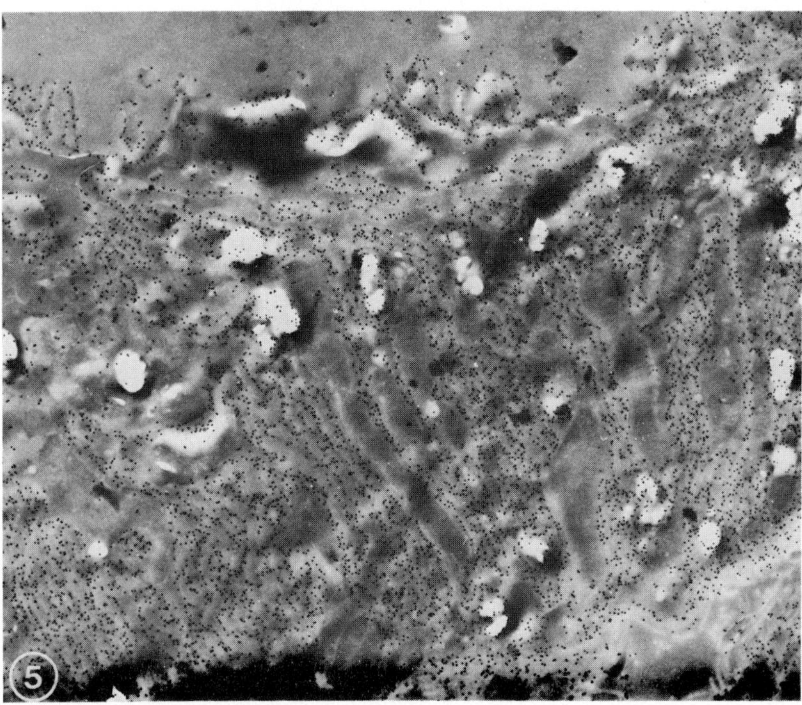

Fig. 5. Electron micrograph of toad kidney embedded in glutaraldehyde cross-linked BSA. The sections were processed as described in Fig. 4. In cells from the distal tubule, the basolateral membrane and its invaginations are intensely labelled with gold particles. There is also labelling of the apical membrane and the apical vesicles. Mitochondria and the cytosol are devoid of gold particles. Section stained with lead citrate and uranyl acetate. Magnification = ×12,150.

tracellular antigen detection. This is mainly due to the impermeability of the cell membrane to antibodies, leading to false-negative results or to resort to techniques using solvents or detergents to allow penetration of antibodies with the consequent loss of fine structure. In post-embedding staining procedures, such diffusion restrictions are not encountered, but the problem has been the supporting matrix used to obtain thin sections. This has been overcome in three different ways: glutaraldehyde cross-linked serum albumin embedding [16, 24, 25]; ultracryotomy [14, 18, 26] and plastic embedding [27]. We have tested the streptavidin- or avidin-biotin bridge technique on BSA-embedded tissue sections and on Lowicryl thin sections of toad kidney.

3.3.1. Localisation of the (Na^+, K^+)ATPase α-subunit in kidney tubular cells

An indirect procedure was applied in which the rabbit α-subunit antibody was reacted with biotinylated sheep anti-rabbit IgG antibodies and visualised with 10 nm gold-streptavidin complexes. Specific binding of α-subunit antibodies on thin

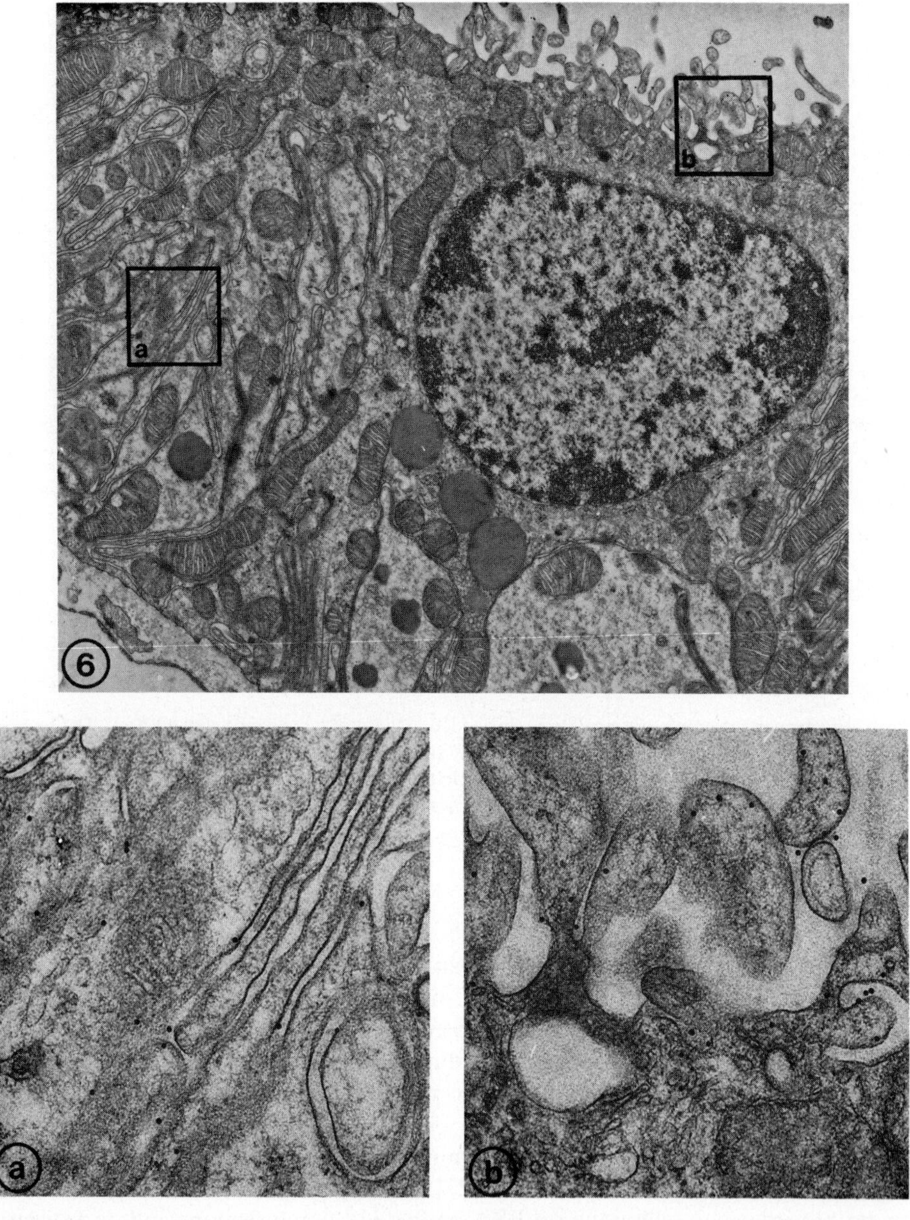

Fig. 6. Electron micrograph of toad kidney embedded in Epon. The sections were processed as described in Fig. 4. The fine structure of distal tubular cells is well preserved. The labelling density, however, is drastically reduced when compared to labelling on BSA-embedded tissue. Gold particles are associated with the infoldings of the basal cell plasma membrane (inset a: Magnification = ×46,400). Gold particles are also decorating the inner or the outer aspect of the apical plasma membrane (inset b: Magnification = ×40,800). The sections were stained with uranyl acetate and lead citrate. Magnification = ×9,600.

sections of BSA-embedded toad kidney tissue indicate that α-subunit antigenic determinants are associated both with the basolateral and apical cell surface membrane. In the proximal tubular cells (Fig. 4) intracellular vesicles mainly situated in the apical cytoplasm of the tubular cells are also decorated with gold particles. The biological significance of the labelling of apical membranes and intracellular vesicles will be discussed elsewhere and our data confirm earlier observations reported by Kyte [28] on mammalian kidney. In the distal tubule, deep invaginations of the basal plasma membrane which frequently surround mitochondria are heavily labelled (Fig. 5). It appears that the α-subunit antibody also reacts with antigenic determinants present in the interstitial space. This labelling disappears when antibodies reacting with the ectoplasmic domain of the α-subunit are removed from the polyclonal antibody preparation. Because of the insufficient staining properties of BSA-embedded tissue sections, it is difficult to correlate immunological labelling with the underlying subcellular structures. In order to identify unequivocally the intracellular organelles associated with the α-subunit antigenic determinants, we have labelled conventional Epon sections in the same procedure described above. The sections were not etched with hydrogen peroxide prior to immunological labelling. As shown in Fig. 6, the labelling density was drastically reduced. One might expect organic solvents and epoxy resins to diminish antigenicity or to restrict antigenic determinant availability (see Chapter 3, pp. 29–36). Recently, we have tested the avidin-biotin bridge technique on sections from Lowicryl-embedded kidney tissue. In this case, the α-subunit antigenic determinants were visualised using ferritin-avidin conjugates. As seen in Fig. 7, the labelling density on the basolateral membrane of distal tubular cells is similar to that obtained in BSA-embedded tissue; however, the fine structural preservation and the staining properties of Lowicryl sections make it the ideal system for the detection of intracellular antigens.

3.3.2. Localisation of (Na^+, K^+)ATPase subunits on viable amphibian leukocytes

The reaction sequence in this system was designed to resemble the conditions of an enzyme immunoassay (ELISA) and to eliminate the need for gold- or ferritin-streptavidin conjugates. The substitution of biotinyl-BSA-gold for streptavidin-gold, or avidin-ferritin was suitable since localisation of α-subunit was specific on cells of low antigen density, such as erythrocytes. Such a system is economical since readily available macromolecules such as bovine serum albumin can be derivatised with biotin via linker arms and then bound to colloidal gold. Thus visualising reagents have to be prepared only once and can be used to detect any biotinylated ligand, including antibodies, hormones, lectins, or macromolecules mediating biological functions. Biotinylated-BSA-gold complexes are stable over several weeks provided polyethylene glycol (0.2%) is added to the preparation. We have used this detection system to localise the α-subunit of (Na^+, K^+)ATPase on toad leukocytes. The fixed cells, incubated with saturating concentrations of biotinylated α-subunit antibodies (via a 7-carbon spacer arm: B-ω-AC-NHS), were then reacted with streptavidin followed by biotinyl-BSA-gold (using B-ω-AC-NHS). The pre-

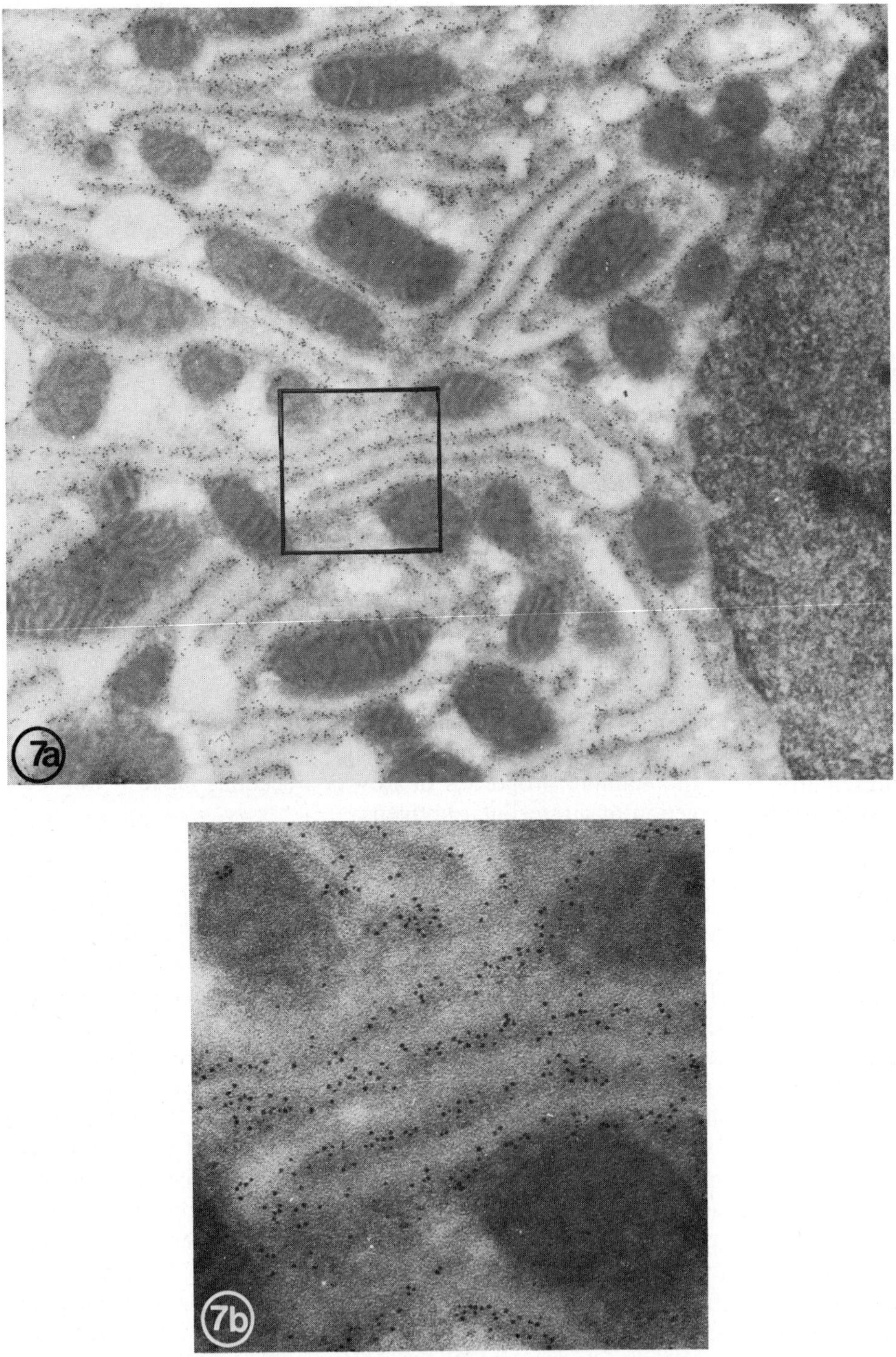

7a

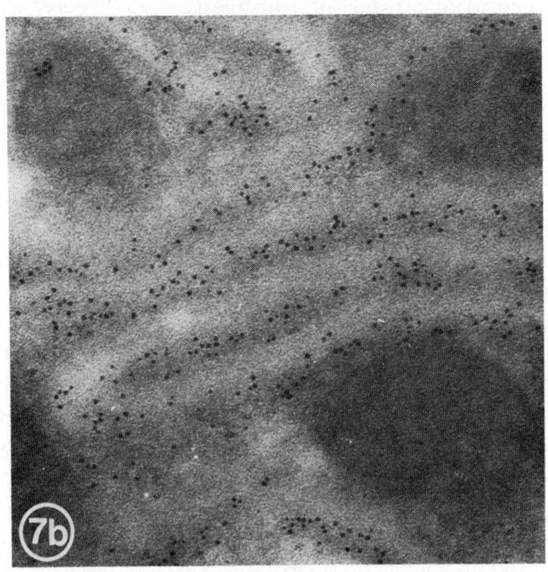

sence of the spacer arm both on BSA-gold and on the detecting antibody drastically increased the labelling efficiency.

As with all particle-tagged immunological reagents, results obtained with the (strept)avidin-biotin bridge technique are amenable to quantification by morphometric analysis of particle density on the section surface. This requires, however, that the antigens are available for interaction with immunological reagents at least at the section surface regardless of their distribution in cells. It is also conditional on the fact that tissue processing does not destroy immunogenicity, and immunological reagents interact with antigens in a stoichiometric relation.

A further advantage of the streptavidin-biotin bridge technique is that binders (i.e. hormone receptors) can be purified to homogeneity using biotinylated ligands and insolubilised streptavidin [23]. Therefore, the same biotinylated ligand can serve first to characterise the binding and biological properties of the derivatised ligand in a fluid phase, second to purify the binding sites by affinity chromatography, and finally to correlate these biochemical parameters with the subcellular, cellular, or tissue distribution of the binder. Finally, the streptavidin-biotin bridge technique has been applied to enzyme-linked immunoassay (ELISA) and to immunoaffinity partitioning (Papermaster and Flanagan, manuscript in preparation).

4. Acknowledgements

We thank Liliane Racine, Barbara Schneider, Lawrence Altman and Eugenia Farinon for their excellent and skillful assistance. We are grateful to Dr. J. Brunner (University of Zürich) for providing us with biotinylated tetra-β-alanine. We acknowledge Dr. R. Solari for revising the manuscript, A. Cesco for her secretarial assistance, and P. Dubied for photographic and graphic work. This work was supported by a Grant 3.413.0.83 from the Swiss National Science Foundation, by USPHS Grant EY-03239, The Veterans Administration, and a grant from Amersham International plc.

5. References

1. Bayer, E.A., Rivnay, B. and Skutelsky, E. (1979) On the mode of liposome-cell interactions biotin-conjugated lipids as ultrastructural probes. Biochem. Biophys. Acta 550, 464–473.

Fig. 7. Electron micrograph of toad kidney embedded in Lowicryl. The tissue was processed as described in section 2, and sections were incubated first with affinity-purified rabbit IgG anti-α-subunit, then with biotinylated sheep IgG anti-rabbit IgG and finally with ferritin-avidin. Sections through distal tubular cells. The tissue fine structure is more readily distinguishable. Ferritin grains are associated with the infoldings of the basal cell membrane and their density is comparable to that observed on sections from BSA-embedded tissue. As shown in the inset (Fig. 7a: Magnification = ×67,890) staining over cytosol and mitochondria is low. The sections were stained with uranyl acetate and lead citrate. Magnification = ×27,900.

2. Heitzmann, H. and Richards, F.M. (1974) Use of the avidin-biotin complex for specific staining of biological membranes in electron microscopy. Proc. Natl. Acad. Sci. USA 71, 3537–3541.
3. Manning, J.E., Hershey, N.D., Broker, T.R., Pelegrini, M., Mitchell, H.K. and Davidson, N. (1975) A new method of *in situ* hybridization. Chromosoma 53, 107–117.
4. Papermaster, D.S., Schneider, B.G., Zorn, M.A. and Kraehenbuhl, J.P. (1978) Immunocytochemical localization of opsin in outer segments and Golgi zones of frog photoreceptor cells. An EM analysis of cross-linked albumin-embedded retinas. J. Cell Biol. 77, 196–210.
5. Green, N.M. (1975) Avidin. Adv. Prot. Res. 29, 85–133.
6. Hoffmann, K., Wood, S.W., Brinton, C.C., Montibeller, J.A. and Finn, F.M. (1980) Iminobiotin affinity columns and their application to retrieval of streptavidin. Proc. Natl. Acad. Sci. USA 77, 4666–4668.
7. Wooley, D.W. and Longsworth, L.G. (1942) Isolation of an antibiotin factor from egg white. J. Biol. Chem. 142, 285–290.
8. Chaiet, L. and Wolf, F.J. (1964) The properties of streptavidin, a biotin-binding protein produced by *Streptomycetes*. Arch. Biochem. Biophys. 106, 1–5.
9. Bayer, E.A. and Wilchek, M. (1980) The use of avidin-biotin complex as a tool in molecular biology. Meth. Biochem. Anal. 26, 1–45.
10. Costello, S.M., Felix, R.T. and Giese, R.W. (1979) Enhancement of immune cellular agglutination by use of an avidin-biotin system. Clin. Chem. 25, 1572–1580.
11. Brandtzaeg, P. (1973) Conjugates of immunoglobulin G with different fluorochromes. I. Characterization by anionic-exchange chromatography. Scand. J. Immunol. 2, 273–290.
12. Frens, G. (1973) Controlled nucleation for the regulation of the particle size in monodisperse gold suspension. Nature (London) 241, 20–22.
13. Geoghegan, W.D. and Ackerman, G.A. (1977) Adsorption of horseradish peroxidase, ovomucoid and anti-immunoglobulin to colloidal gold for the indirect detection of concanavalin A, wheat germ agglutinin and goat anti-human immunoglobulin G on cell surfaces at the electron microscopic level: A new method, theory and application. J. Histochem. Cytochem. 25, 1187–1200.
14. Tokuyasu, K.T. (1973) A technique for ultracryotomy of cell suspensions and tissues. J. Cell Biol. 57, 551–556.
15. Girardet, M., Geering, K., Frantes, J.M., Geser, D., Rossier, B.C., Kraehenbuhl, J.P. and Bron, C. (1981) Immunochemical evidence for a transmembrane orientation of both the (Na^+, K^+)-ATPase subunits. Biochemistry 20, 6684–6691.
16. Kraehenbuhl, J.P., Racine, L. and Jamieson, J.D. (1977) Immunocytochemical localization of secretory proteins in bovine pancreatic exocrine cells. J. Cell Biol. 72, 406–415.
17. Altman, L., Schneider, B.G. and Papermaster, D.S. (1983) Rapid (4 hr) method for embedding tissues in Lowicryl for immunoelectron microscopy. J. Cell Biol. 97, 309A (Abstract).
18. Tokuyasu, K.T. (1978) A study of positive staining of ultrathin frozen sections. J. Ultrastruct. Res. 63, 287–307.
19. Jörgensen, P.L. (1974) Purification and characterization of ($Na^+ + K^+$)-ATPase. III. Purification from the outer medulla of mammalian kidney after selective removal of membrane components by sodium dodecylsulphate. Biochem. Biophys. Acta 356, 36–52.
20. Geering, K. and Rossier, B.C. (1979) Purification and characterization of ($Na^+ + K^+$)-ATPase from toad kidney. Biochem. Biophys. Acta 566, 157–170.
21. Solari, R. and Kraehenbuhl, J.P. (1984). Biosynthesis of the IgA antibody receptor—a model for the transepithelial sorting of a membrane glycoprotein. Cell 36, 61–71.
22. Häuptle, M.T., Suard, Y.M.L., Bogenmann, E., Reggio, M., Racine, L. and Kraehenbuhl, J.P. (1983) Effect of cell shape change on the function and differentiation of rabbit mammary cells in culture. J. Cell Biol. 96, 1425–1434.
23. Häuptle, M.T., Aubert, M., Djiane, J. and Kraehenbuhl, J.P. (1983) Binding sites for lactogenic and somatogenic hormones from rabbit mammary gland and liver. J. Biol. Chem. 258, 305–314.
24. McLean, J.B. and Singer, S.J. (1970) A general method for the specific staining of intracellular antigens with ferritin-antibody conjugates. Proc. Natl. Acad. Sci. USA 65, 122–128.
25. Kraehenbuhl, J.P. and Jamieson, J.D. (1972) Solid phase conjugation of ferritin to Fab fragments for use in antigen localization on thin sections. Proc. Natl. Acad. Sci. USA 69, 1771–1775.

26. Tokuyasu, K.T. and Singer, S.J. (1976) Improved procedure for immunoferritin labeling of ultrathin frozen sections. J. Cell Biol. 71, 894–899.
27. Sternberger, L.A. (1974) Immunocytochemistry. Prentice Hall, Inc., Englewood Clifts, N.J.
28. Kyte, J. (1976) Immunoferritin determination of the distribution of $(Na^+ + K^+)$ATPase over the plasma membranes of renal convoluted tubules. I. Distal segment. J. Cell Biol. 68, 287–318.

Immunolabelling for Electron Microscopy (Polak/Varndell, eds)

CHAPTER 9

The protein A-gold technique for antigen localisation in tissue sections by light and electron microscopy

Jürgen Roth

Biocenter, Department of Electron Microscopy, University of Basel, CH-4056 Basel, Switzerland

CONTENTS

1. INTRODUCTION

Protein A is a cell wall constituent produced by most strains of *Staphylococcus aureus* [1]. From the viewpoint of immunochemistry the most interesting property of protein A is its unique ability to interact with immunoglobulins, notably immunoglobulin G (for review see Goding [2]). All applications of protein A in immunocytochemistry depend on its high affinity and binding to immunoglobulins, which is rapid and reaches saturation level in about 30 min. The K_D values for human and rabbit IgG are similar to antigen-antibody reactions [3–5]. This interaction is a pseudoimmune reaction and takes place mainly at the CH_2 and CH_3 domains of the Fc region of IgG [6]. There are numerous reports on variation in binding to various IgG subclasses and interaction with other immunoglobulin classes [7–9]. In addition, there is also variation in reactivity of protein A with immunoglobulins from different animal species [10, 11]. Despite this, protein A has been shown to be a useful reagent for binding immunoglobulins from several animal species, a property which renders this protein of general applicability in immunocytochemistry. Biberfeld and coworkers [10] initially applied fluorescence-labelled protein A as a second layer reagent in the indirect immunofluorescence technique. Later, protein A conjugated either to ferritin [12] or peroxidase [13] was introduced for antigen localisation at the light and electron microscope level. In 1971, Faulk and Taylor [14] introduced a new marker, colloidal gold, in immunocytochemistry. Since then, colloidal gold coated with different classes of biological macromolecules has been applied for light and electron microscopical cytochemistry (for review see Roth [15]). Romano and Romano [16] were the first to prepare a protein A-gold (pAg) complex which was used in a pre-embedding technique for the localisation of antigens present at the cell surface. In contrast, our main interest was to explore the possibility of immunolocalisation of intracellular antigens in thin sections from resin-embedded material. We were able to show the usefulness of the pAg complex for post-embedding antigen localisation by light and electron microscopy ([17–19] for review). Subsequently, the pAg technique was successfully adapted for the staining of ultrathin frozen sections [20]. This brief review deals with the principle of the pAg technique and cites examples of its application for the localisation of different intracellular antigens. Other relevant reviews on colloidal gold as a marker and on the pAg technique have appeared [15, 21, 22] and should be consulted for further information.

2. Tissue processing

Chemical fixation, organic solvent dehydration and embedding in resin adversely affects antigenicity. This general problem is clearly and widely recognised and appropriate conditions have to be devised for each particular antigen.

2.1. Fixation

Principally, we use two standard conditions for fixation by immersion or vascular perfusion:

a) different concentrations of glutaraldehyde (4–0.5%) or
b) mixtures of paraformaldehyde (4–2%) and glutaraldehyde (2–0.1%); fixation time varies between 30 min and 2 h [15, 21].

Most antigens do not tolerate further fixation with osmium tetroxide. However, Bendayan and Zollinger [23] showed that the antigenicity of certain proteins can be restored after osmication by treatment of the thin sections with sodium metaperiodate (Fig. 3). In addition to the above mentioned fixatives, Bouin's fluid or Carnoy's solution can be recommended for light microscopy [24].

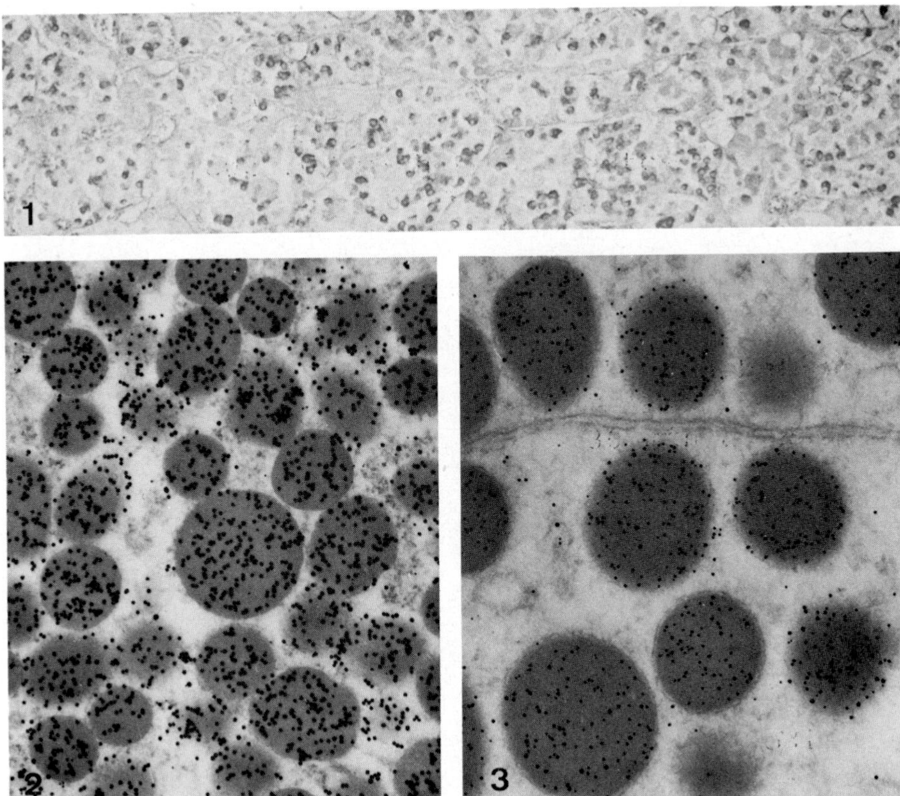

Figs. 1–3. Section from paraffin-embedded human pituitary gland stained for growth hormone (Fig. 1). The positive cells appear black in the photograph but when investigated by bright field transmission microscopy they are pink. Thin sections from Epon-embedded human pituitary adenoma fixed only in glutaraldehyde (Fig. 2) or in glutaraldehyde and osmium tetroxide (Fig. 3) and stained for growth hormone. A dense labelling with 14 nm gold particles is seen over the secretory granules despite no etching of the thin sections. The sections from osmicated tissue were treated with 1% periodic acid for 4 min before labelling. Again, the secretory granules are heavily stained (with 7 nm gold particles). Note the well preserved plasma membrane and secretory granule membrane.

2.2. *Embedding*

For light microscopy, cryostat sections of fixed tissues or sections from paraffin-embedded material can be processed [24] (Figs. 1, 4, 9 and 10). Semithin sections from Epon [25] or low temperature Lowicryl K4M (unpublished data) embedded material provide superior resolution (Figs. 5–7). For electron microscopy, routine embedding with Epon 812 allows in many instances successful antigen localisation on thin sections (Figs. 2 and 8) ([15, 21, 26, 27] for review). In the meantime, several investigations with the low temperature embedding procedure using

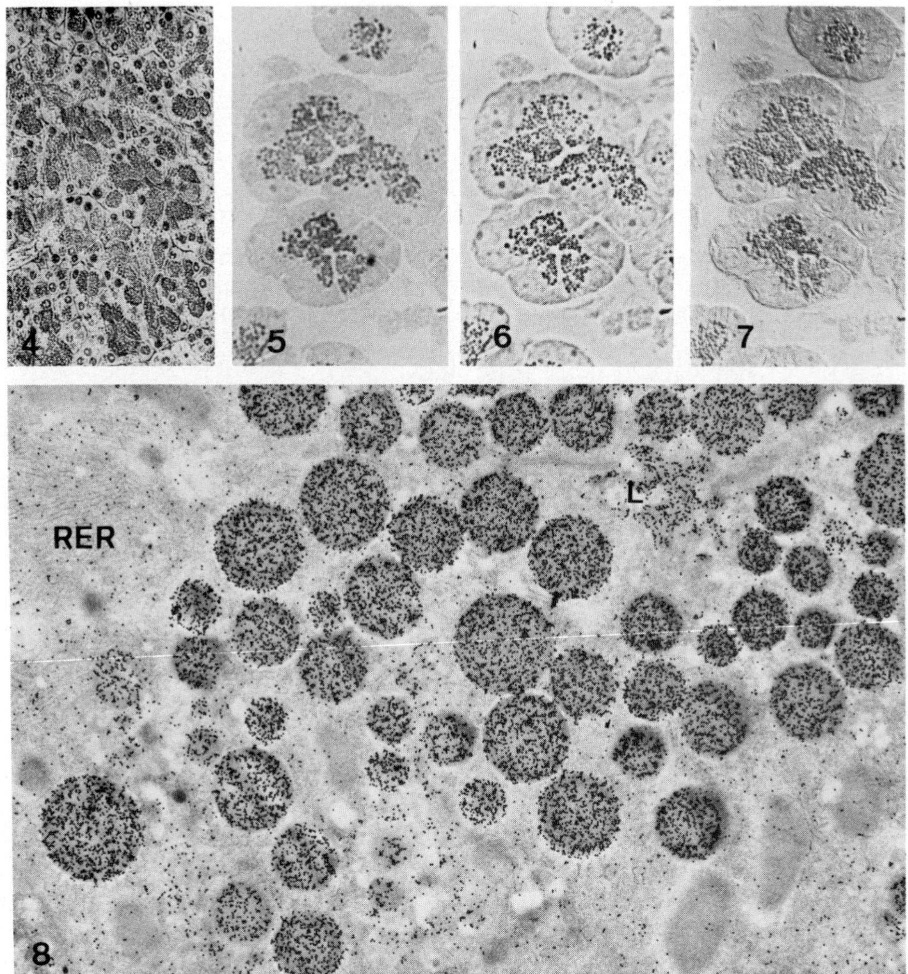

Figs. 4–8. Rat pancreas stained for amylase with protein A-gold complexes prepared from 14 nm gold particles. In the paraffin section (Fig. 4) an intense staining is seen over granular structures in the apex of the acinar cells. This staining pattern can be better appreciated on semithin sections (about 0.5 μm) from Lowicryl K4M embedded tissue when viewed by bright field (Fig. 5), phase contrast (Fig. 6) or Nomarski interference contrast microscopy (Fig. 7). Figure 8 illustrates the localisation of amylase with the same protein A-gold complex as used for light microscopy on a thin section from glutaraldehyde-fixed and Epon-embedded pancreatic tissue. An intense labelling is present over zymogen granules and the contents of the acinar lumen (L). The rough endoplasmic reticulum (RER) is also labelled.

Lowicryl K4M [28] have clearly demonstrated that beside superior structural preservation and drastically lowered background staining [29] more intense specific staining can be obtained (Figs. 11 and 12) [30, 31]. Some antigens [32, 33] and lectin-binding sites could only be demonstrated in Lowicryl K4M embedded tissues.

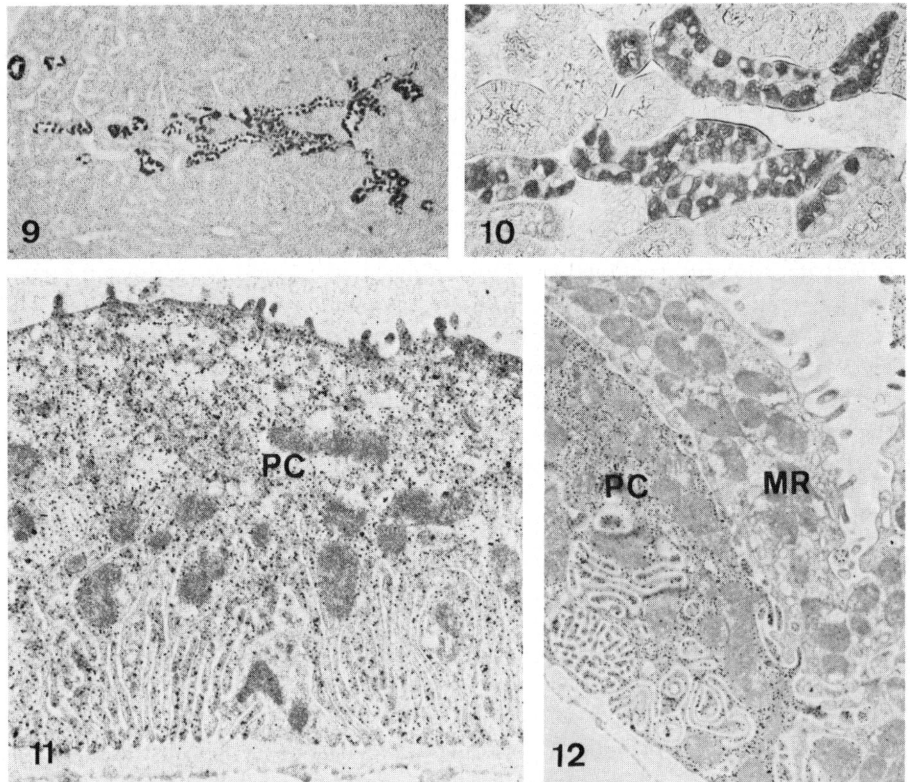

Figs. 9–12. Rat kidney stained for vitamin D-dependent calcium binding protein. In paraffin sections (Figs. 9 and 10) many positive tubular profiles are present in the cortex which were identified as distal convoluted tubules, connecting segments and cortical collecting ducts. At a higher magnification as shown in Fig. 10, the presence of positive and negative cells in distal convoluted tubules is seen. On thin sections of low temperature Lowicryl K4M embedded kidney (Figs. 11 and 12), the principal cells (PC) are positive for calcium binding protein whereas the intercalated (mitochondria-rich) cells (MR) are unstained.

3. Cytochemical procedure

3.1. Preparation of protein A-gold complexes

Colloidal gold can be prepared according to various recipes ([15, 21] for review; see also Chapters 2, 11 and 15). Monodisperse sols of small particle size (down to 14 nm) can be obtained with Frens' procedure [34] or sols of 3 nm particles may be manufactured using a modified Zsigmondy recipe [19, 35]. Other protocols [36, 37] yield heterodisperse sols which after complex formation with protein A can be sized to homogeneity by density-gradient centrifugation [38]. The formation of protein-colloidal gold complexes is pH dependent [39] and, therefore, the colloidal gold sols are adjusted to pH 5.9–6.2 for complex formation with protein A. For stabilisation of 10 ml colloidal gold, about 60 μg protein A are needed for 3 nm particles, about 50 μg protein A for heterodisperse particles prepared according to the methods of Zsigmondy and Thiessen [36] and Stathis and Fabrikanos [37], and approximately 30 μg protein A for 14 nm particles [34]. Further details can be found in [15, 21] and in Chapter 11, pp. 129–142.

3.2. Incubation of sections

For light microscopy, sections are deparaffinised or treated to remove the resin [40]. Thin sections for electron microscopy are never conventionally etched with hydrogen peroxide. In the case of osmium tetroxide fixation, thin sections are treated with oxidising reagents, such as sodium metaperiodate [23]. The sections on glass slides or nickel grids are initially treated with 0.5–1% ovalbumin in phosphate-buffered saline (PBS) for 5 min and incubated with appropriately diluted antibodies for 2 h at room temperature. After two rinses with PBS, appropriately diluted protein A-gold is applied for 1 h at room temperature. The thin sections are then washed and finally counterstained with uranyl acetate and lead acetate. A more detailed description can be found in [15, 21].

4. Application of the protein A-gold technique for post-embedding antigen localisation

The reliability of the protein A-gold technique has been demonstrated by the successful localisation of different classes of proteins such as several secretory proteins and notably polypeptides in various cellular compartments; of enzymes in mitochondria, peroxisomes and granules of eosinophilic leukocytes; of cytoskeletal elements; of extracellular space components; of cytosolic proteins; of membrane integral proteins, etc. ([15] for review).

The light and electron microscopical demonstration of growth hormone in human pituitary is illustrated in Figs. 1–3. Growth hormone was detected in paraffin sections from formaldehyde-fixed tissue and also in Epon thin sections from non-osmicated and osmicated tissue. Our immunocytochemical studies on exocrine pancreatic enzymes confirmed established biochemical data on secretion [41] by demonstrating the presence of several enzymes in the rough endoplasmic reticulum, Golgi apparatus, condensing vacuoles, zymogen granules and acinar lumen [26] (Figs. 4–8). Data were obtained suggesting the involvement of the whole Golgi stack in processing of the secretory products [29]. Double labelling studies showed the presence of different enzymes in the same cellular compartments [19, 42]. Quantitative evaluation of the labelling revealed differences in labelling intensity for the various enzymes and cellular compartments [26].

The immunocytochemical localisation of the vitamin D-dependent calcium binding protein [43] in its main target, duodenum and kidney [33, 44, 45] allowed more definite conclusions about its function in intestinal calcium absorption and renal calcium reabsorption. At the light microscopical level (Figs. 9 and 10), calcium-binding protein was found in certain cells in the distal convoluted tubules, parts of the collecting duct system and the segments connecting these tubular regions. By electron microscopy (Figs. 11 and 12), the immunoreactive cell type was identified as the principal cell. In this cell type, calcium-binding protein was found throughout the cytosol and the nuclear euchromatin but not in preferential association with membranes. This subcellular localisation of calcium-binding protein was also found in intestinal absorptive cells but as shown by quantitative immunocytochemistry the antigen accumulates in the apical and basal region of these cells [33]. These findings suggest that the vitamin D-dependent calcium binding protein may play a role in intracellular calcium regulation and/or movement of calcium across the epithelial cells.

First investigations on the topology of glycosylation steps have already yielded highly interesting results. Galactosyltransferase has been localised in the Golgi apparatus where it is present in a distinct subcompartment consisting of 2–3 Golgi cisternae which are also positive for thiamine pyrophosphatase [32]. Galactose residues as visualised with the sequence *Ricinus communis* lectin I, antibody, protein A-gold complex were detected also in *trans* Golgi cisternae [46].

Collectively, these data demonstrate the high degree of functional specialisation within the Golgi apparatus. On the other hand, they clearly show the great potential of modern immunocytochemical procedures such as the protein A-gold technique in cell biology research.

5. Acknowledgements

It is a great pleasure to thank Mrs A.-K. Beilstein and E. Oesch for technical assistance. The work summarised in this chapter received support from the Swiss National Science Foundation, Grant No. 3.443–0.83.

6. References

1. Verwey, E.F. (1940) A type-specific antigenic protein derived from the Staphylococcus. J. Exp. Med. 71, 635–644.
2. Goding, J.W. (1978) Use of staphylococcal protein A as an immunological reagent. J. Immunol. Meth. 20, 241–253.
3. Langone, J.J. (1982) Use of labelled protein A in quantitative immunochemical analysis of antigens and antibodies. J. Immunol. Meth. 51, 3–22.
4. Kronvall, G. and Frommel, D. (1970) Definition of staphylococcal protein A reactivity for human immunoglobulin G fragments. Immunochemistry 7, 124–127.
5. Myhre, E.B. and Kronvall G. (1980) Immunochemical aspects of Fc-mediated binding of human IgG subclasses to group A, C and G streptococci. Mol. Immunol. 17, 1563–1573.
6. Forsgreen, A. and Sjöquist, J. (1966) Protein A from *Staphylococcus aureus*. Pseudo-immune reaction with human γ-globulin. J. Immunol. 97, 822–827.
7. Harboe, M. and Fölling, J. (1974) Recognition of two distinct groups of human IgM and IgA based on different binding to staphylococci. Scand. J. Immunol. 3, 471–482.
8. Hjelm, H. (1975) Isolation of IgG_3 from normal human sera and from a patient with multiple myeloma by using protein A-Sepharose 4B. Scand. J. Immunol. 4, 633–640.

9. Vidal, M. and Conde, F.R. (1980) Studies of the IgM and IgA contamination obtained by eluting IgG from protein A-Sepharose column with pH steps. J. Immunol. Meth. 35, 169–172.
10. Biberfeld, B., Ghetie, V. and Sjöquist, J. (1975) Demonstration and assaying of IgG antibodies in tissues and on cells by labelled staphylococcal protein A. J. Immunol. Meth. 6, 249–259.
11. Goudswaard, J., van der Donk, J.A., Noardzig, A., van Dam, R.H. and Vaerman, J.-P. (1978) Protein A reactivity of various mammalian immunoglobulins. Scand. J. Immunol. 8, 21–28.
12. Bächi, T., Dorval, G., Wigzell, H. and Binz, H. (1977) Staphylococcal protein A in immunoferritin techniques. Scand. J. Immunol. 6, 241–246.
13. Dubois-Dalcq, M., McFarland, H. and McFarlin, D. (1977) Protein A-peroxidase: A valuable tool for the localization of antigens. J. Histochem. Cytochem. 24, 1201–1206.
14. Faulk, W.P. and Taylor, G.M. (1971) An immunocolloid method for the electron microscope. Immunochemistry 8, 1081–1083.
15. Roth J. (1983) The colloidal gold marker system for light and electron microscopic cytochemistry. In: G. R. Bullock and P. Petrusz (Eds.) Techniques in Immunocytochemistry, Vol. II. Academic Press, London and New York, pp. 217–284.
16. Romano, E.L. and Romano, M. (1977) Staphylococcal protein A bound to colloidal gold: a useful reagent to label antigen-antibody sites in electron microscopy. Immunochemistry 14, 711–715.
17. Roth, J., Bendayan, M. and Orci, L. (1978) Ultrastructural localization of intracellular antigens by the use of protein A-gold complex. J. Histochem. Cytochem. 26, 1074–1081.
18. Roth, J., Bendayan, M. and Orci, L. (1980) FITC-protein A-gold complex for light and electron microscopic immunocytochemistry. J. Histochem. Cytochem. 28, 55–57.
19. Roth, J. (1982) The preparation of protein A-gold complexes with 3 nm and 15 nm gold particles and their use in labelling multiple antigens on ultrathin sections. Histochem. J. 14, 791–801.
20. Geuze, H.J., Slot, J.W., van der Ley, P.A. and Scheffer, R.C.T. (1981) Use of colloidal gold particles in double labelling immunoelectron microscopy of ultrathin frozen tissue sections. J. Cell Biol. 89, 653–665.
21. Roth, J. (1982) The protein A-gold (pAg) technique—a qualitative and quantitative approach for antigen localization on thin sections. In: G. R. Bullock and P. Petrusz (Eds.) Techniques in Immunocytochemistry, Vol. I. Academic Press, London and New York, pp. 107–133.
22. Horisberger, M. (1981) Colloidal gold: a cytochemical marker for light and fluorescent microscopy and for transmission and scanning electron microscopy. Scanning Electron Microsc. II, 9–31.
23. Bendayan, M. and Zollinger, M. (1983) Ultrastructural localization of antigenic sites on osmium-fixed tissues applying the protein A-gold technique. J. Histochem. Cytochem. 31, 101–109.
24. Roth, J. (1982) Applications of immunocolloids in light microscopy. Preparation of protein A-silver and protein A-gold complexes and their application for localization of single and multiple antigens in paraffin sections. J. Histochem. Cytochem. 30, 691–696.
25. Roth, J., Brown, D. and Orci, L. (1983) Regional distribution of N-acetyl-D-galactosamine residues in the glycocalyx of glomerular podocytes. J. Cell Biol. 96, 1189–1196.
26. Bendayan, M., Roth, J., Perrelet, A. and Orci, L. (1980) Quantitative immunocytochemical localization of pancreatic secretory proteins in subcellular compartments of the rat acinar cell. J. Histochem. Cytochem. 28, 149–160.
27. Roth, J. Ravazzola, M., Bendayan, M. and Orci, L. (1981) Application of the protein A-gold technique for electron microscopic demonstration of polypeptide hormones. Endocrinology 108, 247–253.
28. Carlemalm, E., Garavito, M. and Villiger, W. (1982) Resin development for electron microscopy and an analysis of embedding at low temperature. J. Microsc. (London) 126, 123–143.
29. Roth, J., Bendayan, M., Carlemalm, E., Villiger, W. and Garavito, M. (1981) Enhancement of structural preservation and immunocytochemical staining in low temperature embedded pancreatic tissue. J. Histochem. Cytochem. 29, 663–671.
30. Bendayan, M. and Shore, G.C. (1982) Immunocytochemical localization of mitochondrial proteins in the rat hepatocyte. J. Histochem. Cytochem. 30, 139–147.
31. Craig, S. and Goodchild, D.J. (1982) Postembedding immunolabelling. Some effects of tissue preparation on the antigenicity of plant proteins. Eur. J. Cell Biol. 28, 251–256.

32. Roth, J. and Berger, E.G. (1983) Immunocytochemical localization of galactosyltransferase in Hela cells: codistribution with thiamine pyrophosphatase in trans Golgi cisternae. J. Cell Biol. 93, 223–229.
33. Thorens, B., Roth, J., Perrelet, A., Norman, A.W. and Orci, L. (1982) Immunocytochemical localization of the vitamin D-dependent calcium binding protein in chick duodenum. J. Cell Biol. 94, 115–122.
34. Frens, G. (1973) Controlled nucleation for the regulation of the particle size in monodisperse gold suspensions. Nature Phys. Sci. 241, 20–30.
35. Zsigmondy, R. (1905) Zur Erkenntnisse der Kolloide. G. Fischer, Jena.
36. Zsigmondy, R. and Thiessen, P.A. (1925) Das kolloidale Gold. Akademische Verlagsgesellschaft, Leipzig.
37. Stathis, E.C. and Fabrikanos, A. (1958) Preparation of colloidal gold. Chem. Indust. 27, 860–861.
38. Slot, J.W. and Geuze, H.J. (1981) Sizing of protein A-colloidal gold probes for immunoelectron microscopy. J. Cell Biol. 90, 533–536.
39. Geoghegan, W.D. and Ackerman, G.A. (1977) Adsorption of horseradish peroxidase, ovomucoid and anti-immunoglobulin to colloidal gold for the indirect detection of Concanavalin A, wheat germ agglutinin and goat anti-human immunoglobulin G on cell surfaces at the electron microscopic level: a new method, theory and application. J. Histochem. Cytochem. 25, 1187–1200.
40. Maxwell, M.H. (1978) Two rapid and simple methods used for the removal of resins from 1.0 μm thick epoxy sections. J. Microsc. (London) 112, 253–255.
41. Palade, G. (1975) Intracellular aspects of the process of protein synthesis. Science 189, 347–358.
42. Bendayan, M. (1982) Double immunocytochemical labeling applying the protein A-gold technique. J. Histochem. Cytochem. 30, 81–85.
43. Norman, A.W., Roth, J. and Orci, L. (1982) The vitamin D endocrine system: steroid metabolism, hormone receptors, and biological response (calcium binding proteins). Endocr. Rev. 3, 331–336.
44. Roth, J., Thorens, B., Hunziker, W., Norman, A.W. and Orci, L. (1981) Vitamin D-dependent calcium-binding protein: immunocytochemical localization in chick kidney. Science 214, 197–200.
45. Roth, J., Brown, D., Norman, A.W. and Orci, L. (1982) Localization of the vitamin D-dependent calcium-binding protein in mammalian kidney. Am. J. Physiol. 243, F243–F252.
46. Griffiths, G., Brands, R., Burke, B., Louvard, D. and Warren, G. (1982) Viral membrane proteins acquire galactose in trans Golgi cisternae during intracellular transport. J. Cell Biol. 95, 781–792.

Immunolabelling for Electron Microscopy (Polak/Varndell, eds)

CHAPTER 10

Labelled antigen detection methods

Lars-Inge Larsson

Unit of Histochemistry, Frederik den V's vej 11, DK-2100 Copenhagen 0, Denmark

CONTENTS

1. Definition of the problem

Unspecific immunocytochemical staining may result from both immunological and non-immunological phenomena. The latter (including endogenous enzyme activity or autofluorescence, and non-immunological binding of antibodies or conjugates to tissue) are usually detected by second-level controls (including omission of the various antibody layers and substitution of the primary antiserum with pre-immune sera), as they are caused by the properties of the tissue material rather than by the properties of one particular antiserum [1].

Notwithstanding the seriousness of non-immunological unspecificity, most errors seem, however, to be caused by immunological unspecificity. Such unspecificity is caused by two factors:

a) immunological cross-reactivity, and
b) contaminating antibodies.

As the antigen-combining site on an IgG molecule is of limited size it will react only with a limited region of most antigens. In the case of peptides and proteins the size of such a region corresponds to 3–8 (usually 3–4) amino acid residues. These residues may either be covalently linked (continuous antigenic site) or may be brought together by the conformation of the antigen (discontinuous antigenic site) [2]. Assuming that nature has the 20 common amino acids to combine in various constellations (which, admittedly, is an oversimplification since acetylations, sulphations etc. of individual amino acids occur post-translationally), a tripeptide unit can be built in 8,000 (20^3) ways, a tetrapeptide unit in 160,000 (20^4) ways etc.

Although, of course, this naive calculation is an underestimate, it illustrates that the presence of a particular tri- or tetrapeptide unit in only one specific protein or peptide is an exception rather than a rule. This is particularly well illustrated by hormone and enzyme families that have evolved through gene duplication mechanisms.

Thus, an antiserum against a peptide or a protein need not, and most often, is not, specific for that peptide or protein, but may be highly specific for the small part of the antigen it binds to. Realisation of this fact has caused the introduction of terms like "-like immunoreactivity". A way of dealing with this problem is region-specific immunocytochemistry, using sets of antibodies directed against multiple antigenic regions of the peptide to be examined [3]. This, of course, allows detection of a larger part of the peptide sequence. As the probability of coincidence of multiple antigenic regions in the same compartment drastically decreases with the number of antigenic regions detected, region-specific immunocytochemistry gives promise of the ultimate immunocytochemical goal—the detection of only one molecular species [4].

Polyclonal (e.g. rabbit or guinea pig) antisera are heterogenous and contain antibody subpopulations that differ in region-specificity and avidity. Moreover, they may contain contaminating antibodies either induced by previous infections/immunisations, representing autoantibodies or being evoked by unknown contaminants in the immunogen preparation. Thus, polyclonal antisera are often not immediately suited for critical region-specific immunocytochemistry as the common methods of immunocytochemistry (immunofluorescence, immunoperoxidase, peroxidase-antiperoxidase and avidin-biotin-peroxidase techniques) indiscriminately will detect all antibodies that bind to a tissue section, whether they are specific or not.

The problem may be dealt with in three ways:

a) Removal of the unwanted antibodies. In the simplest case, which actually holds true for many antisera, contaminating antibodies may be of much lower titre than the desired antibodies. Hence, simple dilution may sometimes suffice. The importance of not selecting only a standard dilution of all antisera but to make endpoint dilutions can therefore not be overstated. Moreover, if the nature of the contaminating (or cross-reactive) antibody is known it can be removed by adsorption against the contaminating (or cross-reactive) antigen, preferably coupled to a solid support like cyanogen bromide-activated Sepharose® beads [5]. This, however, is rarely the case. The last resort, hence, is to couple the antigen under investigation to a solid support, bind the desired antibody population and wash away the undesired rest, whereafter the desired antibodies are eluted. The problem with such affinity-purification is that the desired antibodies most often are not completely eluted. Hence, antibodies of high avidity can not be eluted under non-denaturing conditions and affinity-purified antisera are therefore enriched in low-avidity antibodies, which although useful for biochemical purposes, become dislodged during immunocytochemical staining [6].

b) Monoclonal antibodies. Production of antibodies from a single clone of hybridoma cells is an extremely useful approach for obtaining a homogeneous population of IgG or IgM molecules. Bemoaned problems with monoclonals include their supposedly low avidity and the fact that they sometimes only react with the native, unfixed antigen. However, these are early technical troubles which already are on their way to being overcome. Particularly the method of screening antibodies must be as immunocytochemical-like as possible if good staining results are to be expected [5].
c) Labelled antigen detection techniques including the radioimmunocytochemical (RICH) [7] and gold-labelled antigen detection (GLAD) [8] techniques, which are unique immunocytochemical methods since they detect only the desired specific antibodies that bind to tissue sections.

2. Advantages and disadvantages with labelled antigen detection techniques

IgG molecules each possess two equivalent antigen-combining sites. If one of these two sites is allowed to bind pure labelled antigen and the other site is allowed to bind tissue-antigen the IgG molecule will form a bridge between the tissue and the labelled antigen, the latter marking the site of the reaction between tissue-antigen and antibody (Fig. 1).

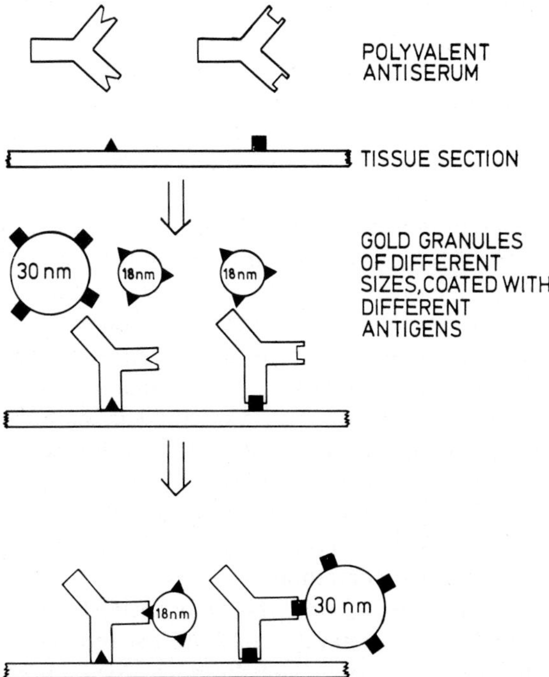

Fig. 1. Principle of the GLAD method and its use for immunocytochemical double-staining

In the first application of this method, the radioimmunocytochemical (RICH) method [7], ^{125}I-labelled antigen was mixed with IgG molecules in such proportions that a preponderance of antigen-antibody complexes binding ^{125}I-labelled antigen with only one of their two combining sites were formed (easily accomplished in the zone of slight antibody surplus; for detailed instructions see [7]). The remaining site was, hence, free to combine with tissue-bound antigen when such "RICH complexes" were applied to tissue sections. Subsequently, the site of reaction was detected by autoradiography.

The RICH method works very well at the light microscopical level and has been used for the localisation of numerous peptide antigens, including pancreatic polypeptide [7], gastrin [1] and ACTH [9]. The dilutions used in this method surpassed those used in the peroxidase-antiperoxidase PAP method [7]. With one gastrin antiserum (No. 2604), which does not cross-react with the related hormone cholecystokinin (CCK) in a radioimmunoassay (RIA) system, but which, when used in conventional immunocytochemical systems, fully cross-reacts with this peptide, particularly interesting results were obtained. Thus, as in the RIA system, RICH complexes made from ^{125}I-labelled synthetic human gastrin I and antibody 2604 reacted only with gastrin cells and not with CCK cells [1]. This fortuitous behaviour has tentatively been ascribed to the selection by the tracer of a unique high-specificity class of antibodies only reacting with gastrin (for a full discussion see [1, 5]).

Although specificity and sensitivity criteria of this new method surpassed those of previous techniques, the precision of the RICH method at the electron microscopical level is as low as with all autoradiographic methods.

An alternative label to ^{125}I, was therefore sought for electron microscopical investigations. Among several possible markers we selected colloidal gold particles [10–12] since these were highly electron dense, and which therefore allowed optimal contrasting of sections. The gold particles could also be made in several sizes, thereby allowing double-staining experiments to be performed on single tissue sections. The new method, similar to the RICH method in principle, was named the GLAD (Gold-Labelled Antigen Detection) technique [8].

The GLAD method involves on-grid staining of ultrathin (resin or ultracryotome) sections with the appropriate antisera. Subsequently, gold granules to which the appropriate antigen (protein or peptide-protein conjugate) has been adsorbed, are applied. As IgG molecules predominantly react with antigens in ultrathin sections with only one of their two antigen-combining sites [8], free antigen-combining sites sticking out from the sections will bind antigen on the gold granules, thus the site of antigen-antibody reaction will be marked [8]. Subsequently, the grids may be refixed, contrasted and observed in the electron microscope. Dilutions used in the GLAD method are generally on par with those of the peroxidase-antiperoxidase (PAP) method.

One of the major advantages of both GLAD and RICH methods is that, unlike all other immunocytochemical methods, only the desired specific antibodies are detected. Unspecific antibodies may well bind to tissue, but being unable to recognise the specific labelled antigen will remain undetected [8]. This specificity discrimination also allows the pre-treatment of the sections with 10% solutions of normal serum from the same species as that donating the primary antiserum [8]. This is not possible with conventional immunocytochemical techniques, which will detect all antibodies of a given species that bind to tissue, but forms an attractive way of blocking unspecific IgG binding in the GLAD (and RICH) technique. Another advantage of this specificity discrimination is that it allows for very simple double-staining experiments. Thus, two antisera directed against two different antigens may be mixed, applied to the ultrathin sections, which subsequently are exposed to a mixture of small gold granules coated with one of the antigens and larger gold granules coated with the other antigen (Fig. 1). The two sizes of gold particles will thus separately mark the two antigen populations present in the tissue [5, 8].

A particular advantage of the GLAD method is that it readily accommodates region-specific immunocytochemistry. Thus, many antisera even if directed only against one single antigen contain a subpopulation of antibodies reacting with different epitopes of that antigen. By coupling different peptide fragments to gold granules of different sizes it is possible to select for particular region-specific antibody subpopulations [8]. Thus, it has proven possible to use separately no less than three distinct region-specific subpopulations in one single antiserum. A first requirement for this, however, is a firm knowledge of what antibody subpopulations are present in a particular antiserum. This is most conveniently tested by use of the paper cytochemical model system developed in 1979 [13]. In this method, strips of filter papers on which 2 μl droplets of peptide or protein solutions have been immobilised are immunocytochemically stained. The strips provide a direct read-out of the specificity and cross-reactivity of different sera and this knowledge can subsequently be directly transferred to tissue-staining experiments. A subsequently developed test system employing nitrocellulose filter instead of cellulose may well be used instead [14]. Despite claims in the literature [15], however, this latter system seems to be less sensitive than the cellulose system (compare data in references [13] and [15]). Moreover, the nitrocellulose system is limited to only those antigens that can attach non-covalently to the substrate.

Disadvantages with the GLAD method are mainly associated with the amount of labour involved. Thus, every antigen to be studied has to be coupled to gold particles. Large molecular weight antigens can be coupled directly to the gold particles, whereas smaller peptide antigens have first to be conjugated to a carrier protein like albumin in order to be adsorbable. Due to the amount of labour, it is therefore recommended to use the GLAD (and RICH) methods in cases of troublesome antisera or when specific double-staining experiments are to be performed [7, 8]. Full methodological details have been published previously [1, 5] and will, for reasons of space, not be repeated here.

3. References

1. Larsson, L.-I. (1981) Peptide Immunocytochemistry. Prog. Histochem. Cytochem. Vol. 13, No. 4, pp. 1–85.

2. ATASSI, M.Z. and SMITH, J.A. (1978) A proposal for the nomenclature of antigenic sites in peptides and proteins. Immunochemistry 15, 609–610.
3. LARSSON, L.-I. and REHFELD, J.F. (1977) Characterization of antral gastrin cells with region-specific antisera. J. Histochem. Cytochem. 25, 1317-1321.
4. LARSSON, L.-I. (1981) Immunochemistry of secretory peptides. Immunochemical dissections of peptide diversity. J. Histochem. Cytochem. 29, 1032–1042.
5. LARSSON, L.-I. (1983) Methods for immunocytochemistry of neurohormonal peptides. In: A. Björklund and T. Hökfelt (Eds.) Handbook of Chemical Neuroanatomy, Vol. 1. Elsevier, Amsterdam, pp. 147–209.
6. STERNBERGER, L.A. (1979) Immunocytochemistry, 2nd Edn. John Wiley & Sons, Inc. New York.
7. LARSSON, L.-I. and SCHWARTZ, T.W. (1977) Radioimmunocytochemistry—A novel immunocytochemical principle. J. Histochem. Cytochem. 25, 1140–1146.
8. LARSSON, L.-I. (1979) Simultaneous ultrastructural demonstration of multiple peptides in endocrine cells by a novel immunocytochemical method. Nature (London) 282, 743–746.
9. RAPPAY, G.Y., KÁRTESZI, M. and MAKARA, G.B. (1979) ACTH radioimmunocytochemistry (RICH) on rat anterior pituitary cells. Histochemistry 59, 207–213.
10. FRENS, G. (1973) Controlled nucleation for the regulation of the particle size in monodisperse gold suspensions. Nature (Phys. Sci.) 241, 20–22.
11. FAULK, W.P. and TAYLOR, C.M. (1971) An immunocolloid method for the electronmicroscope. Immunochemistry 8, 1081–1083.
12. HORISBERGER, M. (1979) Evaluation of colloidal gold as a cytochemical marker for transmission and scanning electronmicroscopy. Biol. Cell. 36, 253–258.
13. LARSSON, L.-I. (1981) A novel immunocytochemical model system for specificity and sensitivity screening of antisera against multiple antigens. J. Histochem. Cytochem. 29, 408–410.
14. HEBRINK, P., VAN BUSSEL, F.J. and WARNAAR, S.O. (1982) The antigen spot test (AST): a highly sensitive assay for detection of antibodies. J. Immunol. Methods 48, 293–298.
15. BOSMAN, F.T., CRAMER-KNIJNENBURG, G. and VAN BERGEN HENEGOUW, J. (1983) Efficiency and sensitivity of indirect immunoperoxidase methods. Histochemistry 77, 185–194.

Immunolabelling for Electron Microscopy (Polak/Varndell, eds)

CHAPTER 11

Gold markers for single and double immunolabelling of ultrathin cryosections

Jan W. Slot and Hans J. Geuze

Centre for Electron Microscopy, School of Medicine, University of Utrecht, Nicolaas Beetsstraat 22, 3511 HG Utrecht, The Netherlands

CONTENTS

1. INTRODUCTION

Electron-dense particles are very useful for immunolabelling ultrathin sections. Alternatives like radiolabels combined with autoradiography or enzyme markers, give a less precise antigen localisation and need more complicated labelling procedures which are, particularly in ultrathin cryosections, not compatible with good preservation of ultrastructural detail (see Chapter 7, pp. 83–93). One of the first, and initially the most successful electron-dense immunomarker was the iron storage protein ferritin, introduced by Singer in 1959 [1]. An attractive alternative to ferritin is colloidal gold, originally introduced into immunochemistry by Faulk and Taylor [2]. There is increasing interest in the application of gold particles for electron microscopy, probably because of their high electron density, the availability of many different particle sizes and the simple procedure by which they may be coupled to macromolecules.

In this chapter we deal with the preparation of immunogold conjugates and their use in immunolocalisation studies on ultrathin cryosections. We use gold markers in indirect labelling procedures. First, specific immunoglobulins are allowed to react with the antigen to be studied in the section. Subsequently the marker is bound to the specific immunoglobulin. To achieve this the marker is complexed with a marker carrier, a protein with affinity to the specific immunoglobulin (Ig). Two marker carriers are used. The first is protein A (pA), a staphylococcal cell wall protein with specific affinity for the Fc portion of IgG from many mammalian species. The second is an immunoglobulin, which we will call the carrier Ig, which is directed against the primary specific immunoglobulin. According to the carrier used, we will indicate the methods as the pA-gold and the Ig-gold method.

2. pA-GOLD METHOD

Protein A was first used in combination with gold markers by Romano and Romano [3]. Later, Roth and co-workers emphasised the usefulness of pA-gold for intracellular localisation studies. They performed many immunogold studies with sections from resin-embedded tissue [4]. We found protein A-gold a suitable marker for ultrathin cryosections and developed a double labelling procedure using these reagents [5] (Fig. 1).

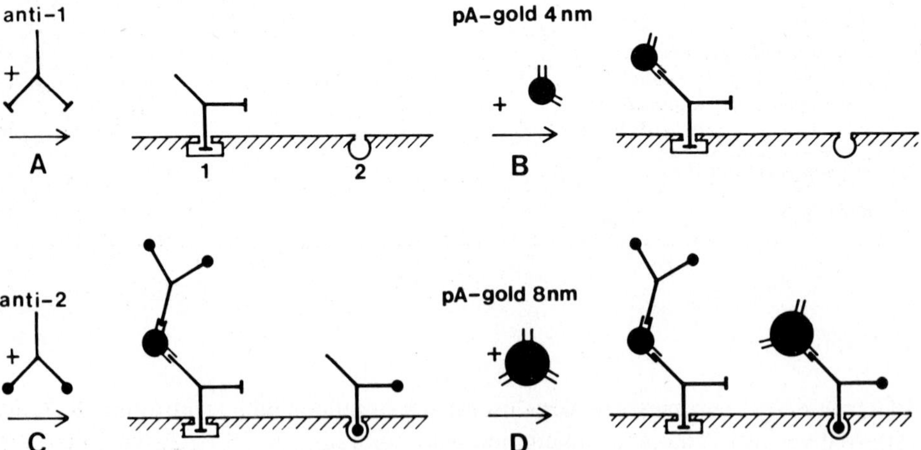

Fig. 1. (A–D) The four successive incubations for pA-gold double labelling of antigen 1 and 2. Anti-2 which is captured by free immunoglobulin binding sites on protein A during (C) is not marked by the second pA-gold. The 4 and 8 nm markers used in (B) and (D) are interchangeable, provided that free protein A is added during the last minutes of incubation (B).

2.1. Sizing of pA-gold by gradient centrifugation

In the procedure to prepare pA-gold, protein A is added in excess to the colloidal gold particles so that some free protein remains after conjugation. This is usually washed away by repeated centrifugation. We replaced these washings by centrifugation over a continuous density gradient [6]. This has three advantages:

a) Uniformly sized samples can be collected from the gradient. This is very important for double labelling experiments. For high resolution localisation in cryosections we found the best choice was pA-gold preparations of 4 nm and 8 nm with non-overlapping size ranges, which were isolated from conjugates of protein A with gold sols prepared using white phosphorus and ascorbic acid, respectively (Fig. 2). Smaller markers are not readily observed in the high contrast sections obtained after uranyl-methylcellulose treatment. Markers of larger size are less sensitive [6].
b) During the centrifugation procedures undertaken to remove free protein A, a tightly packed portion of the pellet cannot normally be resuspended and thus significant amounts of the conjugate are lost. The gradient purification proce-

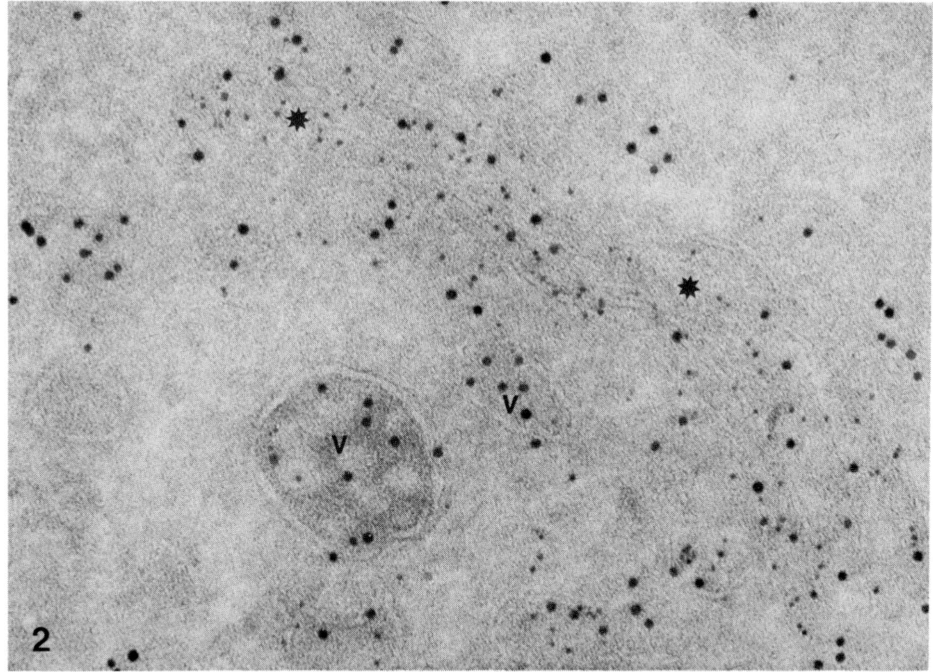

Fig. 2. Rat liver parenchymal cell. Double labelled according to the procedure in Figure 1: (A) anti-ASGP receptor (B) 4 nm pA-gold (C) anti-albumin (D) 8 nm pA-gold. Although ASGP-receptor and albumin occur together in the entire Golgi system, the receptor occurs more frequently in the cisternae (asterisk) than in secretory vesicles containing albumin and VLDL particles (V). Magnification = ×150,000.

dure is a much milder method to remove unbound pA, since the preparations are pelleted only once, thus less of the conjugate is lost.

c) Small aggregates formed during the conjugation procedure move relatively fast over the gradient. The larger ones are pelleted, and smaller complexes accumulate towards the bottom fractions of the gradient. The upper part of the gradient is largely devoid of aggregates. The abovementioned 4 nm and 8 nm preparations are collected from this upper part. Usually these are also the most concentrated fractions which indicates the efficiency of the procedure.

2.2. Double labelling

We introduced a double labelling procedure using pA-gold of different sizes for the co-localisation of the membrane associated glycoprotein GP2 and the secretory protein amylase in rat pancreas [5]. Later, the technique was refined by the introduction of uniformly sized 4 nm and 8 nm pA-gold preparations [6]. In Figure 2 these markers are used together for the co-localisation of albumin and the asialoglycoprotein (ASGP)-receptor in rat liver parenchymal cells. Relative to the distribution of albumin, the receptor seems to occur more frequently in the cisternae of the Golgi system rather than in the secretory vesicles, suggesting a difference in drainage of albumin and the membrane-bound receptor from the Golgi stack [7]. The method was also successfully used to trace the uptake process of ASGP by liver cells and the *c*ompartment was defined where we observed *u*ncoupling of *r*eceptor and its *l*igand, termed CURL [8]. More recent studies, using the same technique, revealed that other receptor-ligand complexes are also taken up in CURL and that ligands with different destinations in the cell are sorted in CURL [20].

The high resolution attained in these studies was possible, not only because of the small and sensitive pA-gold preparations used, but also because of excellent delineation of the ultrastructure in cryosections, which was achieved after Griffiths' modification [9] of Tokuyasu's uranyl-methylcellulose treatment. The enhanced contrast in the sections is compatible with the use of highly electron-dense gold markers, but not with ferritin.

The pA-gold double labelling procedure consists simply of two single immunolabelling steps in sequence (Fig. 1). Usually we observe no significant co-labelling of the first antibody binding sites by the subsequently applied pA-gold complex, regardless of whether the small or the large marker goes first. This is conditional on the grounds that incubation with the first pA-gold reagent is exhausted by the addition of free protein A prior to the application of the second immunoreagents.

Nevertheless, protein A is capable of binding at least two IgG molecules [10], which means that free IgG binding sites are left on the protein A molecules bound to the first specific immunoglobulin. Moreover, unoccupied protein A will occur on the gold marker used in the first staining, particularly when the large marker is used first (8 nm gold particles adsorb approximately 10 protein A molecules). Therefore, many potential Fc binding sites can be expected in the area where the first antigen occurs, when the second specific immunoglobulin is applied. Apparently, if the second specific immunoglobulin, in addition to its specific antigen binding, is captured by these unoccupied protein A molecules the resulting complex is not recognised by the subsequent pA-gold immunoreagent. This is compatible with the assumption that IgG can only bind one protein A molecule [10].

Double labelling can also be achieved with immunoglobulin as marker carrier. By using two specific immunoglobulins from different sources, for instance rabbit and guinea pig, indirect double labelling is possible. Dutton et al. [11] described a method in which goat antibodies against rabbit or guinea pig immunoglobulins were coupled to either ferritin or Imposil, an iron-dextran compound. These particles can be distinguished by way of their different shape and size. Gold markers of different sizes may also be used effectively [12]. In this procedure both labels can be applied simultaneously, which can be particularly advantageous when the two antigens studied occur close together. This is important because in the sequential pA-gold double labelling procedure, the first reaction may hinder the second to some extent. On the other hand, the pA-gold method allows the use of two specific immunoglobulins from the same source (for example, rabbit in Fig. 2).

2.3. Labelling pattern, sensitivity

The monovalency of IgG for protein A suggests that a specific immunoglobulin molecule bound to an antigen is visualised by a single marker particle. This results mostly in a one-to-one distribution of gold particles when 8 nm pA-gold is used (Fig. 3). The 4 nm marker can be found distributed in small clusters of up to 4 particles, which is probably dependent on the size of the antigen molecule and on the number of specific antibodies bound (Fig. 4A).

Protein A-gold probes are sensitive immunoreagents which are used to localise many substances in cryosections. The 4 nm pA-gold tends to give a stronger reaction in terms of number of gold particles than the 8 nm marker. In part this may be due to less sterical hindrance and/or repulsion forces in the case of the 4 nm probe, so that more than one particle can bind to one antigen molecule. In addition, however, the smaller probe may penetrate deeper into the sections so that more antigen becomes accessible. For instance, we observed that, under mild fixation conditions, the 4 nm complex can penetrate into the cytosolic material in cryosections and label antigens like the coat protein clathrin [13]. Using 8 nm pA-gold, we found a rather poor labelling for clathrin, presumably because that probe did not penetrate. All these factors together (sterical hindrance, repulsion, penetration) probably cause the rapid decrease in sensitivity with increasing sizes of gold particles [6].

Based on a comparison of pA-gold immunoglobulin-ferritin it has been suggested that ferritin is more efficient as an immunomarker than colloidal gold [14]. However, this may be much more a matter of difference in reaction mechanism of both marker carriers. We also observed a more intense reaction with immunoglobulin-ferritin than with 4 nm pA-gold. This is illustrated in Figure 4A, B, for the case of amylase in rat pancreas. However, in areas with a moderate reaction (Fig. 5) it appeared that ferritin was often grouped in small clusters of up to 10 particles,

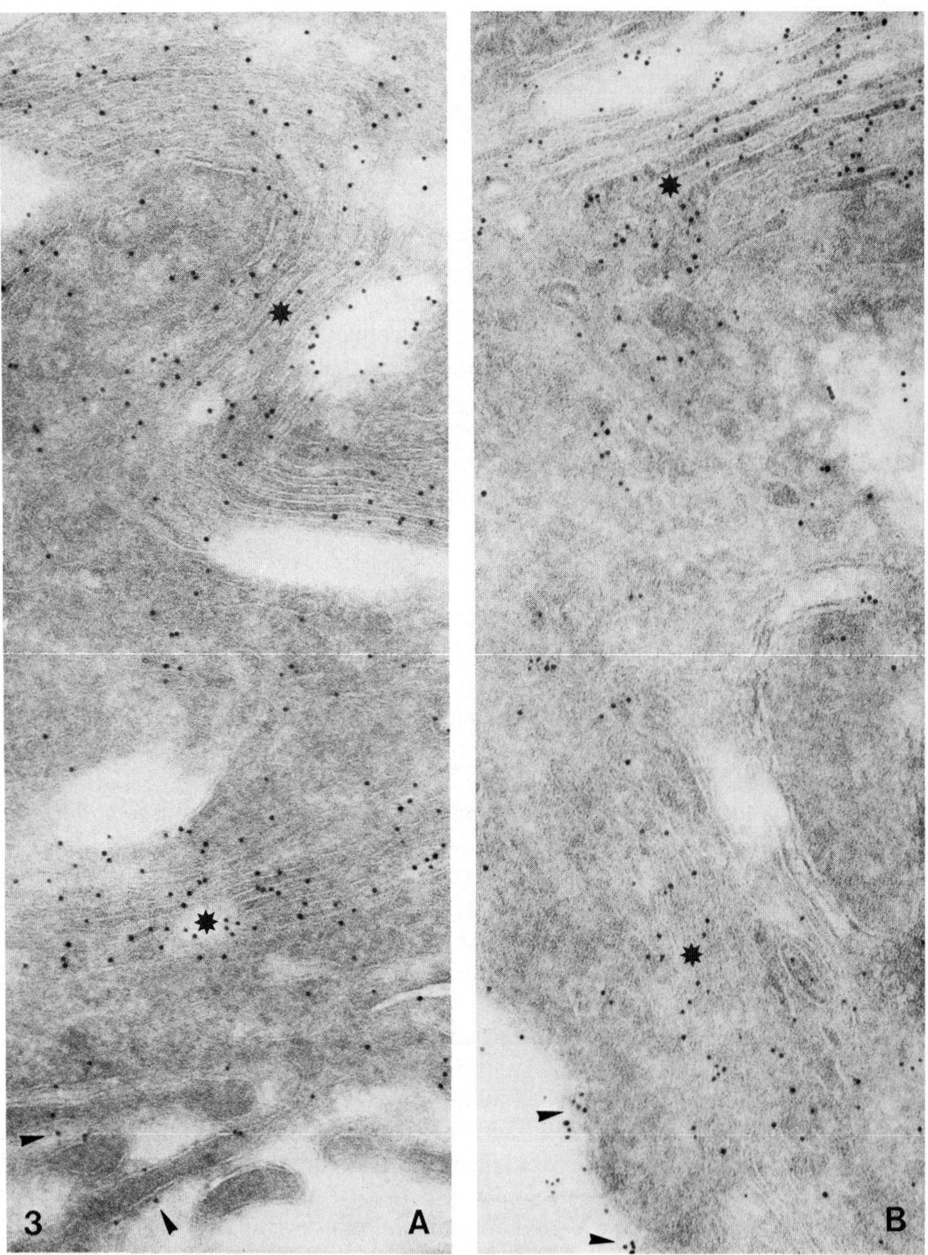

Fig. 3. Human duodenal absorptive cells. Immunolabelling of the receptor for polymeric IgA (secretory component (SC) with 8 nm pA-gold (A) and Ig-gold (B). SC is distributed all through the Golgi cisternae (asterisk) and along the lateral cell membrane (arrowheads). In (B) the label is often grouped in clusters of 2–4 particles, a phenomenon rarely seen with pA-gold. Magnification = ×75,000.

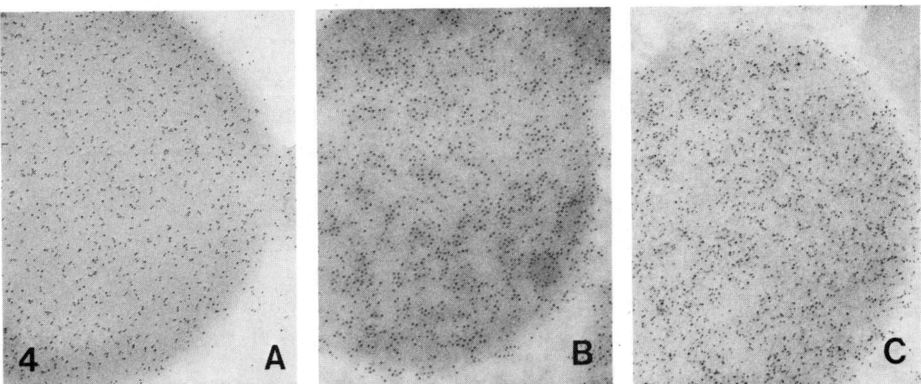

Fig. 4. Rat pancreatic zymogen granules. Immunolabelling of amylase with 4 nm pA-gold (A), immunoglobulin-ferritin (B) or Ig-gold (C). The number of particles in (A) is less than in (B) and (C), but the particles in (A) are distributed more homogeneously. Sections A and B are not stained with uranyl. Magnification = ×52,000.

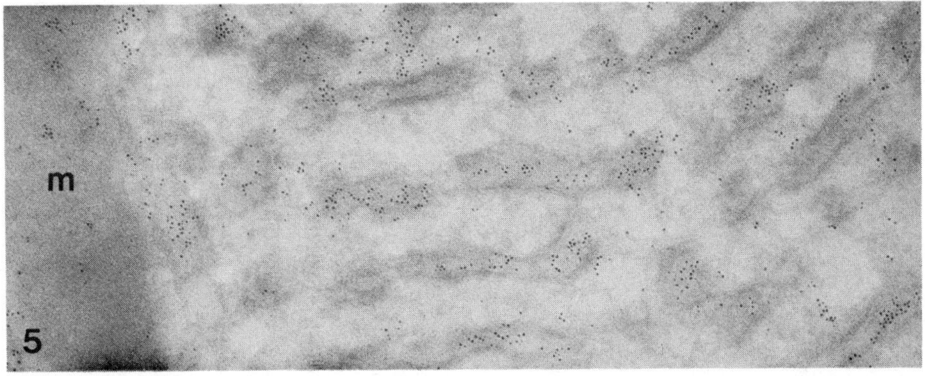

Fig. 5. Same section as in Figure 4B. The low amylase in the rough endoplasmic reticulum (RER) allows to notice the clustered appearance of ferritin particles which are probably grouped around one amylase molecule. Some mitochondrial (m) labelling is also visible, probably because of inadequate cross-linkage of amylase and leakage. Magnification = ×60,000.

indicating that, in contrast to pA-gold, the specific immunoglobulin molecule can bind more than one immunoglobulin-ferritin complex. Therefore, a signal amplification rather than efficiency enhancement underlies the more intense labelling with immunoglobulin-ferritin.

3. Ig-GOLD METHOD

Gold markers were first used in immunocytochemistry bound to immunoglobulins [2]. The preparation of stable Ig-gold is not as simple as the procedure to make pA-gold. In general, the correct pH for binding proteins to gold is slightly more alkaline than the isoelectric point of the protein [15]. Polyvalent immunoglobulin preparations consist of a mixture of molecules with a wide range of isoelectric points, which makes adjustment to the correct pH difficult. Nevertheless, several attempts to bind immunoglobulin fractions to gold have been described [15, 16]. Recently, a method which seems more generally applicable to immunoglobulin fractions from various sources was worked out by De Mey and co-workers [17]. They dialysed affinity-purified immunoglobulin preparations to pH 9 at low ionic strength before coupling to gold sols with the same pH. We used their coupling procedure, made homogeneous Ig-gold fraction by gradient centrifugation in the same way as that described for pA-gold [6] and used the probes to label cryosections.

3.1. Preparation of Ig-gold

As a carrier we used goat anti-rabbit immunoglobulin. Initially, immunoglobulin preparations were routinely dialysed against PBS and stored in the freezer following affinity purification. Though clear after thawing, such preparations (approximately 1 mg/ml) become cloudy during dialysis against 2 mM Borax. We centrifuged the precipitate down (30 min × 50,000 g) and the supernatant was used to make the Ig-gold complex. This complex was stable in the presence of high salt concentration, but it moved rather fast on the gradient when compared to pA-gold preparations. On grids the preparation appeared to consist mainly of small clumps of gold particles. The gradient fractions mainly differed in number of particles in the clumps rather than in particle size (Fig. 6A, B). However, when freshly affinity-purified immunoglobulin fractions from the same serum were dialysed directly from the high salt in which it was eluted (3 M KCNS) to 2 mM Borax, the solution stayed clear. Clumps of particles were only scarce in Ig-gold complexes made from these immunoglobulin preparations (Fig. 6C, F), and gradient centrifugation yielded uniformly sized fractions in the upper part of the gradient (Fig. 6D, G). Small clumps still occurred in the lower part (Fig. 6E, H).

These observations illustrate that the procedure for making Ig-gold is rather critical. Particularly for making uniformly sized probes special measures are required to prevent the immunoglobulin from aggregation. When dialysing from PBS to 2 mM Borax, the ionic strength of the immunoglobulin solution may drop to a critical value, while the pH still shifts over the range of isoelectric points, where the protein is sensitive to aggregation. This may explain why some immunoglobulin precipitated. On the other hand, when we started from the non-buffered immunoglobulin solution in 3 M KCNS the high pH was probably reached during dialysis while the immunoglobulin was still in high salt concentration, a condition that prevented immunoglobulin from aggregation and precipitation. This sensitivity to aggregation may vary for immunoglobulin fractions from different sera. For instance, another goat immunoglobulin fraction, directed against mouse immunoglobulin, aggregated even under the conditions that worked well for the goat anti-rabbit immunoglobulin. With anti-mouse immunoglobulin we could achieve

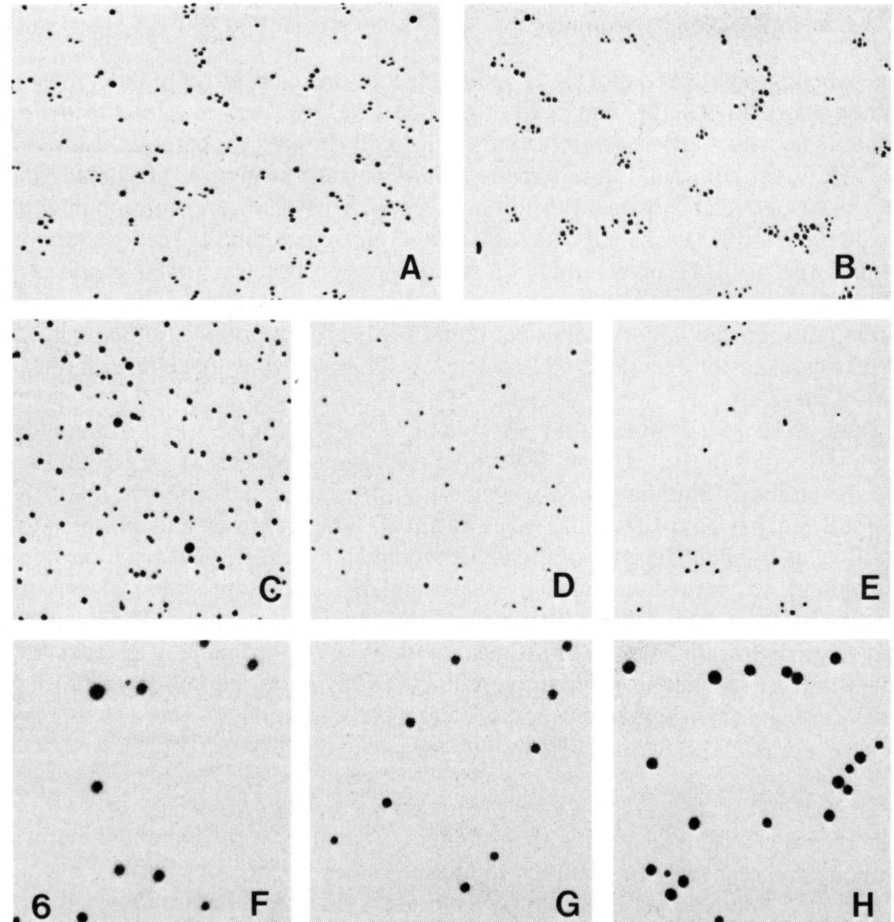

Fig. 6. The composition of Ig-gold preparations visualised by adsorbing them to carbon-formvar coated copper grids. The grids were floated for 15 min on drops of the Ig-gold preparations to which some protein was added (~ 0.1% BSA, to prevent clustering on the grid) and were then washed with distilled water. Upper (A) and lower (B) gradient fractions were prepared from $gold_5$ and an immunoglobulin solution from which precipitate has been removed by centrifugation. The preparation consists mainly of small clumps of gold particles probably formed around small immunoglobulin aggregates which remained in the supernatant. Total preparations of $gold_5$ (C) and $gold_{12}$ (F) with immunoglobulin solutions that did not aggregate are largely composed of single particles with various sizes. Uniform fractions can be collected from these preparations by gradient centrifugation. (D) and (G) show such fractions from $gold_5$ and $gold_{12}$, respectively, taken from the upper part of the gradient. A mixture of larger particles and small heterogeneous clumps occurs in the lower part of the gradients (E and H). Magnification = ×120,000.

good results after diluting the solution in KCNS to 0.3 mg/ml and raising the Borax pH to 9.2.

When starting from good immunoglobulin solutions the gradient sizing method worked well for Ig-gold with the same advantages as mentioned for pA-gold.

3.2. *Labelling pattern, sensitivity*

The labelling characteristics of Ig-gold preparations closely resembled those for immunoglobulin-ferritin. The yield of gold particles was higher than after pA-gold labelling and the particles were arranged in small clusters. Apparently it is not the marker, ferritin or gold, but rather the marker carrier, protein A or immunoglobulin, that causes the observed difference between pA-gold and immunoglobulin-ferritin (section 2.3). The clusters formed by 8 nm Ig-gold on the sections are much smaller (up to 4 particles) than those of 4 nm probes (up to 10 particles per cluster). This is probably because increasing sterical hindrance and repulsion forces between the larger probes counteract cluster formation. Still larger Ig-gold probes may cluster even less, so that the labelling pattern will become identical to that of pA-gold.

There were no indications that pA-gold picked up less of the specific immunoglobulin molecules in the sections. If we take each Ig-gold cluster as tagged on to one specific antibody binding site, our general impression is that labelling intensity of Ig-gold and pA-gold of similar sizes is equal. This suggests that pA-gold is as sensitive as Ig-gold. Because of the clustered labelling, the Ig-gold reaction is more convenient to appreciate, which is particularly important when the antigen concentration is low. On the other hand, the Ig-gold reaction will be saturated earlier at high antigen concentrations, so that, pA-gold labelling gives a better quantitative impression of the antigen distribution. Also, the one-to-one distribution of pA-gold markers allows a more precise localisation.

4. Conclusions

Complexes of colloidal gold with protein A as well as with immunoglobulin are both useful reagents for indirect immunolocalisation in ultrathin cryosections, particularly when particle sizes of 4–8 nm are used. The preparation of the gold complexes is simple. However, the low solubility of immunoglobulin at low ionic strength can cause aggregation of the protein and a clumped dispersion of the ultimate Ig-gold. Gradient centrifugation is a proper method to isolate uniform and unclumped preparations and is an elegant way to remove free protein A or immunoglobulin.

The Ig-gold method gives a higher yield of label than pA-gold. This is, however, not the result of a more efficient recognition of antigens but is rather a result of secondary amplification. By consequence Ig-gold is a convenient marker in cases of low antigen concentration. pA-gold gives a more correct impression of the antigen distribution.

5. Appendix

5.1. *Colloidal gold sols*

Gold sols containing particles with a mean diameter of 5–6 nm ($gold_5$) are prepared by using white phosphorus [18]. 1.5 ml of a 1% $HAuCl_4$ solution and 1.4 ml of

$0.1\,M\,K_2CO_3$ are added to 120 ml of distilled water. Add 1 ml of white phosphorus (Merck, Darmstadt, West Germany) in diethylether and mix well. (The phosphorus solution is prepared by adding 1 part of ether saturated with white phosphorus to 4 parts of ether.) The mixture is left for 15 min at room temperature and then boiled under reflux until the colour turns from brownish to red (about 5 min).

Sols with particle diameters of about 12 nm ($gold_{12}$) are prepared by reduction of $HAuCl_4$ with sodium ascorbate [19]. 1 ml of 1% $HAuCl_4$ and 1 ml of $0.1\,M\,K_2CO_3$ are mixed in 25 ml of distilled water. While stirring, add quickly 1 ml of a 0.7% solution of sodium ascorbate (BDH Chemicals, Ltd., Poole, U.K.). This reaction is performed on ice. Higher temperatures tend to increase the particle size. After addition of ascorbate, the colour immediately becomes purple-red. Then the volume is adjusted to 100 ml with distilled water. Finally, heat the solution until boiling so that the colour becomes red. The amount of carbonate added in both recipes brings the pH of the sols to approximately 6, which is good for preparing pA-gold. For preparing Ig-gold the sol was brought to pH >9 by adding 0.7 ml $0.1\,M\,K_2CO_3$ to 100 ml sol (checked by pH paper).

The amount of carrier-protein (protein A or immunoglobulin) needed to stabilise a sol, which is probably identical to the amount that can bind to that sol, is determined as follows [16]: Mix 0.25 ml gold sol in small glass tubes with various amounts of the protein. Add 0.25 ml of 10% NaCl after 1 min. As long as more protein can be bound, the solution turns blue. The lowest concentration of protein that prevents colour change is taken as the stabilisation point.

5.2. *pA-gold preparation* [6]

Take 30 ml gold sol. Add protein A (Pharmacia), from a 2 mg/ml stock solution in distilled water, in an amount so that the concentration exceeds the stabilisation point by 10%. Add 0.3 ml of 5% polyethylene glycol (Carbowax 20 K) after 1 min, to be sure that the gold particles are stabilised maximally. Then pellet the pA-gold by centrifugation: $gold_5$, 45 min at 125,000 g_{av}; $gold_{12}$, 45 min at 50,000 g_{av}. The pellet is composed of a large loose part and a small tightly packed part. Remove the supernatant without disturbing the pellet and resuspend the loose part of the pellet in the remainder of the supernatant (the small part that could not be removed without taking some of the pellet). This is layered over a 10–30% continuous sucrose or glycerol gradient (volume 10.5 ml, length 8 cm) in PBS pH 7.2. The gradient is centrifuged in a SW 41 rotor (Beckman Instruments) for 45 min at 41,000 rpm ($gold_5$) or for 30 min at 20,000 rpm ($gold_{12}$). From about 1 cm below the top down to the bottom, the gradient is stained red by pA-gold. The upper half of this red zone is collected in successive fractions of about 1 ml. These are largely free of clumps and the average sizes of the particles in the fractions are 4–5 nm ($gold_5$) and 7–9 nm ($gold_{12}$).

For storage, the preparations can be dialysed to 45% glycerol in PBS and kept at $-18°C$. Long-term storage can also be achieved by freezing small samples ($\sim 100\,\mu l$) in lower concentrations of glycerol at $-70°C$ (J. L. M. Leunissen, personal communication).

5.3. *Ig-gold preparation*

Goat anti-rabbit immunoglobulin was complexed to gold largely according to the method of De Mey et al. [17]. 4 ml of antiserum was adsorbed to 2 ml CNBr activated Sepharose gel to which rabbit immunoglobulin was bound, for 2 h at room temperature. Then the gel was washed by PBS and anti-rabbit immunoglobulin was eluted with 4 ml 3 M KCNS. The protein concentration (E^{280}) was adjusted to 1 mg/ml in 3 M KCNS and the solution was dialysed at room temperature against 2 mM Borax, pH 9.

Immunoglobulin preparations easily form aggregates during this procedure. Part of the aggregates precipitate, but the smaller ones stay in solution even after centrifugation (30 min × 50,000*g*). The tendency to form aggregations varies according to the immunoglobulin preparations used. To avoid aggregation the above procedure can be modified by further diluting the immunoglobulin before dialysis or by slightly raising the pH of the Borax solution.

Complexing of immunoglobulin to gold and subsequent purification of homogeneous Ig-gold fractions by gradient centrifugation are identical to the pA-gold procedure except that:

- Coupling occurs at pH > 9. Immediately after coupling the pH is adjusted to 8.2 (HCl).
- Carbowax is replaced by 0.25% bovine serum albumin (BSA). This protein is added as a 10% aqueous solution at pH 8.2.
- PBS in the glycerol or sucrose gradient is replaced by TBS (0.01 M Tris, 0.15 M NaCl, pH 8.2).

Ig-gold can be stored for a long time at 4°C with 0.02% sodium azide included. Probably the probes can also be frozen as mentioned for pA-gold.

5.4. *Immunolabelling of ultrathin cryosections*

Ultrathin cryosections are prepared from glutaraldehyde and/or paraformaldehyde-fixed tissue after infusion with 2.3 M sucrose, according to Tokuyasu (see Chapter 6, pp. 71–82). They are mounted on carbon-coated copper grids, and passed onto drops of the following solutions at room temperature for immunolabelling [5].

1) 2% gelatin in PBS 10 min.
2) 0.02 M glycine in PBS, 10 min. Alternatively one can use 1% borohydrate in PBS.
3) Specific immunoglobulin. Preferably affinity pure preparations, ~25 μg/ml. Drops as small as 5 μl are sufficient. 30 min.
4) PBS. 4 × 1 min.
5) pA-gold or Ig-gold, both diluted just before use with PBS + 1% BSA, to $E^{520} \simeq 0.05$. Drops of 5 μl are sufficient. 30 min.
6) PBS. 4 × 5 min.

For double labelling with pA-gold, free protein A 0.05 mg/ml is added to step 5 during the last few minutes of the incubation. Then steps 3–6 are repeated with a second specific immunoglobulin in step 3 and a pA-gold marker of different size in step 5.

Finally the sections can be uranyl stained and embedded in methylcellulose after Tokuyasu as modified by Griffiths [9]:

7) Distilled water. 3 × a few seconds.
8) 2% uranyl acetate in 0.15 M oxalic acid, brought to pH 7 with 5% NH_4OH. 10 min.
9) Distilled water. 3 × a few seconds.
10) 2–4% uranyl acetate. About 10 min.
11) Distilled water, approximately 1 sec. Be careful, the uranyl is easily extracted. This step can be replaced by 2 or 3 fast passages over drops of 1.5% methylcellulose.
12) 1.5% methylcellulose in distilled water (Tylose MH 300 or Methocel MC 400 cP, Fluka). The solution is prepared by adding 1.5 g methylcellulose to 100 ml distilled water. Stir at 4°C for 24 h or even longer and store in the refrigerator.
13) Pick up the grids within 30 sec from the methylcellulose drop with a wire loop (∅ > ∅ grid). Remove excess of methylcellulose with filter paper, let the grid dry and remove it from the loop.

6. Acknowledgements

We would like to thank Mrs. Ada Baars for her excellent technical assistance, Mr. Jos Berkers for providing part of the illustrations (Fig. 3), Dr. Jane Peppard (Chester Beatty Res. Inst., Sutton, Surrey, U.K.) who gave us the anti-SC preparation and Dr. Allan Schwartz (Sidney Farber Cancer Inst., Boston-Mass., U.S.A.) who provided anti-ASGP-receptor IgG.

7. References

1. Singer, S.J. (1959) Preparation of an electron-dense antibody conjugate. Nature (London) 183, 1523–1524.
2. Faulk, W.P. and Taylor, G.M. (1971) An immunocolloid method for the electron microscope. Immunochemistry 8, 1081–1083.
3. Romano, E.L. and Romano, M. (1977) Staphylococcal protein A bound to colloidal gold: A useful reagent to label antigen-antibody sites in electron microscopy. Immunochemistry 14, 711–715.
4. Roth, J. (1983) The colloidal gold marker system for light and electron microscopic cytochemistry. In: G.R. Bullock and P. Petrusz (Eds.) Techniques in Immunocytochemistry, Vol. 2. Academic Press, London, pp. 217–284.
5. Geuze, H.J., Slot, J.W., Scheffer, R.C.T. and van der Ley, P.A. (1981) Use of colloidal gold particles in double-labeling immunoelectron microscopy of ultrathin frozen tissue sections. J. Cell Biol. 89, 653–665.
6. Slot, J.W. and Geuze, H.J. (1981) Sizing of protein A-colloidal gold probes for immunoelectron microscopy. J.Cell Biol. 90, 533–536.
7. Slot, J.W. and Geuze, H.J. (1983) Immunoelectron microscopic exploration of the Golgi complex. J. Histochem. Cytochem. 31, 1049–1056.

8. Geuze, H.J., Slot, J.W., Strous, G.J.A.M., Lodish, H.F. and Schwartz, A.L. (1983) Intracellular site of asialoglycoprotein receptor-ligand uncoupling: double-label immunoelectron microscopy during receptor-mediated endocytosis, Cell 32, 277–287.
9. Griffiths, G., Brands, R. Burke, B. Louvard, D. and Warren, G. (1982) Viral membrane proteins acquire galactose in trans Golgi cisternae during intracellular transport. J. Cell Biol. 95, 781–792.
10. Mota, G., Gethie, V. and Sjöquist, J. (1978) Characterization of the soluble complex formed by reacting rabbit IgG with protein A of S. aureus. Immunochemistry 15, 639–642.
11. Dutton, A., Tokuyasu, K.T. and Singer, S.J. (1979) Iron-dextran antibody conjugates. General method for simultaneous staining of two components in high resolution immunoelectron microscopy. Proc. Natl. Acad. Sci. USA 76, 3392–3396.
12. Tapia, F.J., Varndell, I.M., Probert, L., De Mey, J. and Polak, J.M. (1983) Double immunogold staining method for the simultaneous ultrastructural localization of regulatory peptides. J. Histochem. Cytochem. 31, 977–981.
13. Slot, J.W. and Geuze, H.J. (1983) The use of protein A-colloidal gold (PAG) complexes as immunolabels in ultrathin frozen sections. In: A.C. Cuello (Ed.) Immunohistochemistry. IBRO, pp. 323–346.
14. Tokuyasu, K.T. (1983) Present state of immunocryoultramicrotomy. J. Histochem. Cytochem. 31, 164–167.
15. Geoghegan, W.D. and Ackerman, G.A. (1977) Adsorption of horseradish-peroxidase, ovomucoid and anti-immunoglobulin to colloidal gold for the indirect detection of Concanavalin A, wheat germ agglutinin and goat anti-human immunoglobulin G on cell surfaces at the electron microscopic level: a new method, theory and application. J. Histochem. Cytochem. 25, 1187–1200.
16. Horisberger, M. and Rosset, J. (1977) Colloidal gold, a useful marker for transmission and scanning electron microscopy. J. Histochem. Cytochem. 25, 295–305.
17. De Mey, J., Moeremans, M. Geuens, G. Nuydens, R. and De Brabander, M. (1981) High resolution light and electron microscopic localization of tubulin with the IGS (immunogold staining) method. Cell Biol. Int. Rep. 5, 889–899.
18. Zsigmondy, R. (1905) Zur Erkenntnis der Kolloide, Jena, Germany.
19. Stathis, E.C. and Fabrikanos, A. (1958) Preparation of colloidal gold. Chem. Ind. (London) 27, 860–861.
20. Geuze, H.J., Slot, J.W., Strous, G.J.A.M., Peppard, J., Lodish, H.F. and Schwartz, A.L. (1984) Intracellular receptor sorting during endocytosis: Comparative immuno-electron microscopy of multiple receptors in rat liver. Cell 37, 195–204.

Immunolabelling for Electron Microscopy (Polak/Varndell, eds)

CHAPTER 12

Double labelling cytochemistry applying the protein A-gold technique

Moïse Bendayan and Heather Stephens

Department of Anatomy, Université de Montréal, Montréal, Québec, Canada

CONTENTS

1. Introduction

The application of cytochemical techniques allows for the in situ morphological localisation of a wide range of macromolecules, taking advantage of their biochemical properties. Moreover, there is considerable interest in simultaneously revealing several proteins in order to compare and correlate their specific morphological distribution. The concomitant detection of various macromolecules can be performed by applying cytochemical techniques on serial sections cut from a tissue block, or, in a more convincing and elegant way, by performing double labelling on the same tissue section using different markers.

The protein A-gold immunocytochemical technique was introduced as an alternative to other methods for the ultrastructural localisation of antigenic sites (for review see references [1, 2]). It is an indirect post-embedding technique based on the unique property of protein A to interact with the Fc region of immunoglobulins. To be applied in electron microscopy, protein A has to be labelled with an electron-dense marker. In many instances colloidal gold has been chosen because of its unique properties.

In particular, the size of the colloidal gold particles can be varied according to the procedure used for their preparation, which renders this marker particularly attractive for double labelling techniques. Indeed, using antibodies directed against two different proteins together with protein A-gold complexes formed by gold particles of two different sizes, two antigenic sites can be revealed simultaneously on the same section. Protein A complexed with colloidal silver can also be used successfully either for the detection of one particular antigen or in combination with protein A-gold in the double labelling approach.

2. Technical principles

The principles of the basic protein A-gold technique, which is a two-step labelling procedure, is described in Fig. 1. The labelling is achieved by incubating the thin

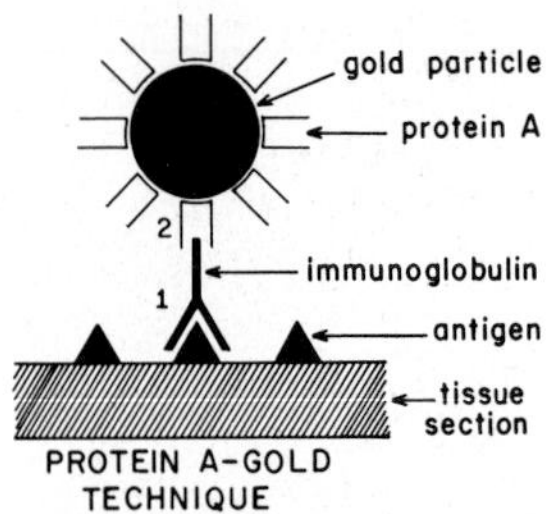

Fig. 1. Diagram illustrating the principles of the protein A-gold technique. It is a two-step post-embedding method carried out on thin tissue sections. In the first step the immunoglobulin G interacts with the antigen exposed at the surface of the section. The antigen-antibody complex is then revealed in the second step by the protein A-gold complex. The molecules of protein A surrounding the gold particles interact with the Fc region of the immunoglobulin. Indirectly, the gold particle enables the localisation of the antigen on the tissue section. (To make the diagram more comprehensible, relative proportions have not been respected.) (Adapted from reference [2].)

sections with the specific antibody followed by the protein A-gold complex. However, for each antigen, conditions of fixation and embedding of the tissue, as well as those of incubation have to be worked out in order to obtain optimal localisation. Application of this technique on pancreatic exocrine tissue, using antibodies directed against secretory proteins has allowed the localisation of various pancreatic enzymes in the different cellular compartments involved in protein secretion [3, 4]: rough endoplasmic reticulum, Golgi apparatus, immature and mature secretory granules (Fig. 2). Furthermore, since the protein A-gold technique allows for quantitative evaluation, the intensity of the labelling obtained for each enzyme could be determined. An increasing gradient was found from the rough endoplasmic reticulum to the Golgi and to the granule which reflects the process of protein concentration occurring along the secretory pathway [3, 4]. These results suggested that the pancreatic enzymes are processed through the same cellular compartments. To demonstrate this, the double labelling approach

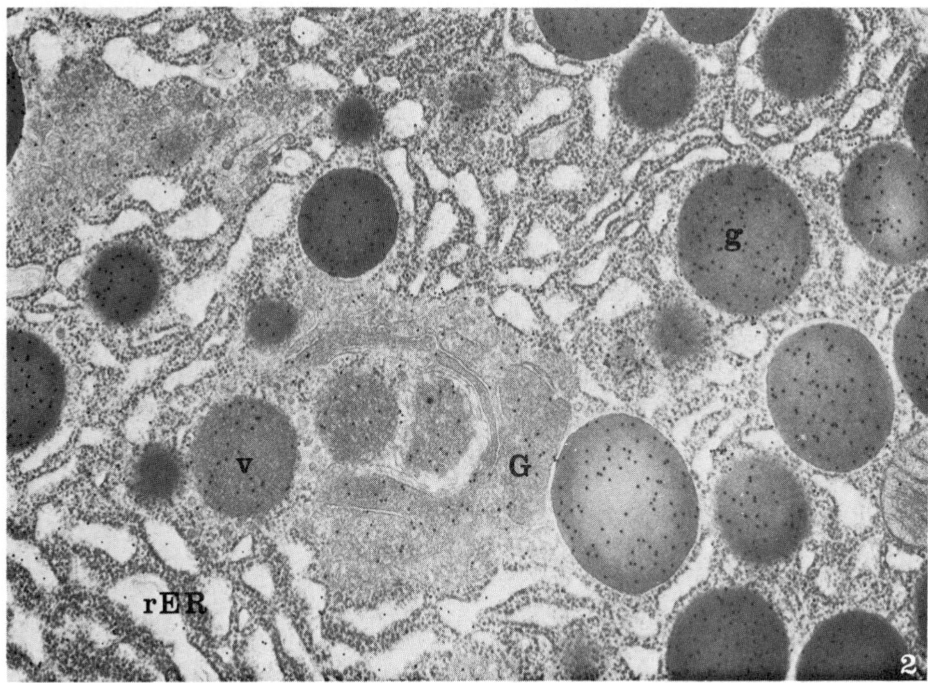

Fig. 2. Localisation of amylase on a pancreatic acinar cell. The labelling by gold particles appears to be present over the rough endoplasmic reticulum (rER), the Golgi cisternae (G), the condensing vacuoles (V) or immature secretory granules and the zymogen or mature secretory granules (g). (1% Glutaraldehyde; 1% osmium tetroxide; Epon; sodium metaperiodate/anti-amylase/protein A-gold.) Magnification = ×23,000.

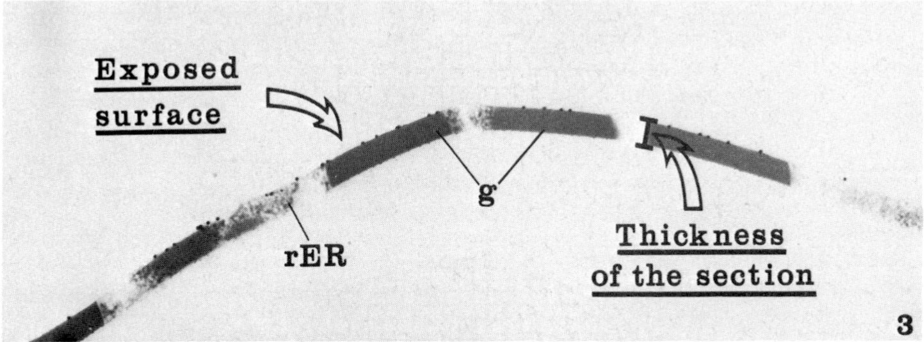

Fig. 3. This figure illustrates a cross-section obtained at a right angle from a previously labelled thin section. The experiment was carried out in the following manner: a conventional thin section of rat pancreatic tissue was labelled by the protein A-gold technique for the demonstration of amylase. The section was then re-embedded in Epon and cut perpendicularly to the labelled surface. The labelling by gold particles appears to be restricted to the surface of the section over the rough endoplasmic reticulum (rER) and the secretory granules (g). No particles are found in the depth of the section demonstrating that only those antigenic sites exposed at the surface of the section can be revealed by the protein A-gold technique (adapted from reference [2]). Magnification = ×31,000.

was developed and applied [5]. The examination of a labelled tissue section cut at a right angle to the original surface, reveals that the labelling occurs exclusively at the surface of the section and that only antigenic sites exposed by the cutting procedure can be detected (Fig. 3). This is particularly important with respect to the double labelling approach.

For the simultaneous labelling of two antigenic sites on the same section protein A-gold complexes formed by gold particles of different sizes can be used. However, we were confronted with a major problem since protein A interacts with the Fab as well as to the Fc region of IgG molecules albeit to a lesser extent; cross-reaction between the first and the second labelling may occur if both labelling procedures are performed on the same side of the tissue section. In order to avoid such cross-reaction, we have developed a technique which makes use of both surfaces of a tissue section allowing for simultaneous, but independent, demonstration of two antigenic sites [5].

2.1. *Procedure for double immunolabelling*

The technique is carried out in the following way. Protein A-gold complexes formed by gold particles of two sizes are first prepared using the appropriate procedure (for review see reference [2]). Protein A-silver complexes can be prepared according to a technique previously described [6]. Ultrathin tissue sections should be mounted on uncoated nickel grids so both faces of the sections (face A and face B, Fig. 4) are exposed and available for labelling. The adhesion of

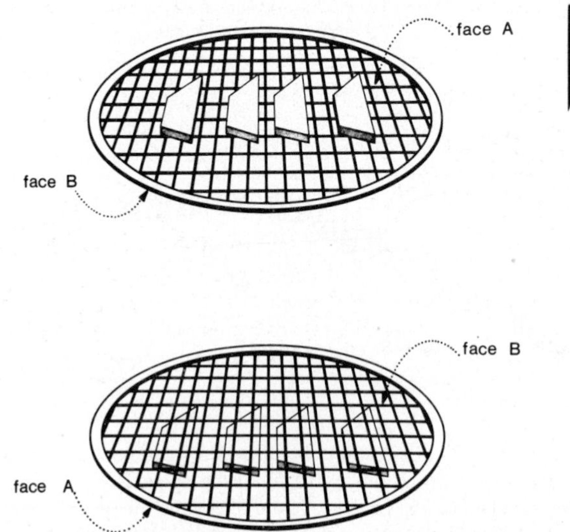

Fig. 4. Diagram illustrating the nomenclature of faces A and B of an electron microscope grid supporting some tissue sections. On face A of the grid the tissue sections have their entire surface exposed, while on face B only the surface of the tissue sections between the bars of the grid is exposed and available for labelling (adapted from reference [5]).

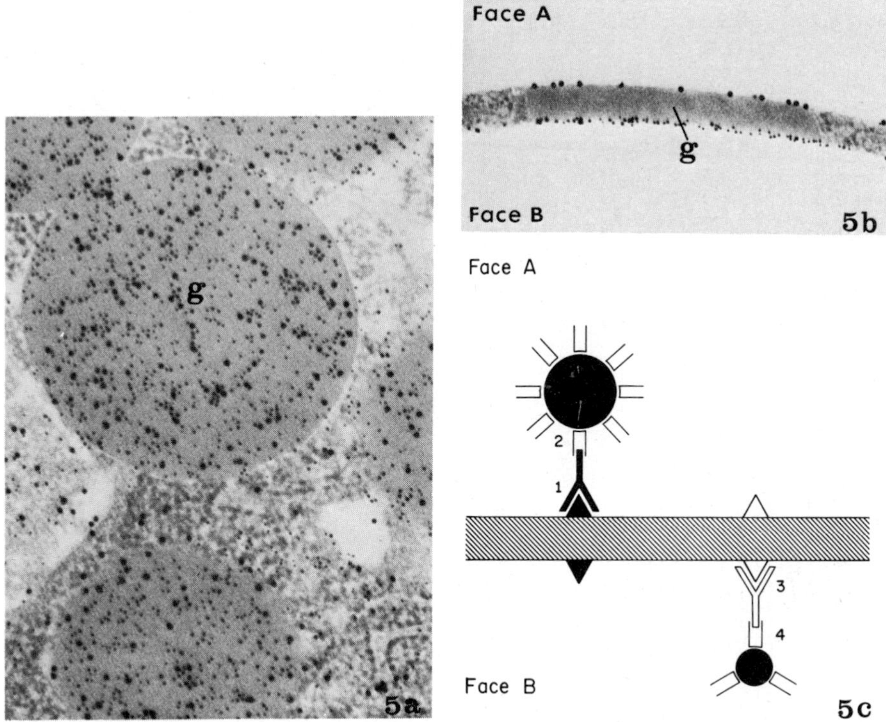

Fig. 5. Double labelling performed using faces A and B of a thin tissue section. (a) Simultaneous localisation of amylase and chymotrypsinogen in a pancreatic acinar cell by the double labelling approach. Both enzymes, amylase (as revealed by the large gold particles on face A) and chymotrypsinogen (as revealed by the small gold particles on face B) are present in the same secretory granule (g). No artifactual co-labelling is observed. (1% Glutaraldehyde; 1% osmium tetroxide; Epon; sodium metaperiodate; face A: anti-amylase/protein A-large gold particles; face B: anti-chymotrypsinogen/ protein A-small gold particles.) Magnification = ×52,000. (b) Cross-section of a thin tissue section double labelled for the demonstration of amylase and chymotrypsinogen as described in Fig. 5a. It appears that the large gold particles revealing amylase antigenic sites are restricted to face A of the tissue section while the small ones, revealing chymotrypsinogen antigenic sites, are restricted to face B. (g) secretory granule. Magnification = ×56,000. (c) Diagram illustrating the principle and the different steps of the double labelling approach using both faces of a thin tissue section. 1) Incubation of face A with the first antibody; 2) incubation with the protein A-gold complex formed by large particles; 3) incubation of face B with a second antibody; 4) incubation with the protein A-gold complex formed with small gold particles.

the sections to the grids can be enhanced by treating the grids with 20% acetic acid for a few seconds, followed by 90% ethanol and rinsing in distilled water (this treatment should be performed prior to the mounting of the tissue sections; [5]).

The labelling of face A of the tissue section is carried out according to the protocol described for the protein A-gold technique [1, 2], using the protein A-gold complex formed with large gold particles. Incubation of the grid for 5 min at room

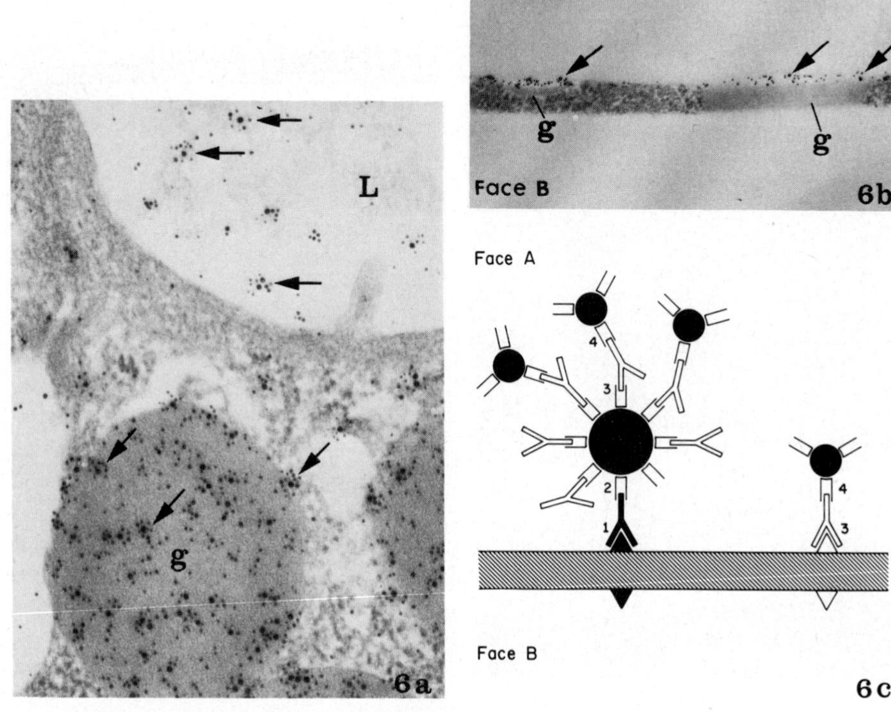

Fig. 6. Double labelling performed on the same face of a tissue section and demonstrating the artifactual co-distribution of the large and small gold particles. In such a protocol, the second antibody interacts with the protein A molecules surrounding the large gold particles thereby inducing an apparent co-labelling. The small gold particles are clearly associated with the large ones (arrows). (L) acinar lumen; (g) secretory granule. (1% Glutaraldehyde; 1% osmium tetroxide; sodium periodate; face A: anti-amylase/protein A-large gold particles/anti-chymotrypsinogen/protein A-large gold particles.) Fig. 6(a) Magnification = ×56,000; (b) Magnification = ×36,000.

temperature on a drop of 0.01 M phosphate-buffered saline containing 1% ovalbumin is then followed by transfer of the grid without rinsing onto a drop of the diluted specific antibody (various periods of time according to the antibody and its dilution). The grid is then rinsed to remove excess unbound antibodies and placed onto a drop of the protein A-gold complex for incubation (30–60 min at room temperature). Finally, the grid is rinsed in buffer and water. In the case of osmium tetroxide post-fixed tissues, pre-treatment of the sections with a saturated solution of sodium metaperiodate for one hour at room temperature should be performed prior to immunolabelling [7]. This allows for the retention of fine ultrastructural preservation and, thus, for better resolution in the results. Once the labelling of face A of the tissue sections is performed, the grids are dried and turned over. The second face of the sections, face B, can then be labelled according to the same protocol using a second antibody and a protein A-gold complex formed with small

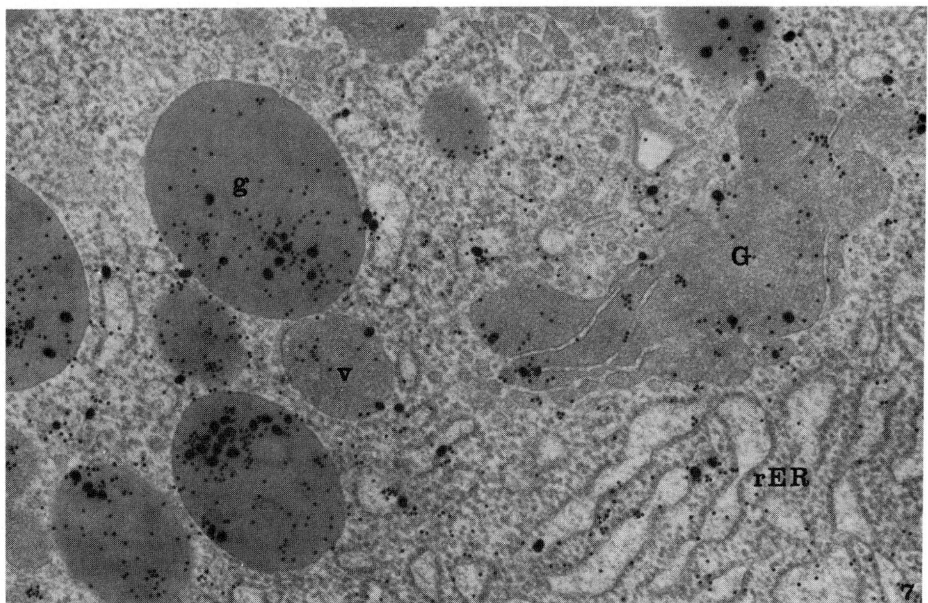

Fig. 7. Simultaneous localisation of carboxypeptidase A (as revealed by the large gold particles, approximately 35 nm) and carboxypeptidase B (as revealed by the small gold particles, approximately 14 nm) over the rough endoplasmic reticulum (rER), the Golgi cisternae (G), the condensing vacuoles (V) and the mature secretory granules (g) of a pancreatic acinar cell. The non-uniform pattern of labelling can be clearly observed over the secretory granules which suggests a compartmentalisation of proteins in the granules [4]. Both types of gold particles were prepared by the citrate method [2]. (1% Glutaraldehyde; 1% osmium tetroxide; Epon; sodium periodate; face A: anti-carboxypeptidase A/protein A-large gold particles; face B: anti-carboxypeptidase B/protein A-small gold particles.) Magnification = ×40,000.

Figs. 7–9. Demonstration of various secretory proteins in pancreatic acinar cells. The double labelling technique has been applied using faces A and B of the thin tissue sections.

gold particles. Care has to be taken not to wet both sides of the grid with the same reagent. The tissue sections can be stained with uranyl acetate and lead citrate before examination in the electron microscope. The "transparency" of the section under the electron beam allows for the simultaneous visualisation of the gold particles present on both sides of the section. The principle and results of this double labelling technique is illustrated in Fig. 5.

3. Discussion

With such an approach, since the labelling of each of the faces of the sections is performed independently, no co-labelling due to artifactual cross-reaction between reagents is observed. This is further and better demonstrated by the examination of

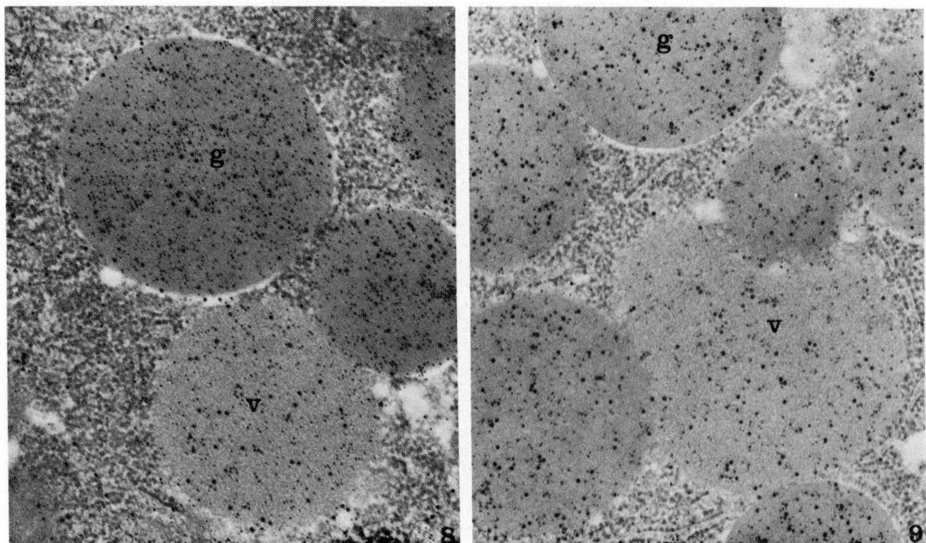

Fig. 8. Simultaneous localisation of trypsinogen (as revealed by the large gold particles, approximately 14 nm) and chymotrypsinogen (as revealed by the small gold particles, approximately 8 nm) in condensing vacuoles (V) and secretory granules (g). The large gold particles were prepared by the citrate method and the small ones by the tannic acid method [2]. (1% Glutaraldehyde; Epon; face A: anti-trypsinogen/protein A-large gold particles; face B: anti-chymotrypsinogen/protein A-small gold particles.) Magnification = ×31,500.

Fig. 9. Simultaneous localisation of amylase (as revealed by the small silver particles, approximately 9 nm) and trypsinogen (as revealed by the large gold particles, approximately 14 nm) in condensing vacuoles (V) and secretory granules (g). The silver colloidal particles were prepared by the method described by Roth [6]. (1% Glutaraldehyde; Epon; face A: anti-trypsinogen/protein A-gold; face B: anti-amylase/protein A-silver.) Magnification = ×31,500.

a double labelled section cut at a right angle (Fig. 5b). In contrast, if both labellings are performed on the same face of the section artifactual co-labelling is systematically obtained (Fig. 6).

The technique was applied for the simultaneous demonstration of various enzymes in the different cellular compartments of the secretory pathway of pancreatic acinar cells [5]. The enzymes were found to be present simultaneously in the same compartments (Fig. 7, 8 and 9). Such results brought support to the proposal that pancreatic secretory proteins are processed by the cells in a parallel manner throughout the rough endoplasmic reticulum-Golgi-granule secretory pathway [5].

A second approach for the simultaneous demonstration of two proteins on the same thin section has been developed by combining pre-embedding cytochemical techniques with the post-embedding protein A-gold immunocytochemical method. First, a given enzyme is localised by established cytochemical techniques in a pre-embedding step and after processing the tissue for electron microscopy, specific

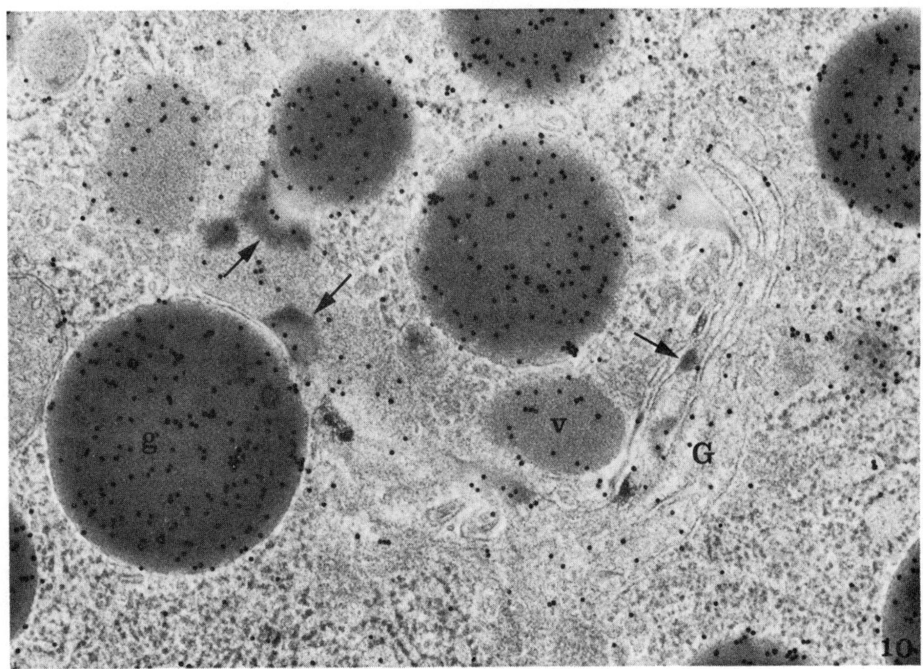

Fig. 10. Simultaneous localisation of thiamine pyrophosphatase (TPPase) and amylase in the Golgi apparatus of the rat pancreatic acinar cell. The lead deposits (arrows) revealing the site of the TPPase are present exclusively in the *trans* cisternae of the Golgi apparatus (G) while the gold particles revealing amylase antigenic sites are present over all Golgi cisternae, condensing vacuoles (V) and secretory granules (g). (1% Glutaraldehyde/TPPase cytochemical reaction/1% osmium tetroxide; Epon; sodium periodate; anti-amylase/protein A-gold.) (Adapted from reference [2].) Magnification = ×40,500.

antigenic sites can be revealed on the tissue thin sections by applying the protein A-gold technique. This approach has been successfully applied for the simultaneous demonstration of different phosphatases and secretory proteins in the Golgi apparatus of pancreatic acinar cells (Figs. 10 and 11) [2, 4] for phosphatases and galactosyltransferase in the Golgi apparatus of Hela cells [8] as well as for acetylcholinesterase and various basal laminae components in neuromuscular junctions (Figs. 12, 13 and 14) ([9] and in preparation). Thus the protein A-gold technique as such, or in combination with other cytochemical techniques, appears as a very versatile and powerful approach in cytochemistry.

4. Acknowledgements

This investigation was supported by grants from the Medical Research Council and the Muscular Dystrophy Association of Canada. The authors are recipients of scholarships from the MRC (M.B.) and the FRSQ (H.S.).

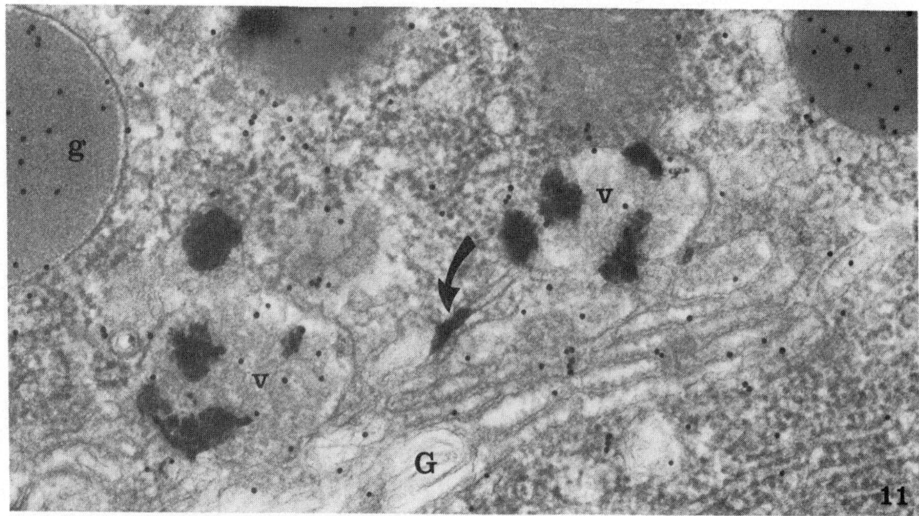

Fig. 11. Simultaneous localisation of acid phosphatase and amylase in the Golgi apparatus of the rat pancreatic acinar cell. The lead deposits revealing the sites of the acid phosphatase are present in the rigid *trans* cisternae (arrow) of the Golgi apparatus (G) and in the condensing vacuoles (V). The gold particles revealing amylase antigenic sites are present over the Golgi cisternae (G), the condensing vacuoles (V) and the secretory granules (g) but are absent over the rigid *trans* cisternae (arrow). (1% Glutaraldehyde/acid phosphatase cytochemical reaction/1% osmium tetroxide; Epon; sodium periodate, anti-amylase/protein A-gold.) (Adapted from reference [4].) Magnification = ×49,000.

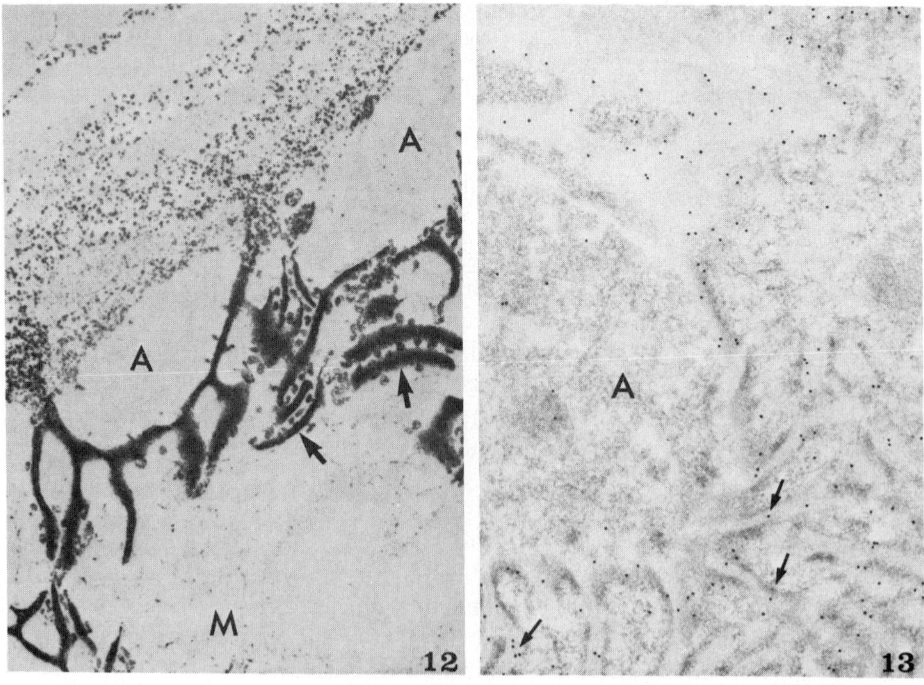

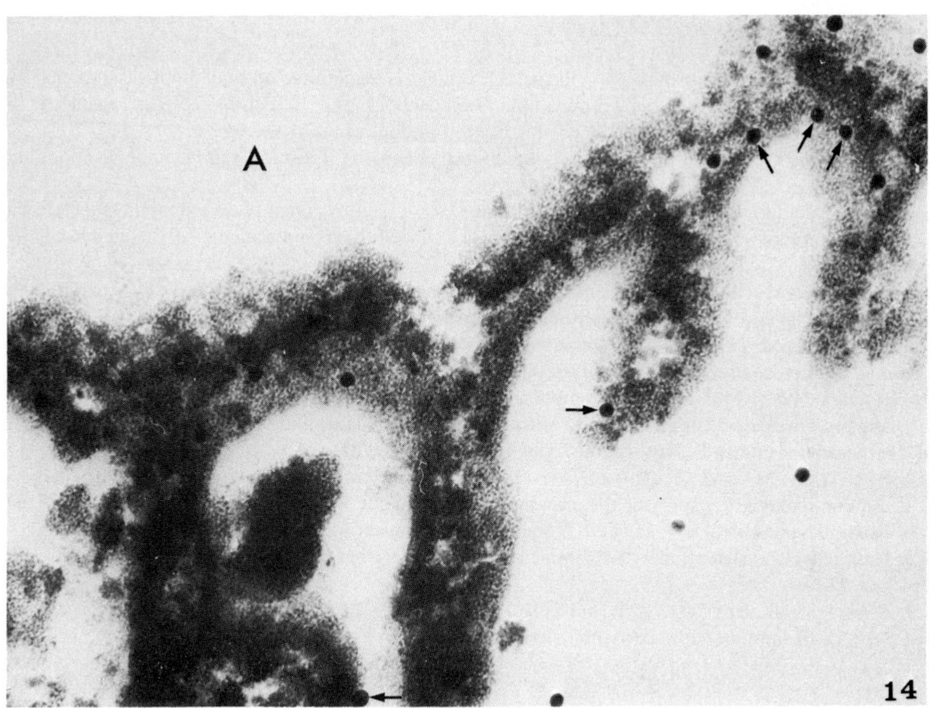

Fig. 14. Simultaneous localisation of acetylcholinesterase and collagen type IV at the level of a neuro-muscular junction. Both proteins appear to be present simultaneously at the level of the synaptic clefts. Furthermore, in some instances the gold particles revealing collagen type IV antigenicity appear to be at the edge of the acetylcholinesterase reaction product (arrows) suggesting the existence of a certain pattern of distribution of these proteins. (4% Paraformaldehyde; acetylcholinesterase cytochemical reaction; Lowicryl K4M; anti-collagen IV/protein A-gold.) Magnification = ×124,500.

Fig. 12. Localisation of acetylcholinesterase (AChE) at the level of a neuro-muscular junction. The electron-dense lead deposits revealing AChE sites are present in the primary synaptic folds underneath the axonal processes (A) and in the secondary synaptic clefts (some identified by arrows). M = muscle cell. (4% Paraformaldehyde; acetylcholinesterase cytochemical reaction; Lowicryl K4M.) Magnification = ×8,000.

Fig. 13. Localisation of collagen type IV at the level of a neuro-muscular junction. The gold particles are present over the basal laminae material in the secondary synaptic clefts (some identified by arrows). (4% Paraformaldehyde; Lowicryl K4M; anti-collagen IV/protein A-gold.) Magnification = ×29,000.

5. References

1. ROTH, J. (1982) The protein A-gold (pAG) technique, qualitative and quantitative approach for antigen localization on thin sections. In: G.R. Bullock and P. Petrusz (Eds.) Techniques in Immunocytochemistry, Vol. I. Academic Press, London, pp. 107–133.
2. BENDAYAN, M. (1984) Protein A-gold immunocytochemistry: Technical approach, applications and limitations. J. Elect. Microsc. Tech. (In press).
3. BENDAYAN, M., ROTH, J., PERRELET, A. and ORCI, L. (1980) Quantitative immunocytochemical localization of pancreatic secretory proteins in subcellular compartments of rat acinar cell. J. Histochem. Cytochem. 28, 149–160.
4. BENDAYAN, M. (1984) Concentration of amylase along its secretory pathway in the pancreatic acinar cell as revealed by high resolution immunocytochemistry. Histochem. J. 16, 85–108.
5. BENDAYAN, M. (1982) Double immunocytochemical labeling applying the protein A-gold technique. J. Histochem. Cytochem. 30, 81–85.
6. ROTH, J. (1982) Applications of immunocolloids in light microscopy. Preparation of protein A-silver and protein A-gold complexes and their application for localization of single and multiple antigens in paraffin sections. J. Histochem. Cytochem. 30, 691–696.
7. BENDAYAN, M. and ZOLLINGER, M. (1983) Ultrastructural localization of antigenic sites on osmium-fixed tissues applying the protein A-gold technique. J. Histochem. Cytochem. 31, 101–109.
8. ROTH, J. and BERGER, E.G. (1982) Immunocytochemical localization of galactosyltransferase in Hela cells. Codistribution with thiamine pyrophosphatase in trans-Golgi cisternae. J. Cell Biol., 93, 223–229.
9. STEPHENS, H., BENDAYAN, M. and GISIGER, V. (1983) Ultrastructural co-localization of collagen type IV or laminin with acetylcholinesterase (AChE) in endplate regions. J. Cell Biol. 97, 234a.

Immunolabelling for Electron Microscopy (Polak/Varndell, eds)

CHAPTER 13

Double immunostaining procedures: Techniques and applications

Ian M. Varndell and Julia M. Polak

Department of Histochemistry, Royal Postgraduate Medical School, Hammersmith Hospital, Du Cane Road, London W12 0HS, U.K.

CONTENTS

1. Introduction

One logical extension of the capacity to localise a single antigen or group of similar antigens at the electron microscopical level is the development of a reliable procedure which enables the discrimination of two or more distinct, but co-existing or neighbouring, antigens. Several contributors to this book have broached the subject of multiple immunolabelling and a variety of marker combinations have been proposed [1–3]. In the overwhelming majority of cases one of the markers employed is the peroxidase-antiperoxidase (PAP) complex reaction product—usually the oxidised polymer of 3,3′-diaminobenzidine. The PAP marker system has been combined with radiolabelled antibodies [1, 4, 5], ferritin-labelled antibodies [6], ferritin-avidin [7], lectin-ferritin [8] and iron-dextran complexes (e.g. Imposil) [9] as well as colloidal gold-labelled immunoreagents (discussed in detail later). Peroxidase-antiperoxidase immunostaining has also been combined with techniques such as anterograde degeneration [10, 11], anterograde autoradiography [12], transmitter uptake autoradiography [2, 13], Golgi impregnation [14] and retrograde peroxidase tracing [15, 16]. Each technique has its advantages in particular test systems but on the whole these are outweighed in general applications when considerable disadvantages, such as diffusion of reaction products, low electron density of markers and difficult visual differentiation of particulate markers, become apparent.

The quest for a reliable multiple immunostaining technique was made considerably easier by the introduction of colloidal gold [17, 18]. Homogeneously-sized gold particle populations can be manufactured (as discussed at length in the Appendix to Chapter 15) and a number of double immunolabelling procedures have now been described utilising two different particle diameters [19–23]. The majority of these have been mentioned in other chapters throughout this book [3, 24, 25]. However, it is worth considering the basic techniques in some detail at this point. Each of the following techniques can be considered to have some advantages in general applications which outweigh the disadvantages, albeit in some cases marginally.

1.1. Direct immunolabelling

Direct marking of antigens with (gold-) labelled antibodies is the simplest, though not the most sensitive [3, 26] means of immunolocalisation (Fig. 1). This method has been revolutionised by the introduction of monoclonal antibodies. Simultaneous or sequential application of two or more monoclonal antibodies linked to

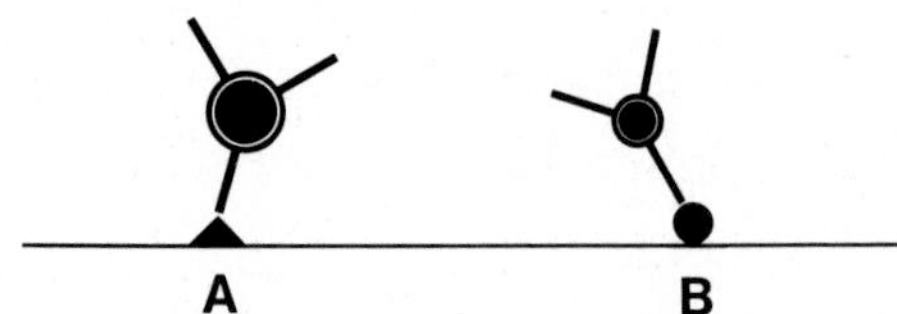

Fig. 1. Diagrammatic representation of immunoglobulin G molecules directly labelled with gold particles of different sizes. The number of immunoglobulin molecules adsorbed to the surface of gold particles is dependent upon several factors which include, size of particle, protein concentration and the class of the immunoglobulin molecules adsorbed. A 5 nm particle may adsorb 2–3 IgG molecules whereas 20 nm particles may bind 10–15 IgG.

different gold particle sizes is used routinely in the investigation of cell surface antigens [27].

Double immunolabelling with the direct marker system necessitates the availability of fairly large amounts of affinity-purified polyclonal or monoclonal antibodies prior to conjugation. Unfortunately, one is seldom in the fortunate position of having sufficient stocks of well characterised antisera to dedicate a considerable volume for gold conjugation. Hopefully as monoclonal antibody technology advances the above statement will cease to hold.

1.2. Indirect immunolabelling—GLAD method

The *G*old-*L*abelled *A*ntigen *D*etection (GLAD) method (Fig. 2) was originally promulgated as a multiple immunolabelling procedure [19]. In this procedure pure

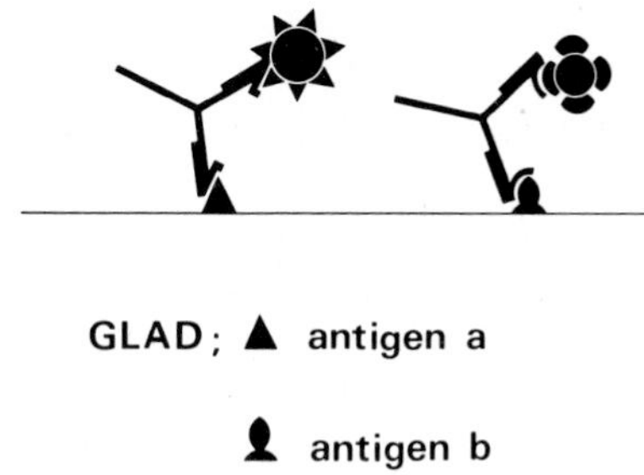

Fig. 2. GLAD (Gold Labelled Antigen Detection) method. Homogeneous populations of two or more gold particle sizes are conjugated to different antigens. Available sites on the IgG, IgM and IgA will link tissue-bound and gold-conjugated antigens.

antigen (purified natural or synthetic) is used in the immunogen and is also conjugated to a single size of gold particles. Antisera raised against the immunogen, and thus directed to parts thereof, are harvested and purified. Selected populations of (region-specific) antisera are then applied to tissue sections and the sites of binding are marked using the gold-linked antigen reagent. This is an attractive technique as the antibody should only recognise the same determinants on the section and on the gold-linked reagent. In practice each antigen to be localised requires a gold-linked reagent. Again, highly purified or synthetic antigens are usually available in small quantities either because of the bulk of the starting tissue and the exhaustive methods involved from extraction, through purification to final characterised form, or prohibitive cost—and frequently both—consequently little is generally left after the raising of antisera. In extensive studies of single antigens the GLAD method could be considered to be the technique of choice. Only the inflexibility of the method precludes its use in general applications.

1.3. Indirect immunolabelling—protein A-gold and immunoglobulin-gold

On the whole, indirect methods of antigen localisation at the electron microscopical level are favoured. Many examples of protein A-gold and immunoglobulin-gold immunocytochemistry are given throughout this book. Discounting monoclonal antibodies, the vast majority of antisera used as primary layer are raised in rabbits. The implication of this is that second layer gold immunoreagents will not discriminate between two rabbit-raised primary antibodies in simultaneous or even sequential procedures unless steps are included to block determinants remaining on the first rabbit immunoglobulin.

Occasionally, either by convention, design or availability some primary antisera are raised in species other than rabbit. We have recently described a double immunolabelling technique based on the availability of primary antisera raised in different species [22]. This technique is described in detail in the Appendix to this chapter.

Two techniques have been described to date in which multiple antisera from the same, or more than one donor species, can be visualised simultaneously.

1.3.1. Selected surface immunolabelling [20]
By careful manipulation of electron microscope grids, one surface of the mounted section may be immunostained without contamination of the other face. Bendayan [20] used this technique to immunostain both faces of a single grid by the sequential application of different antisera which were visualised using two protein A-gold reagents with non-overlapping particle sizes. However, there would appear to be no reason why the protein A-gold complexes could not be replaced by immunoglobulin-gold in a modified version of this procedure. This elegant technique requires great care to ensure that immunoreagents are not transferred from one face of the section to the other. Short incubation times and droplet washing procedures [28, 29] are recommended to obtain optimal results. However, in our hands the tendency appears to be to under-wash the immunostained grids in order to avoid cross-face contamination with the result that background staining is distractingly high.

1.3.2. Double protein A-gold immunolabelling [21]
This involves the sequential application of two separate antibody-protein A-gold procedures with intermediate blocking of potential immunoreactive sites (Fig. 3). The results are generally good but somewhat inconsistent in model systems. We tested this procedure on ultrathin sections of human endocrine pancreas using antisera raised against glicentin and insulin applied sequentially and visualised using two different protein A-gold complexes. The result (Fig. 4) exhibited a high cross-contamination of the first primary antibody-protein A-gold-second primary antibody complex (see Fig. 3) by the second protein A-gold reagent. A possible explanation of this cross-contamination is given in Fig. 5. The original technique described by Roth requires that all available Fc binding sites for protein A on the second primary antibody are occupied. This is unlikely to be the case as the CH_2

and CH_3 domains of immunoglobulins can potentially bind more than one protein A molecule [30], thus some sites on the second primary antibody IgG Fc may be available to react with the second protein A-gold complex. Despite this complication, with carefully planned control immunostaining, and the use of gold particles of small diameter (4–8 nm; see Chapter 11, pp. 129–142), the double protein

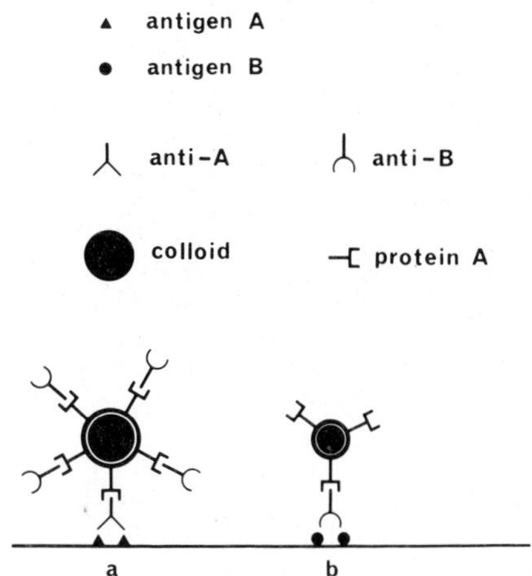

Fig. 3. Protein A-gold double immunocytochemical staining procedure. (a) Antigen A is localised with a specific IgG (anti-A). The Fc component of this antibody is then linked to staphylococcal protein A which is adsorbed to the colloidal particle. Excess antibody to antigen B (anti-B) is then applied which links to specific tissue-bound sites. In addition, the Fc component of unbound anti-B links to available protein A sites on gold particle 1. The application of a second gold particle size (or other colloid such as silver) coated with protein A should adhere to the tissue-bound anti-B Fc regions. The two particle sizes can then be easily distinguished in the electron microscope.

A-gold immunolabelling procedure offers considerable advantages over the other techniques given above on the basis of simplicity, cost of reagents and flexibility.

2. Applications for double immunolabelling

Cellular physiology cannot be viewed in terms of single, isolated events or functions. The localisation of an antigen (for example, a bioactive peptide) within a cell or on the surface of a cell, conveys considerable information about the activity of that cell. Critical interpretation of the ultrastructural appearance may shed some light on the general activity of the cell but little information on the rate of synthesis or secretion, maturity of secretory granules, or the mode and nature of stimulation is gained. Double immunolabelling at least allows a second antigen, in the same cell or tissue to be investigated simultaneously which may provide valuable information

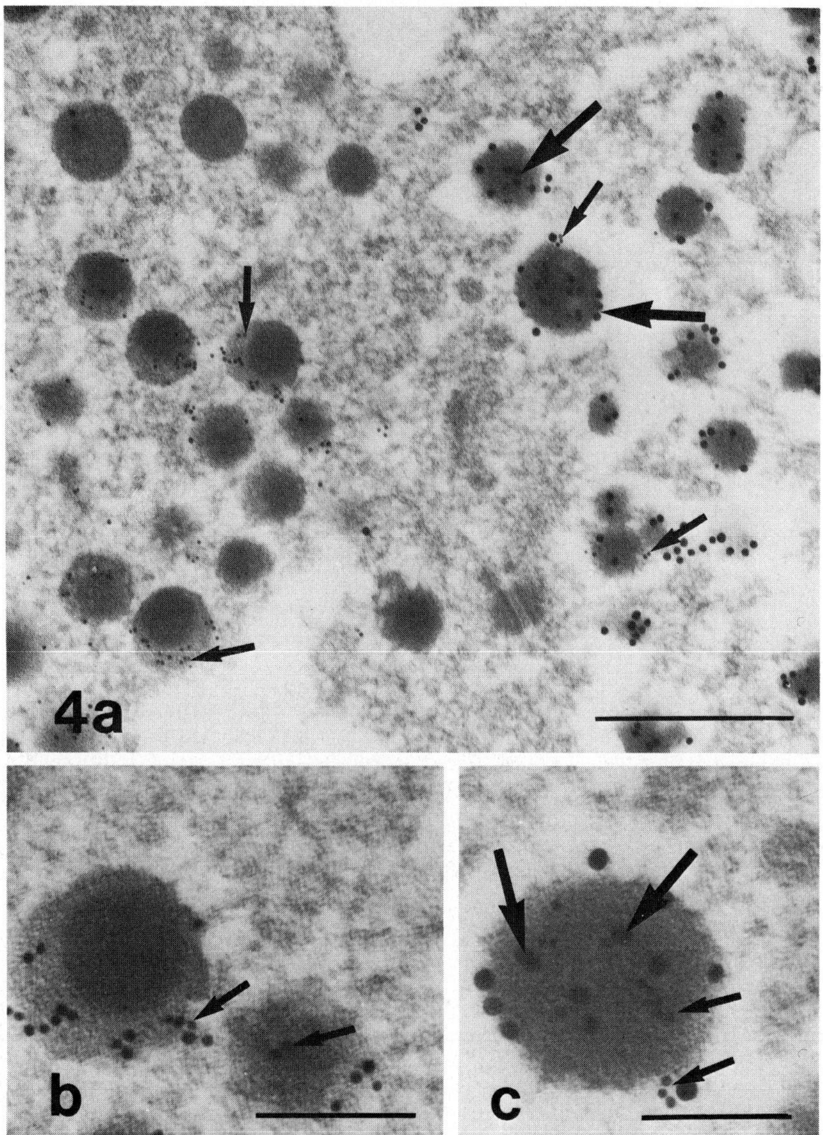

Fig. 4. Double protein A-gold immunostaining of human pancreatic islet. Glicentin immunoreactivity is localised with 10 nm protein A-gold complex (small arrows) both to A cell granules (4a bottom left; 4b) but also to insulin immunoreactive (20 nm protein A-gold, large arrows) B cell granules (4a top right; 4c). A possible mechanism for this cross-contamination is given in Figure 5. Glutaraldehyde fixation. Uranyl acetate and lead citrate counterstains. Scale bars = (a) 500 nm; (b,c) 200 nm.

in the understanding of cellular function. For the sake of simplicity the applications of double immunolabelling have been divided into the cellular demonstration of the basic interactive components—Synthesis (section 2.1), Storage (section 2.2), Stimulus (section 2.3) and Secretion (section 2.4).

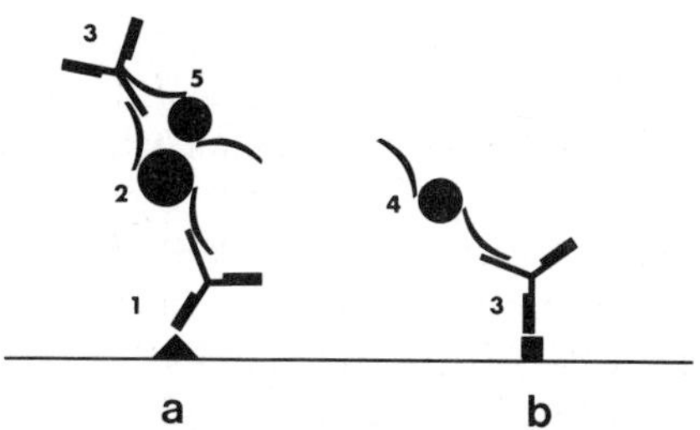

Fig. 5. Diagrammatic representation of possible source of cross-reactivity obtained using double protein A-gold immunocytochemistry. Antigen a is localised by antibody anti-a (step 1). A protein A-gold complex is used to visualise the anti-a antibody (step 2). Protein A is depicted as an elongated protein adsorbed to the surface of the colloidal particle. Antibody anti-b is then added (step 3) to localise tissue-bound antigen b and to saturate all available protein A sites on the first anti-a-protein A-gold complex (steps 1 + 2). A second protein A-gold complex, of different colloid size, is added to visualise the antigen b-antibody anti-b complex (step 4). However, it is possible that some protein A binding sites may be available on the Fc region of the anti-b antibody molecules which are bound into the antigen a-protein A-gold complex. The second protein A-gold may thus react with the first protein A-gold complex (step 5) which would result in an apparent dual localisation for antigen b.

2.1. Synthesis: DNA technology and polyproteins

In peptide-producing endocrine cells and nerve cell bodies, genomic hnRNA transcripts are processed, by putative enzymic cleavage and re-assembly, into cytosolic mRNA encoding the sequence of the peptide precursor molecule. With the advent of DNA technology, cell-free translation systems and other molecular biological procedures, peptide precursor mRNA nucleotide sequences are now being deduced rapidly and accurately. Amino acid sequences can thus be elucidated for the precursor and the presence of peptides existing within that molecule predicted by mapping of trypsin-sensitive dibasic amino acid cleavage sites. Such techniques have revealed so-called ‘polyproteins’—precursor molecules containing two or more bioactive peptide components. Examples such as pro-opiomelanocortin [31] and pro-enkephalin [32] are often quoted but more recently, pro-gastrin [33], pro-glucagon [34, 35], pro-substance P [36] and the calcitonin precursor gene, which is subjected to alternative tissue processing [37, 38] have been disclosed. The calcitonin precursor polyprotein-encoding gene is referred to later in this section.

It is abundantly clear that the vast majority of, if not all, ‘regulatory’ peptides are synthesised (as part of a precursor molecule), processed (cleaved, glycosylated etc.), packaged, and stored in secretory granules within endocrine cells or nerve cell bodies. Release, as a result of specific stimulation, may be endocrine, paracrine or as a neurotransmitter. However, it is equally clear (from biochemical, histochemical and immunocytochemical studies) that not all the bioactive components potentially encoded within the precursor mRNA are ultimately stored, though they may be expressed in a terminally extended (inactive) non-antigenic form. Electron immunocytochemical techniques using antisera recognising only the products of precursor cleavage enable the demonstration of co-existing bioactive molecules.

We have been able to localise ACTH and gamma$_1$-MSH to the same secretory granules in the pars intermedia and pars distalis of mammalian pituitary [39] (Fig. 6), whereas other products of pro-opiomelanocortin cleavage are not detected using our system. Calcitonin has also been found to co-exist with its C-terminal flanking peptide (PDN-21, katacalcin) in C-cells from normal thyroid and in calcitonin-containing cells of medullary carcinomas [40]. Recent studies have shown that calcitonin gene-related peptide (CGRP) also co-exists with calcitonin and katacalcin but only in a sub-population of C-cells from normal and carcinomatous thyroids [41].

The double immunogold staining procedure has been applied to investigate the localisation of precursor-derived molecular forms of gastrin in mammalian gastric antrum [42] and also the distribution of glicentin [43] and pancreatic glucagon in mammalian A cells. These are both reported in some detail in section 2.2.2.

One potential use of double labelling which has so far not been investigated in depth, mainly due to the technical problems which are proving to be insuperable at the present time, is the combination of mRNA hybridisation cytochemistry with ultrastructural immunocytochemistry. Basically, this involves the localisation of mRNA using labelled DNA probes complementary to the mRNA under investigation (cDNA probes), followed by immunostaining of the peptide product, the synthesis of which is directed by the mRNA. cDNA probes can be labelled by the incorporation of tritiated (^{3}H) or biotinylated nucleotides during their manufacture. Unfortunately, there are significant technical problems surrounding the retention of intact mRNA accurately within the cell. Sufficient sites must then be available for the cDNA probe to bind to the mRNA. Because of the nature of the cDNA probe, penetration into the tissue is generally poor. mRNA does not withstand embedding procedures either due to direct denaturation or masking by the components involved, therefore ultrathin frozen sections or tissue slices (Vibratome® or cryostat) would need to be employed for incubation.

Tritiated cDNA probes are localised by autoradiography, however, not only is a prolonged exposure time required but the site of silver grain deposition is not absolutely precise, due to distance and angle of the radiation and also to grain size. Biotinylated cDNA probes offer attractive advantages in that they can be localised using avidin-biotin-peroxidase or streptavidin-gold.

The products of mRNA translation may then be localised using conventional electron immunocytochemical procedures.

2.2. Storage: Sub-cellular storage sites

Peptide-containing endocrine cells and nerve terminals are distinguished from other cell types and from each other largely by the presence of electron-dense secretory granules, varying in their shape, size and the form of their limiting membrane. Interspecific homogeneity of secretory granule morphology from the same cell type is often striking, for example, amphibian and reptilian pancreatic insulin-secreting (B) cells are remarkably similar in appearance to those found in mammalian pancreas. However, secretory granules known to contain a common

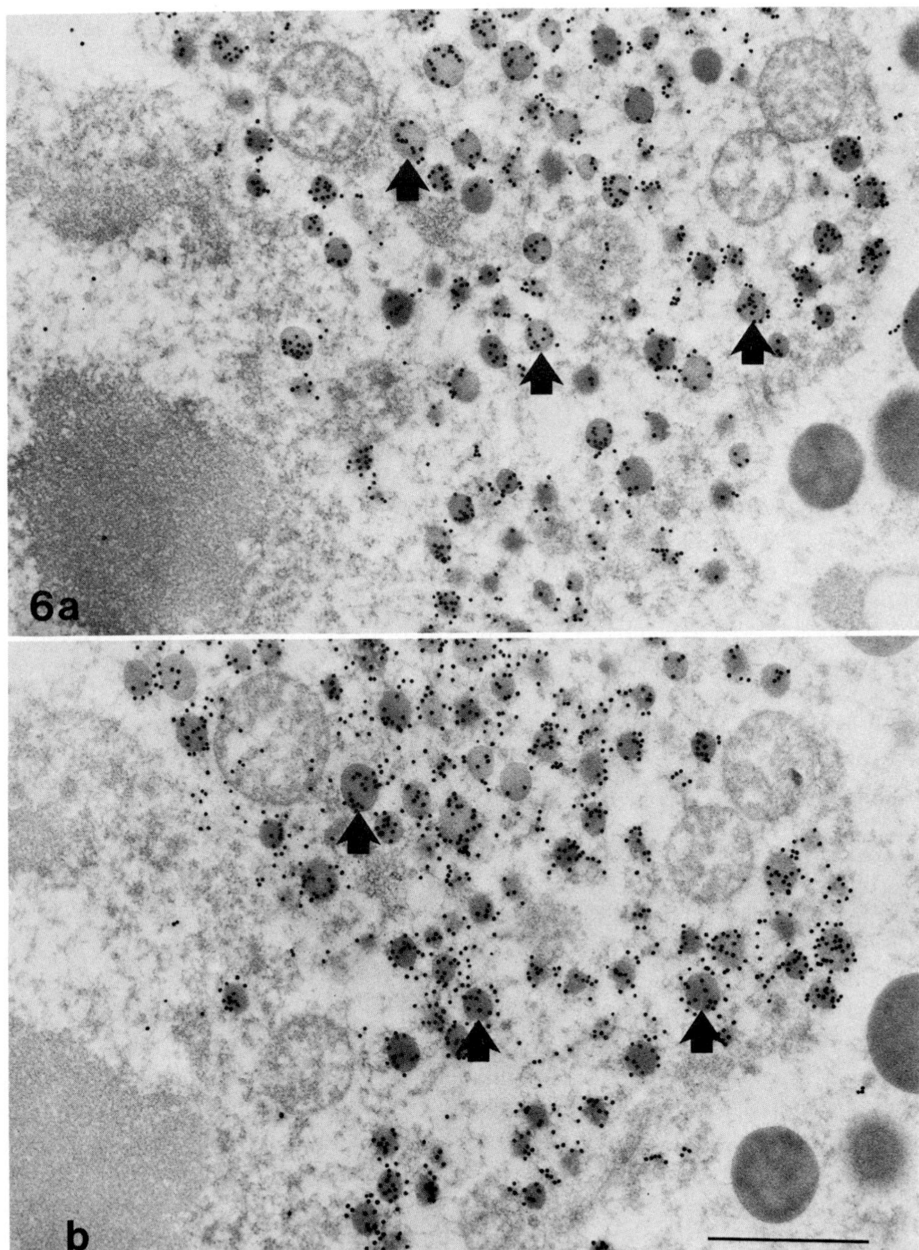

Fig. 6. Adjacent serial ultrathin sections of human pars distalis immunostained for gamma$_1$-MSH (a) and for ACTH (b), demonstrating that both immunoreactivities are co-localised to corticotroph granules. Glutaraldehyde fixation; immunogold staining procedure with 20 nm gold particles. Uranyl acetate and lead citrate counterstains. Scale bar = 500 nm.

peptide from two different organ sites within a single individual may show considerable morphological differences. For example, the chromaffin granules from the adrenal medulla and the carotid body type I cell secretory granules both contain methionine enkephalin, tyrosine hydroxylase and dopamine-β-hydroxylase but are morphologically distinct cell types [44]. It is abundantly clear from such evidence that the morphological appearance of the secretory granules reflects the heterogeneity of the molecular components and is not simply dependent upon the major bioactive agent stored.

Within any one cell type there may also be intergranular differences in morphology (Fig. 7). This may represent the post-translational maturation of a single secretory granule population, or the expression of two or more secretory granule sub-populations containing unrelated products. The application of single and double immunostaining procedures at the electron microscopical level has enabled us to investigate some of these particular problems.

2.2.1. p-Type heterogeneity

Although peptide-containing neurones are morphologically distinct from cholinergic and adrenergic nerve terminals [45, 46] they represent a reasonably homogeneous group, identifiable on the appearance of the electron-dense vesicles they contain [47], collectively referred to as p- (peptidergic-) type. Recently, electron immunocytochemical studies have identified distinct sub-populations of p-type terminals containing morphologically similar vesicles [48, 49]. Image analysis of these terminals has revealed slight morphological differences between the vesicle sub-populations [49], upholding the view that morphological appearance is attributable, at least in part, to peptide content [48–50]. Double immunogold staining procedures have been used to differentiate between morphologically similar vesicles on single ultrathin sections.

Conventional electron microscope studies have revealed that many nerve terminals contain mixed vesicle populations [47]. Such terminals are frequently encountered and commonly contain small (40–60 nm) agranular electron-lucent vesicles in addition to the large (over 100 nm) dense-cored granules. Several studies have concluded that bioactive peptides are localised in the electron-dense secretory granules [48, 51–53] and there is considerable evidence to suggest that the electron-lucent agranular vesicles contain acetylcholine or catecholamines [54]. We have attempted to identify both peptidergic and catecholaminergic components within the mixed vesicle terminals using the double immunogold staining method. However, reliable antibodies to acetylcholine and catecholamines, with the possible exception of serotonin, have not been available until fairly recently [55]. This is partly due to the low immunogenicity of the hapten which is hardly surprising when the ubiquity of these phylogenetically conserved transmitters is considered. Moreover, the amines are extremely susceptible to cross-linking agents, such as glutaraldehyde, with the result that fixation may destroy or at least obscure the antigenic structure. Antisera to the synthesising or converting enzymes (choline acetyltransferase, tyrosine hydroxylase, dopamine-β-hydroxylase and phenylethanolamine-*N*-methyltransferase) are routinely used for the localisation of 'classical'

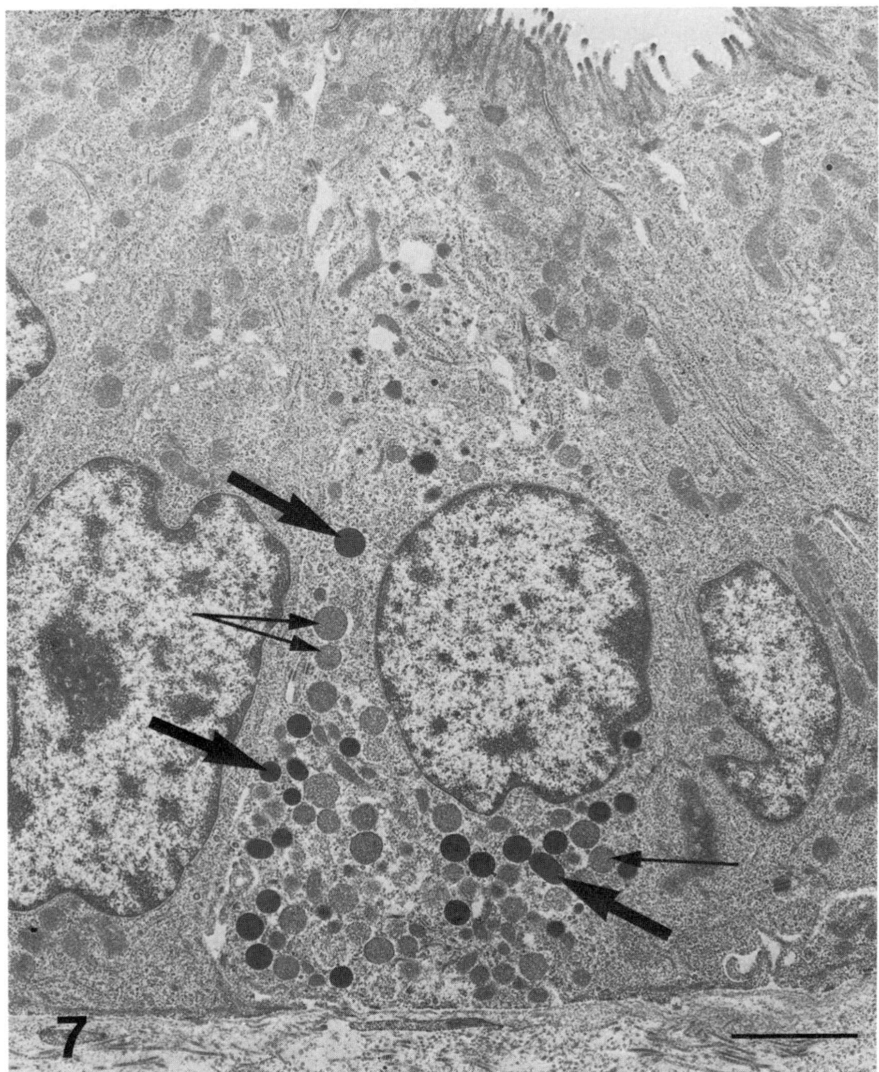

Fig. 7. Somatostatin immunoreactive cell from human colonic mucosa. Two apparently distinct secretory granule types are present in the cytoplasm. Morphological characterisation of the secretory granules by image analysis correlating size, shape and electron density revealed that a single population is present. Glutaraldehyde/osmium tetroxide fixation. Uranyl acetate and lead citrate counterstains. Scale bar = 1 μm.

neurotransmitters [53, 55, 56] but the site of enzyme antigenicity does not necessarily indicate the site of enzyme *activity* or transmitter synthesis and storage.

2.2.2. Intracellular segregation of bioactive antigens

Few examples of intracellular co-existence, rather than co-storage (co-localisation) have been reported from ultrastructural studies. The localisation of amines to small agranular vesicles and of peptides to large electron-dense granules in mixed nerve terminals is now a possibility but the lability of the 'classical neurotransmitter' requires that the site of the converting enzyme(s) is demonstrated.

Using antisera reactive to specific regions of peptides and their precursors we have been able to correlate morphologically-characterised secretory granules with their peptide content. Although our work to date has been largely directed to endocrine cells, the application of the double immunogold staining procedure,

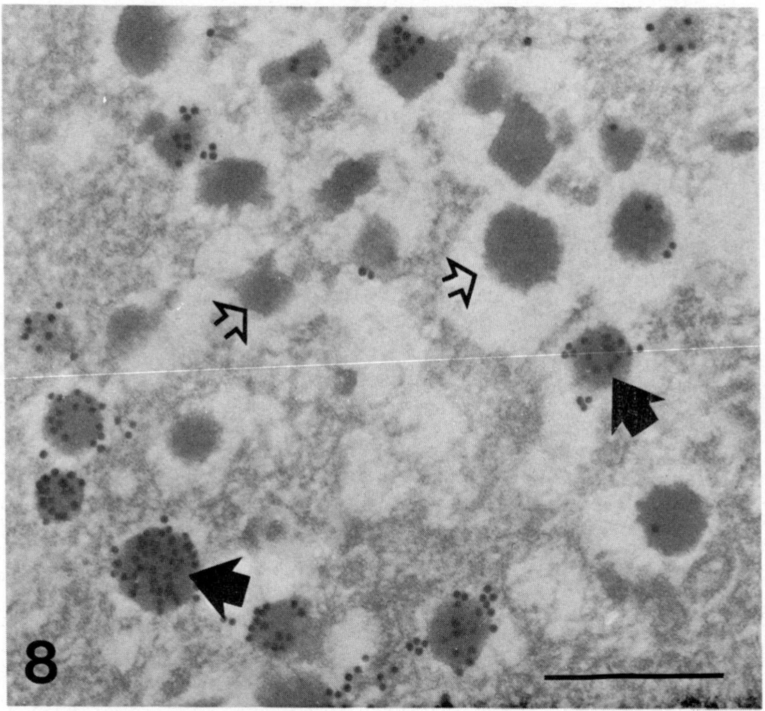

Fig. 8. Intergranular segregation of pro-insulin (C-peptide) antisera raised against C-peptide, but known to cross-react extensively with pro-insulin were applied to this ultrathin section of human pancreas. The homogeneous insulin-immunoreactive secretory granules were immunostained (20 nm gold particles; solid arrows) whereas the differentiated crystalline-cored granules were not immunoreactive (hollow arrows). Glutaraldehyde fixation. Uranyl acetate and lead citrate counterstains. Scale bar = 250 nm.

combined with image analysis techniques may be used in the same way to investigate neural systems. We have been able to localise gastrin and pro-gastrin to morphologically-distinguishable secretory granules in mammalian antral G-cells [42] and recently we have used the double immunogold staining procedure to show the segregation of insulin and pro-insulin in a sub-population of secretory granules from human pancreatic B cells (Fig. 8). Unfortunately, the absence of immuno-staining cannot necessarily be taken to mean that the antigen is not present. An induced (enzyme-mediated, fixation etc.) conformational change could lead to a 'negative' result by non-recognition of the antiserum or loss of one product from the site of storage.

One further example of intracellular co-existence is the presence of γ,γ-enolase (14-3-2 protein, nerve-specific protein, *N*euron-*S*pecific *E*nolase [NSE] [57]) in both neural and endocrine components of the diffuse neuroendocrine system of most organs, including that of the gastroenteropancreatic system and adrenal [58], lung [59, 60] and skin [61]. Comparative investigations into the presence of γ,γ-enolase and regulatory peptides/amines indicate that all peptide/amine-containing cells and neurons also contain this highly acidic glycolytic enzyme [62]. Most of the research currently undertaken on the co-existence of γ,γ-enolase with peptides has been conducted at the light microscope level largely due to the technical problems encountered with the conservation of enzyme antigen integrity at the ultrastructural level. The first convincing ultrastructural localisation of 'nerve-specific protein' was by Langley and co-workers [63] in rat cerebellum, who described its dispersion throughout the cytoplasm, in particular associated with mitochondrial membranes and microtubules. This demonstration was performed using a modification of the pre-embedding PAP method. We have used a similar technique to localise γ,γ-enolase in endocrine cells. However, direct correlation of γ,γ-enolase with peptide or amine content by electron immunocytochemistry has not been reported to date. Recently, we have been able to immunostain both γ,γ-enolase and regulatory peptide antigens using the double immunogold staining procedure on single ultrathin sections of freeze-dried rat pancreatic tissue (see Chapter 16). Our findings corroborate those of Langley [63] with respect to the cytoplasmic localisation of γ,γ-enolase.

2.2.3. Antigen co-localisation

The localisation of two peptide products to a single secretory granule has been referred to earlier in this chapter (see section 2.1).

2.2.3.1. Topographic segregation. Much interest was generated by Ravazzola and Orci's [64] demonstration that pancreatic glucagon and its immediate precursor glicentin [65] are topographically segregated within individual pancreatic A cell granules. Subsequently we were able to demonstrate this segregation using an immunogold procedure [43] (Fig. 9). Recently we have investigated rat pancreatic B cells processed according to the quick-freeze, freeze-dry, vapour fixation method, using the double immunogold staining method. As reported in Chapter 16, we have been able to show an intragranular segregation of insulin and pro-insulin (C-peptide) molecules in a sub-population of secretory granules. The functional significance of such a segregation is not evident however.

2.2.3.2. Co-localisation of peptides derived from a common precursor. The examples of ACTH with gamma$_1$-MSH, and calcitonin with its C-terminal flanking peptide (PDN-21, katacalcin) have been cited earlier in this chapter. In addition to these may be added sub-populations of mammalian antral G cell secretory granules which co-store gastrin-17 and the *N*-terminally extended molecular form, gastrin-34 [42] (Fig. 10). Insulin is also co-stored with its precursor in the majority of B cell secretory granules. The pancreatic B cell thus appears to exhibit intergranular segregation, topographic segregation of insulin and pro-insulin (C-peptide), and the co-localisation of insulin and pro-insulin (C-peptide) in granules in different stages of maturation.

2.2.3.3. Co-localisation of peptides derived from separate precursors. We have recently investigated three examples of apparently unrelated peptides being localised to single secretory granules. Neurotensin-like immunoreactivity has been localised to a sub-population of enkephalin- and noradrenalin-containing chromaffin cells in the cat adrenal medulla [66]. Although the observation of neurotensin-like immunoreactivity has been restricted to the cat adrenal medulla,

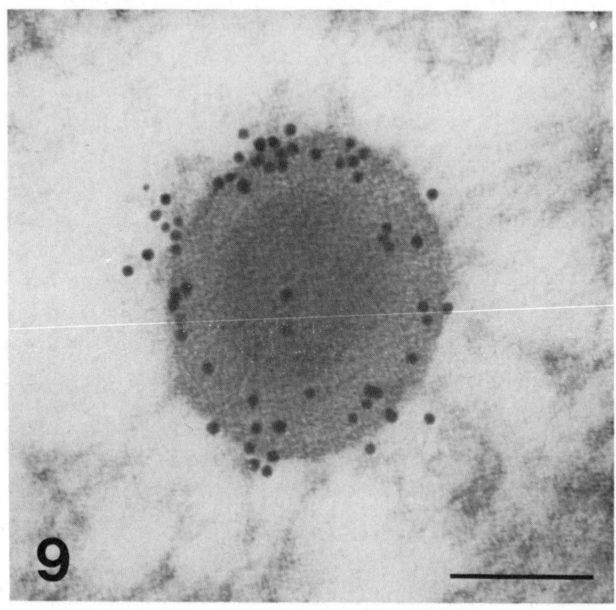

Fig. 9. Glicentin immunoreactivity restricted to the halo of an alpha granule from a human pancreatic glucagon-containing A cell. The core is largely unstained. Glutaraldehyde fixation. Uranyl acetate and lead citrate counterstains. Scale bar = 100 nm.

the recently discovered 36 amino acid peptide *n*euro*p*eptide t*y*rosine (NPY) [67, 68] has been found to co-exist with enkephalin in a large sub-population of noradrenalin-containing chromaffin cells from a wide range of mammals, including mouse, rat, cat, horse and man [69]. Double immunogold staining procedures have revealed the homogeneity of enkephalin and NPY antigenic sites throughout the noradrenergic granules. It should be emphasised that not all the noradrenergic granules co-store enkephalin and/or NPY.

For some years it has been known that colonic enteroglucagon-containing (EG) cells expressed a pancreatic polypeptide (PP)-like determinant [70] which could be localised immunocytochemically. Extracts of colonic mucosa revealed no true PP immunoreactivity [71]. As the full amino acid sequence of neither enteroglucagon

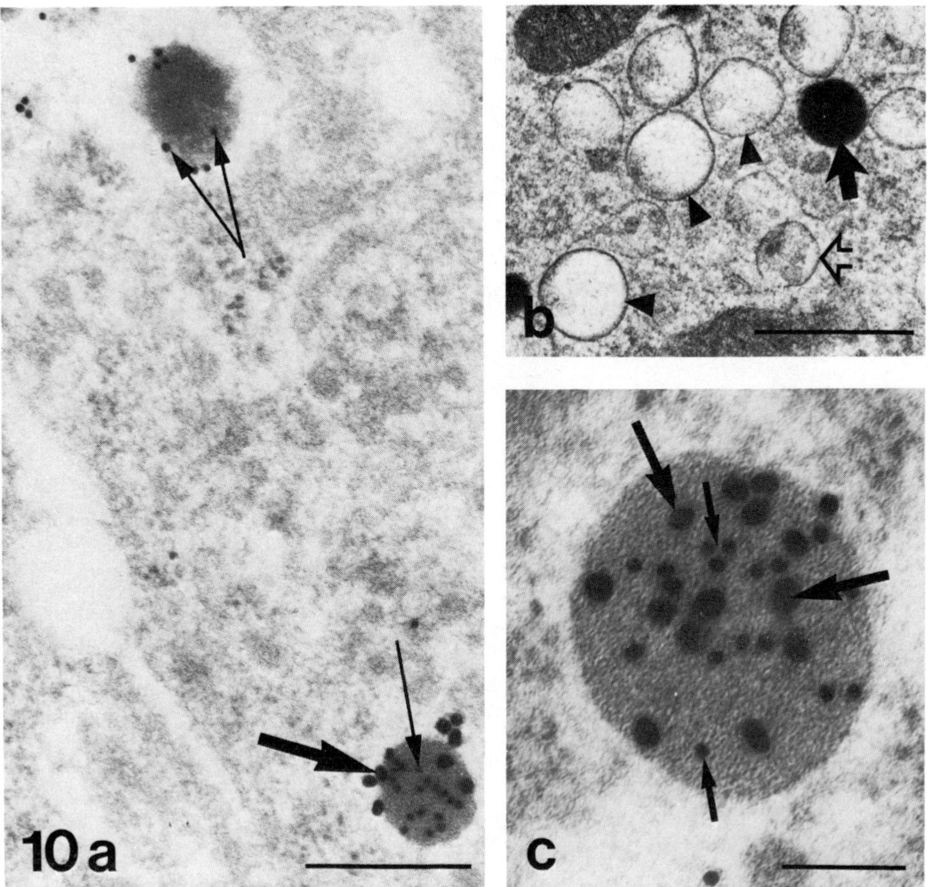

Fig. 10. Gastrin-17 and gastrin-34 immunoreactivity localised to secretory granules from a human gastric antral G cell. (a,c) Double immunogold staining procedure localising G 17 (20 nm gold particles; small arrows) to electron-dense (10a bottom right; 10c) and intermediate 'pale-cored' secretory granules (10a top left). G34 is restricted to the electron-dense secretory granule (40 nm gold particles; 10a bottom right; 10c). (b) The conventional electron microscopical appearance of an antral G cell. Electron-lucent granules (arrowheads), pale-cored granules (hollow arrow) and an electron-dense granule (solid arrow) are visible. Glutaraldehyde fixation. Uranyl acetate and lead citrate counterstains. Scale bars = (a) 300 nm; (b) 500 nm; (c) 100 nm.

nor pancreatic polypeptide precursor was known it was impossible to determine whether the PP-like immunoreactant represented a true co-localisation of unrelated products, co-localisation of different regions of the same precursor or was merely due to antisera cross-reactivity. With the discovery of *p*eptide with tyrosine (PYY), a 36 amino acid molecule which bears striking similarities to PP and NPY described above, by Tatemoto and co-workers [72, 73] and the recent elucidation of pro-glucagon structure [35] we have been able to resolve the problem. PYY has

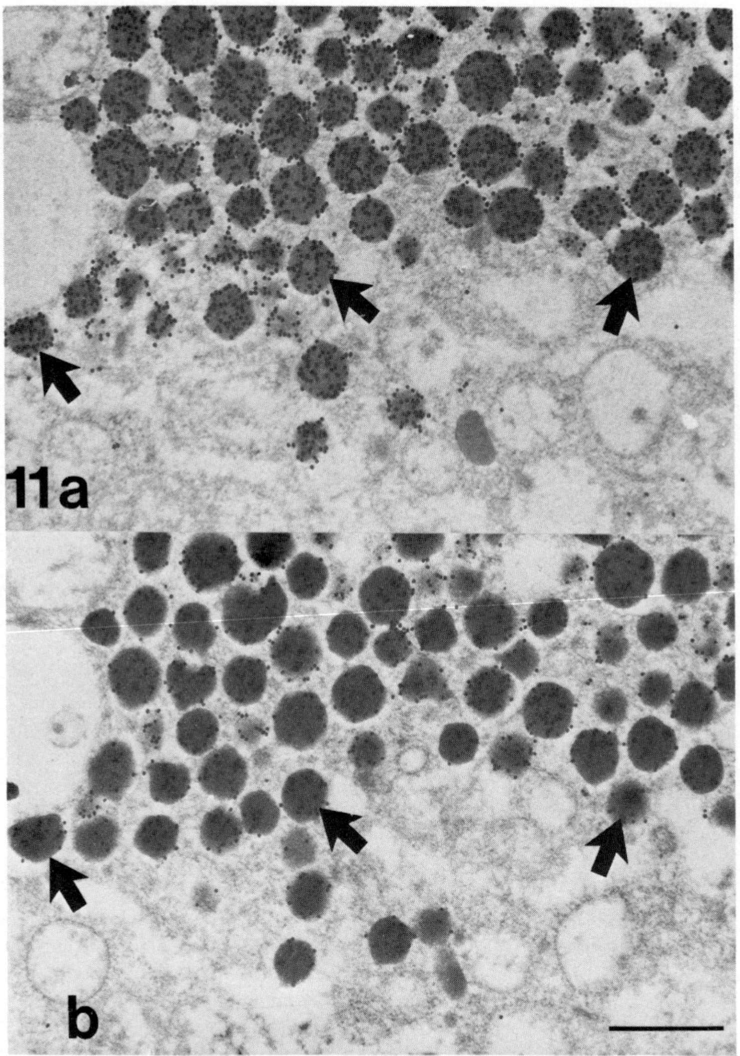

Fig. 11. Ultrathin serial sections of adult human colon immunostained with antisera to glicentin (a) and PYY (b). Arrows indicate identical secretory granule profiles. Glutaraldehyde fixation. Uranyl acetate and lead citrate counterstains. Scale bar = 500 nm.

been found to co-exist in the large majority of, if not all, glucagon/glicentin-containing (EG) cells in the mammalian colon [50]. Figure 11 reveals the co-localisation of both antigens on serial ultrathin sections. A developmental and phylogenetic study has revealed the co-existence of both antigens in glucagon-containing cells from the pancreas and gut of many species from fish to primate, including man. This highlights the close ontogenetic relationship between EG and pancreatic A cells.

2.3. Stimulus: Cell and nerve interactions

At the simplest level a peptide-producing endocrine cell is stimulated to synthesise a product under direct (innervation; paracrine) or indirect (circulating factor) receptor-mediated control. Initiation and rate of secretion, and hence duration of storage, is also controlled by similar signals. It is still not known whether all endocrine cells of one type are controlled in the same way, although the observed differential rates of secretion would suggest multiple stimuli. Double immunogold staining methods will undoubtedly play an important role in the understanding of cell-nerve, cell-cell and nerve-nerve relationships. This work is currently under investigation in our laboratory.

2.3.1. Receptors

The use of antigens adsorbed to small (3–5 nm) gold particles (see GLAD method, section 1.2) may be used in a pre-embedding procedure to localise receptor sites at the ultrastructural level. After processing the application of a post-embedding immunogold staining procedure using gold particles of a larger size in order to reveal an intracellular antigen could provide significant information concerning the mode of stimulation of any particular endocrine cell or nerve cell body.

One exciting development in this field has been the introduction of anti-idiotypic sera [74, 75] in order to localise receptor sites in post-embedding procedures. The use of such antisera in combination with antisera generated against, for example, a secretory protein, would enable a simple double immunogold procedure to be performed. This may be one way to investigate the kinetics of stimulus-storage-secretion.

2.4. Secretion

With slight modifications to the present procedures, the interaction of cytoskeletal proteins with cellular components such as organellar membranes, secretory granules, and enzymes such as neuron-specific enolase (NSE) could be investigated in detail.

It is well documented that pharmacological doses of drugs such as reserpine will deplete monoamine levels whilst substance P is unaffected. Double immunogold staining procedures have enabled relative levels of amine and peptide to be compared prior to, and after, treatment with specific blockers and releasers.

3. Conclusion

Several techniques are currently available for the simultaneous ultrastructural demonstration of two or more tissue-bound antigens. We have attempted to review such techniques in this chapter and to provide a range of applications, either established, new or purely speculative. In our hands the use of immunoglobulins from two different species visualised by the sequential or stimultaneous application of two colloidal gold particle sizes adsorbed with immunoglobulins from a third species offers a simple technique with which to demonstrate dual localisation of antigens in single tissue sections. The details of the procedure are given in the Appendix to this chapter.

This technique has been used to date to demonstrate co-localisation of some antigens in endocrine cells and in nerve terminals. Problems of co-existence of peptide molecular forms and amines with peptides have been, and will continue to be, resolved using this methodology. In conclusion, we now have an important tool at our disposal for the morphological investigation of co-existence and co-localisation.

4. Appendix

In this Appendix we have attempted to describe all the stages involved from tissue preparation to electron immunocytochemistry for the simultaneous ultrastructural visualisation of multiple antigens using the double immunogold staining procedure.

4.1. Tissue preparation

Small blocks of tissue removed at surgery or immediately post-mortem are fixed by immersion in a neutral-buffered aldehyde-based solution for up to 2 h at 4°C. We routinely use either 1% formaldehyde (prepared from its para- polymer) plus 2% glutaraldehyde in 0.075 M sodium phosphate buffer (pH 7.3) or 2.5% glutaraldehyde (ultra-pure) in 0.1 M phosphate buffer (pH 7.2). Cacodylate buffers can also be used. Half of the blocks are rinsed in buffer alone (containing 0.1 M sucrose) for 30–60 min and osmicated (1% osmium tetroxide in Millonig's buffer, or sodium phosphate buffer, pH 7.2 for 1 h at 4°C), whereas the other half is merely rinsed in buffer. We have been successful in localising most regulatory peptide and amine-converting enzyme antigens at the electron microscope level using this fixation schedule.

Following fixation the blocks are dehydrated via an ascending series of alcohols and infiltrated with Araldite epoxy resin following treatment with the intermediate solvent propylene oxide. We have found that epoxy resins such as Araldite, Epon and Spurr possess good electron immunocytochemical qualities whereas the less toxic acrylic polymer resins have not been successful in our hands. Most regulatory peptide and amine-converting enzyme antigens have been localised in heat-cured (60°C, 18–24 h) epoxy resin-embedded tissue; however, a few antigens, notably vasoactive intestinal peptide (VIP) are heat-labile in the presence of resin components. This can be overcome by polymerisation of the resin at room temperature (or below) using ultraviolet irradiation (10 days to 3 weeks; [48]).

Sections of non-osmicated tissue showing silver to silver-grey interference colours (60–100 nm) are collected on cleaned, uncoated 300-mesh nickel or gold grids and allowed to dry overnight. Osmicated tissue can be sectioned and mounted conventionally.

4.2 Double immunogold staining procedure

All incubations are carried out in 60 × 15 μl microtest plates; each well holds 15–20 μl.

a) The grid-mounted ultrathin sections are 'etched' in a 10% hydrogen peroxide solution for 10 min at room temperature. This step is believed to permeabilise the resin (see Chapter 3) thereby aiding antibody penetration. Recent reports indicate that this step may not be essential, particularly if the sections are cut from 'under-cured' (12–18 h; 60°C) resin blocks.
b) Wash thoroughly in microfiltered (0.45 μm pore size) distilled water.
c) Drain grids and place into droplets of normal goat serum (NGS; 1:30 dilution in antiserum diluent, see below) for at least 30 min at room temperature.
d) Drain NGS from the grids using fibre-free absorbent paper and incubate in droplets of primary antisera. We normally use a mixture of two antisera—one raised in a rabbit and one raised in a guinea pig. The antisera should not cross-react and should be directed to different regions of the same molecule, to different molecular forms or to different molecular species. The mixture should be diluted so that each antiserum is present at its pre-determined optimal titre. The antiserum diluent used routinely is PBS containing 0.1% bovine serum albumin (BSA; Sigma fraction; V, globulin free) and 0.01% sodium azide. pH 7.2. Incubation in primary antisera is usually carried out for 1 h at room temperature. This can be varied depending upon the dilution, avidity and affinity of the antisera.
e) After thorough washing in 50 mM Tris buffer (pH 7.2) and 50 mM Tris buffer containing 0.2% BSA for 3 × 15 min with agitation in each case the grids are placed into droplets of 50 mM Tris buffer containing 1% BSA pH 8.2 for 5 min.
f) Incubate the grids in a mixture of gold-labelled anti-primary species IgG at optimal titres for 1 h at room temperature. Our system employs 20 nm gold-labelled anti-guinea pig IgG with 10- or 40 nm gold-labelled goat anti-rabbit IgG. The three gold particle sizes are readily distinguishable by electron microscopy. A protocol for the preparation of gold particles and the techniques for the adsorption of immunoglobulins to gold is given in the Appendix to Chapter 15. The gold sol should be diluted with the pH 8.2 Tris-1% BSA buffer before centrifugation at 2,000 × *g* for 20 min to remove micro-aggregates accumulating with storage of the gold solution.
g) Wash in copious volumes of Tris-0.2% BSA buffer, Tris buffer and distilled water ('jet'-washing and beaker washing) 3 × 15 min each. Finally rinse the grids in microfiltered distilled water, dry by draining and counterstain for conventional electron microscopy. (M)ethanolic heavy metal stains may be employed.
h) Controls for the specificity of the reactions are essential and our routine procedures are as follows: (i) using non-immune rabbit and/or guinea pig serum as first layer; (ii) absorption controls with the respective antigens, antigenic fragments and homologous molecular species; (iii) omitting one or

both of the first-layer antisera; (iv) running simultaneous single immunogold staining method for one or other of the antigens on an adjacent section.

5. References

1. Priestley, J.V. (1984) Pre-embedding ultrastructural immunocytochemistry: Immunoenzyme techniques. In: J. M. Polak and I. M. Varndell (Eds.) Immunolabelling for Electron Microscopy. Elsevier Science Publ., Amsterdam, pp. 37–52.
2. Pelletier, G. and Morel, G. (1984) Immunoenzyme techniques at the electron microscopical level. In: J. M. Polak and I. M. Varndell (Eds.) Immunolabelling for Electron Microscopy. Elsevier Science Publishers, Amsterdam, pp. 83–93.
3. Horisberger, M. (1984) Electron-opaque markers: A review. In: J.M. Polak and I.M. Varndell (Eds.) Immunolabelling for Electron Microscopy. Elsevier Science Publishers, Amsterdam, pp. 17–26.
4. Cuello, A.C., Milstein, C. and Priestley, J.V. (1980) Use of monoclonal antibodies in immunocytochemistry with special reference to the central nervous system. Brain Res. Bull. 5, 575–587.
5. Cuello, A.C., Priestley, J.V. and Milstein, C. (1982) Immunocytochemistry with internally labelled monoclonal antibodies. Proc. Natl. Acad. Sci. USA 79, 665–669.
6. Morgan, C. (1972) Use of ferritin-conjugated antibodies in electron microscopy. Int. Rev. Cytol. 32, 291–326.
7. Heitzmann, H. and Richards, F.M. (1974) Use of the avidin-biotin complex for specific staining of biological membranes in electron microscopy. Proc. Natl. Acad. Sci. USA 71, 3537–3541.
8. Nicholson, G.L. and Singer, S.I. (1971) Ferritin-conjugated agglutinins as specific saccharide stains for electron microscopy: Application to saccharides bound to cell membranes. Proc. Natl. Acad. Sci. USA 68, 942–945.
9. Marshall, P.R. and Rutherford, D. (1972) Physical investigation on colloidal iron-dextran complexes. J. Colloid Interface Sci. 37, 390–402.
10. Hunt, S.V., Kelly, J.S. and Emson, P.C. (1980) The electron microscopic localisation of methionine enkephalin within the superficial layers (I and II) of the spinal cord. Neuroscience 5, 1871–1890.
11. Leranth, C.S. and Frotschner, M. (1983) Commissural afferents to the rat hippocampus terminate on VIP-like immunoreactive non-pyramidal neurons. An E.M. immunocytochemical degenerative study. Brain Res. 276, 357–361.
12. Sumal, K., Blessing, W.W., Joh, T.H., Reis, D.J. and Pickel, V.M. (1983) Synaptic interaction of vagal afferents and catecholaminergic neurons in the rat tractus solitarius. Brain Res. 277, 31–40.
13. Basbaum, A.I., Glazer, E.J. and Lord, B.A.P. (1982) Simultaneous ultrastructural localization of tritiated serotonin and immunoreactive peptides. J. Histochem. Cytochem. 30, 780–784.
14. Somogyi, P., Freund, T.F., Wu, J.-Y. and Smith, A.D. (1983) The section-Golgi impregnation procedure. 2. Immunocytochemical demonstration of glutamate decarboxylase in Golgi-impregnated neurons and in their afferent synaptic boutons in the visual cortex of the cat. Neuroscience 9, 475–490.
15. Ruda, M.A. (1982) Opiates and pain pathways: Demonstration of enkephalin synapses on dorsal horn projection neurons. Science 215, 1523–1525.
16. Priestley, J.V. and Cuello, A.C. (1983) Substance P immunoreactive terminals in the spinal trigeminal nucleus synapse with lamina I neurons projecting to the thalamus. In: P. Skrabanek and D. Powell (Eds.) Substance P Dublin 1983. Boole Press, Dublin, pp. 251–252.
17. Feldherr, C.M. and Marshall, J.M. (1962) The use of colloidal gold for studies of intracellular exchange in amoeba, *Chaos chaos*. J. Cell Biol. 12, 640–645.
18. Faulk, W.P. and Taylor, C.M. (1971) An immunocolloid method for the electron microscope. Immunochemistry 8, 1081–1083.

19. Larsson, L.-I. (1979) Simultaneous ultrastructural demonstration of multiple peptides in endocrine cells by a novel immunocytochemical method. Nature (London) 282, 743–744.
20. Bendayan, M. (1982) Double immunocytochemical labelling applying the protein A-gold technique. J. Histochem. Cytochem. 30, 81–85.
21. Roth, J. (1982) The preparation of protein A-gold complexes with 3 nm and 15 nm gold particles and their use in labelling multiple antigens on ultrathin sections. Histochem. J. 14, 791–801.
22. Tapia, F.J., Varndell, I.M., Probert, L., De Mey, J. and Polak, J.M. (1983) Double immunogold staining method for the simultaneous ultrastructural localisation of regulatory peptides. J. Histochem. Cytochem. 31, 977–981.
23. Varndell, I.M. and Polak, J.M. (1983) The use of immunogold staining procedures in the demonstration of neurochemical coexistence at the ultrastructural level. In: N. Osborne (Ed.) Dale's Principle and Communication Between Neurones. Pergamon Press, Oxford, pp. 179–200.
24. Roth, J. (1984) The protein A-gold technique for antigen localisation in tissue sections by light and electron microscopy. In: J. M. Polak and I. M. Varndell (Eds.) Immunolabelling for Electron Microscopy. Elsevier Science Publishers, Amsterdam, pp. 113–121.
25. Bendayan, M. and Stephens, H. (1984) Double labelling cytochemistry applying the protein A-gold technique. In: J. M. Polak and I. M. Varndell (Eds.) Immunolabelling for Electron Microscopy. Elsevier Science Publishers, Amsterdam, pp. 143–154.
26. De Waele, M., De Mey, J., Moeremans, M., De Brabander, M. and Van Camp, B. (1983) Immunogold staining method for the detection of cell surface antigens with monoclonal antibodies. In: G. R. Bullock and P. Petrusz (Eds.) Techniques in Immunocytochemistry, Vol. 2. Academic Press, London, pp. 1–23.
27. De Waele, M. (1984) Haematological electron immunocytochemistry. Detection of cell surface antigens with monoclonal antibodies. In: J. M. Polak and I. M. Varndell (Eds.) Immunolabelling for Electron Microscopy. Elsevier Science Publishers, Amsterdam, pp. 267–288.
28. Tokuyasu, K.T. (1984) Immuno-cryoultramicrotomy. In: J.M. Polak and I.M. Varndell (Eds.) Immunolabelling for Electron Microscopy. Elsevier Science Publishers, Amsterdam, pp. 71–81.
29. Beesley, J.E. (1984) Recent advances in microbiological immunocytochemistry. In: J.M. Polak and I. M. Varndell (Eds.) Immunolabelling for Electron Microscopy. Elsevier Science Publishers, Amsterdam, pp. 289–303.
30. Kronvall, G. and Frommel, D. (1970) Definition of staphylococcal protein A reactivity for human immunoglobulin G fragments. Immunochemistry 7, 124–127.
31. Nakanishi, S., Inoue, A., Kita, T., Nakamura, M., Chang, A.C.Y., Cohen, S.N. and Numa, S. (1979) Nucleotide sequence of cloned cDNA for bovine corticotropin-β-lipotropin precursor. Nature (London) 278, 423–427.
32. Dandekar, S. and Sabol, S.L. (1982) Cell-free translation and partial characterization of mRNA coding for enkephalin-precursor protein. Proc. Natl. Acad. Sci. USA 79, 1017–1021.
33. Boel, E., Vuust, J., Norris, F., Norris, K., Wind, A., Remfeld, J.F. and Marcker, K.A. (1983) Molecular cloning of human gastrin cDNA: Evidence for evolution of gastrin by gene duplication. Proc. Natl. Acad. Sci. USA 80, 2866–2869.
34. Bell, G.J., Santerret, R.F. and Mullenbach, G.T. (1983) Hamster preproglucagon contains the sequence of glucagon and two related peptides. Nature (London) 302, 716–718.
35. Lopez, L.C., Frazier, M.L., Su, C.-J., Kumar, A. and Saunders, G.F. (1983) Mammalian pancreatic preproglucagon contains three glucagon-related peptides. Proc. Natl. Acad. Sci. U.S.A. 80, 5485–5489.
36. Nawa, H., Hirose, T., Takashima, H., Inayama, S. and Nakanishi, S. (1983) Nucleotide sequences of cloned cDNAs for two types of bovine brain substance P precursor. Nature (London) 306, 32–36.
37. Amara, S., Jonas, V., Rosenfeld, M.G., Ong, E.S. and Evans, R.M. (1982) Alternative RNA processing in calcitonin gene expression generates mRNAs encoding different polypeptide products. Nature (London) 298, 240–244.
38. Rosenfeld, M.G., Mermod, J.J., Amara, S.G., Swanson, L.W., Sawchenko, P.E., Rivier, J., Vale, W.W. and Evans, R.M. (1983) Production of a novel neuropeptide encoded by the calcitonin gene via tissue-specific RNA processing. Nature (London) 304, 129–135.

39. Ali-Rachedi, A., Ferri, G.-L., Varndell, I.M., Van Noorden, S., Schot, L.P.C., Ling, N., Bloom, S.R. and Polak, J.M. (1983) Immunocytochemical evidence for the presence of gamma$_1$-MSH-like immunoreactivity in pituitary corticotrophs and ACTH-producing tumours. Neuroendocrinology 37, 427–433.
40. Ali-Rachedi, A., Varndell, I.M., Facer, P., Hillyard, C.J., Craig, R.K., MacIntyre, I. and Polak, J.M. (1983) Immunocytochemical localisation of katacalcin, a calcium-lowering hormone cleaved from the human calcitonin precursor. J. Clin. Endocrinol Metab. 57, 680–682.
41. Sabate, I., Polak, J.M., Bloom, S.R., Varndell, I.M. Major, J.M., Mulderry, P.M., Stolarsky, L., Evans, R.M. and Rosenfeld, M.G. (1984) Colocalization and expression of calcitonin and CGRP in thyroid gland and medullary carcinoma. Nature (London) (in press).
42. Varndell, I.M., Harris, A., Tapia, F.J., Yanaihara, N., De Mey, J., Bloom, S.R. and Polak, J.M. (1983) Intracellular topography of immunoreactive gastrin demonstrated using electron immunocytochemistry. Experientia 39, 713–717.
43. Varndell, I.M., Tapia, F.J., Probert, L., Buchan, A.M.J., Gu, J., De Mey, J., Bloom, S.R. and Polak, J.M. (1982) Immunogold staining procedure for the localisation of regulatory peptides. Peptides 3, 259–272.
44. Varndell, I.M., Tapia, F.J., De Mey, J., Rush, R.A., Bloom, S.R. and Polak, J.M. (1982) Electron immunocytochemical localization of enkephalin-like material in catecholamine-containing cells of the carotid body, the adrenal medulla and in pheochromocytomas of man and other mammals. J. Histochem. Cytochem. 30, 682–690.
45. Taxi, J. (1965) Contribution a l'etude des connexions des neurones moteurs du systeme nerveux autonomie. Ann. Sci. Nat. Zool. Biol. Anim. 12, 413–674.
46. Baumgarten, H.G., Holstein, A.F. and Owman, C.H. (1970) Auerbach's plexus of mammals and man—electron microscopical identification of three different types of neuronal process in myenteric ganglia of the large intestine from rhesus monkey, guinea-pigs and man. Z. Zellforsch. Mikrosk. Anat. 106, 376–397.
47. Cook, R.D. and Burnstock, G. (1976) The ultrastructure of Auerbach's plexus in the guinea-pig. 1. Neuronal elements. J. Neurocytol. 5, 171–194.
48. Probert, L., De Mey, J. and Polak, J.M. (1981) Distinct subpopulations of enteric p-type neurones contain substance P and vasoactive intestinal polypeptide. Nature (London) 294, 470–471.
49. Probert, L., De Mey, J. and Polak, J.M. (1983) Ultrastructural localization of four different neuropeptides within separate populations of p-type nerves in the guinea pig colon. Gastroenterology 85, 1094–1104.
50. Ali-Rachedi, A., Varndell, I.M., Adrian, T.E., Gapp, D.A., Van Noorden, S., Bloom, S.R. and Polak, J.M. (1984) Peptide YY (PYY) immunoreactivity is co-stored with glucagon-related immunoreactants in endocrine cells of the gut and pancreas. Histochemistry (in press).
51. Pickel, V.M., Joh, T.H., Reis, D.J., Leeman, S.E. and Miller, R.J. (1979) Electron-microscopic localization of substance P and enkephalin in axon terminals related to dendrites of catecholaminergic neurons. Brain Res. 160, 387–400.
52. Floor, E., Grad, E. and Leeman, S.E. (1982) Synaptic vesicles containing substance P purified by chromatography on controlled pore glass. Neuroscience 7, 1647–1655.
53. Cuello, A.C., Jessell, T.M., Kanazawa, I. and Iversen, L.L. (1977) Substance P: Localization in synaptic vesicles in rat central nervous system. J. Neurochem. 29, 747–751.
54. Nagy, A., Baker, R.R., Morris, S.J. and Whittaker, V.P. (1976) The preparation and characterization of synaptic vesicles of high purity. Brain Res. 109, 285–309.
55. Verhofstad, A.A.J., Steinbusch, H.W.M., Joosten, H.W.J., Penke, B., Varga, J. and Goldstein, M. (1983) Immunocytochemical localization of noradrenaline, adrenaline and serotonin. In: J. M. Polak and S. Van Noorden (Eds.) Immunocytochemistry: Practical Applications in Pathology and Biology. John Wright and Sons, Bristol, pp. 143–168.
56. Sabban, E., Goldstein, M., Bohn, M.C. and Black, I.B. (1982) Development of the adrenergic phenotype: Increase in adrenal messenger RNA coding for phenylethanolamine-N-methyltransferase. Proc. Natl. Acad. Sci. U.S.A. 79, 4923–4927.
57. Marangos, P.J. and Zomzely-Neurath, C. (1976) Determination and characterisation of neuron-specific protein (NSP) associated enolase activity. Biochem. Biophys. Res. Commun. 68, 1309–1316.

58. Schmechel, D.E., Marangos, P.J. and Brightman, M.W. (1978) Neuron-specific enolase is a molecular marker for peripheral and central neuroendocrine cells. Nature (London) 276, 834–836.
59. Wharton, J., Polak, J.M., Cole, G.A., Marangos, P.J. and Pearse, A.G.E. (1981) Neuron-specific enolase as an immunocytochemical marker for the diffuse neuroendocrine system in human fetal lung. J. Histochem. Cytochem. 29, 1359–1364.
60. Sheppard, M.N., Johnson, N.F., Cole, G.A., Bloom, S.R., Marangos, P.J. and Polak, J.M. (1982) Neuron specific enolase (NSE) immunostaining: A useful tool for the light microscopical detection of endocrine cell hyperplasia in adult rats exposed to asbestos. Histochemistry 74, 505–513.
61. Gu, J., Polak, J.M., Tapia, F.J., Marangos, P.J. and Pearse, A.G.E. (1981) Neuron-specific enolase in the Merkel cells of mammalian skin. Am. J. Pathol. 104, 63–68.
62. Polak, J.M. and Bloom, S.R. (1982) Regulatory peptides—new aspects. In: E.D. Williams (Ed.) Current Endocrine Topics. Praeger, Eastbourne, pp. 101–147.
63. Langley, O.K., Ghandour, M.S., Vincendon, G. and Gombos, G. (1980) An ultrastructural immunocytochemical study of nerve-specific protein in rat cerebellum. J. Neurocytol. 9, 783–798.
64. Ravazzola, M. and Orci, L. (1980) Glucagon and glicentin immunoreactivity are topologically segregated in the alpha-granule of the human pancreatic A cell. Nature (London) 284, 66–68.
65. Thim, L. and Moody, A.J. (1981) The primary structure of porcine glicentin (proglucagon). Regulatory Peptides 2, 139–150.
66. Terenghi, G., Polak, J.M., Varndell, I.M., Lee, Y.C., Wharton, J. and Bloom, S.R. (1983) Neurotensin-like immunoreactivity in a subpopulation of noradrenaline-containing cells of the cat adrenal gland. Endocrinology 112, 226–233.
67. Tatemoto, K. (1982) Neuropeptide Y: Complete amino acid sequence of the brain peptide. Proc. Natl. Acad. Sci. U.S.A. 79, 5485–5489.
68. Tatemoto, K., Carlquist, M. and Mutt, V. (1982) Neuropeptide Y—a novel brain peptide with structural similarities to peptide YY and pancreatic polypeptide. Nature 296, 659–662.
69. Varndell, I.M., Polak, J.M., Allen, J.M., Terenghi, G. and Bloom, S.R. (1984) Neuropeptide tyrosine (NPY) immunoreactivity in norepinephrine-containing cells and nerves of the mammalian adrenal gland. Endocrinology (in press).
70. Buffa, R., Capella, C., Fontana, P., Usellini, L. and Solcia, E. (1978) Types of endocrine cells in the human colon and rectum. Cell Tissue Res. 192, 227–240.
71. Adrian, T.E., Bloom, S.R., Bryant, M.G., Polak, J.M., Heitz, Ph. and Barnes, A.J. (1976) Distribution and release of human pancreatic polypeptide. Gut 17, 940–944.
72. Tatemoto, K. (1982) Isolation and characterisation of peptide YY (PYY), a candidate gut hormone that inhibits pancreatic exocrine secretion. Proc. Natl. Acad. Sci. U.S.A. 79, 2514–2518.
73. Lundberg, J.M., Tatemoto, K., Terenius, L., Hellstrom, P.M., Mutt, V., Hökfelt, T. and Hamberger B. (1982) Localisation of peptide YY (PYY) in gastrointestinal endocrine cells and effects on intestinal blood flow and motility. Proc. Natl. Acad. Sci. U.S.A. 79, 4471–4475.
74. Wassermann, N.H., Penn, A.S., Freimuth, P.I., Treptow, N., Wentzel, S., Cleveland, W.L. and Erlanger, B.F. (1982) Anti-idiotypic route to anti-acetylcholine receptor antibodies and experimental myasthenia gravis. Proc. Natl. Acad. Sci. U.S.A. 79, 4810–4814.
75. Cleveland, W.L., Wassermann, N.H., Sarangarajan, R., Penn, A.S. and Erlanger, B.F. (1983) Monoclonal antibodies to the acetylcholine receptor by a normally functioning auto-anti-idiotypic mechanism. Nature 305, 56–57.

Immunolabelling for Electron Microscopy (Polak/Varndell, eds)

CHAPTER 14

Freeze-fracture cytochemistry

Pedro Pinto da Silva

Section of Membrane Biology, National Cancer Institute, Frederick Cancer Research Facility, Frederick, MD 21701, U.S.A.

CONTENTS

1. INTRODUCTION

1.1. Freeze-fracture

When frozen biological membranes are fractured they are split along their apolar matrix into two unequal 'halves' [1]. The lipid molecules which comprise the bilayer membrane continuum are separated into two monolayered structures that comprise also the peripheral membrane proteins that are adsorbed at either surface. The process of fracture of integral membrane proteins, in particular transmembrane proteins, is not easy to predict. Conceivably, it depends on the relative expression and anchoring of these proteins at either surface, their possible association into oligomeric forms, and also, the number and disposition of peptide chains across the apolar matrix of the membrane.

1.2. *Freeze-etching*

The experimental approach that led to the proof of membrane splitting—'freeze-etching' [1], provided the first method for the cytochemical characterisation of freeze-fractured membranes. Cells, frozen in distilled water, were freeze-fractured and the preparations allowed to sublime ('etch') at −100°C for 1 min or more. Etched preparations of surface-labelled erythrocyte ghosts revealed that the distribution of antigens [2], as well as of virus and lectin receptors [3, 4] and anionic sites [5], was similar and continuous to that of the membrane particles seen on the fracture faces. These experiments proved crucial features of the chemistry and topology of membrane proteins. The particles were shown to represent the sites of proteins intercalated across the apolar matrix of the bilayer continuum with their antigens and receptors exposed at the outer surface [2, 4, 5].

While etching was a powerful technique it had severe limitations: a) the samples had to withstand freezing without appreciable damage; b) the integral membrane proteins visualised as membrane particles had to be capable of experimentally induced translational reorganisation into patterns (generally aggregates, for example, as induced by low pH in erythrocyte ghost membranes (see reference [6]) that could be related to those formed by the labelling molecules on the surface of the membrane. Very few systems fulfilled these criteria. The introduction of rapid freezing techniques [7, 8] overcame the first limitations and made it possible to observe the outer surface of membranes exposed by etching. However, in intact cells, the inner (cytoplasmic) surfaces of plasma and intracellular membranes are covered by non-etchable cytoplasm. Therefore, rapid freezing techniques were not adequate to address questions related to the partition of membrane components during freeze-fracture of intact (i.e. unlysed) cells. In a few membranes, this problem was approached with the splitting of frozen cells attached by poly-lysine to copper followed by fracture, isolation, and analysis of membrane halves [9, 10].

2. Fracture-label: Direct labelling of freeze-fractured membranes

We have developed alternative methods, collectively termed 'fracture-label', that allow direct, in situ, labelling of freeze-fractured plasma and intracellular membranes [11–13] as well as of cross-fractured cytoplasm. Fracture-label is a relatively simple procedure: cells and tissues are fixed (conventional glutaraldehyde fixation), impregnated in glycerol, frozen, freeze-fractured, and then thawed. Surprisingly, ice crystal damage (see Chapter 16, pp. 235–248), in particular that caused by thawing, is not detectable (Figs. 1 and 4a and c) [11–19]). The split membranes can be labelled and the distribution and partition of membrane components with each membrane half can be assessed.

We have now derived two main methods of fracture label: 'thin-section fracture-label' [11, 12] and 'critical point-drying fracture-label' [13].

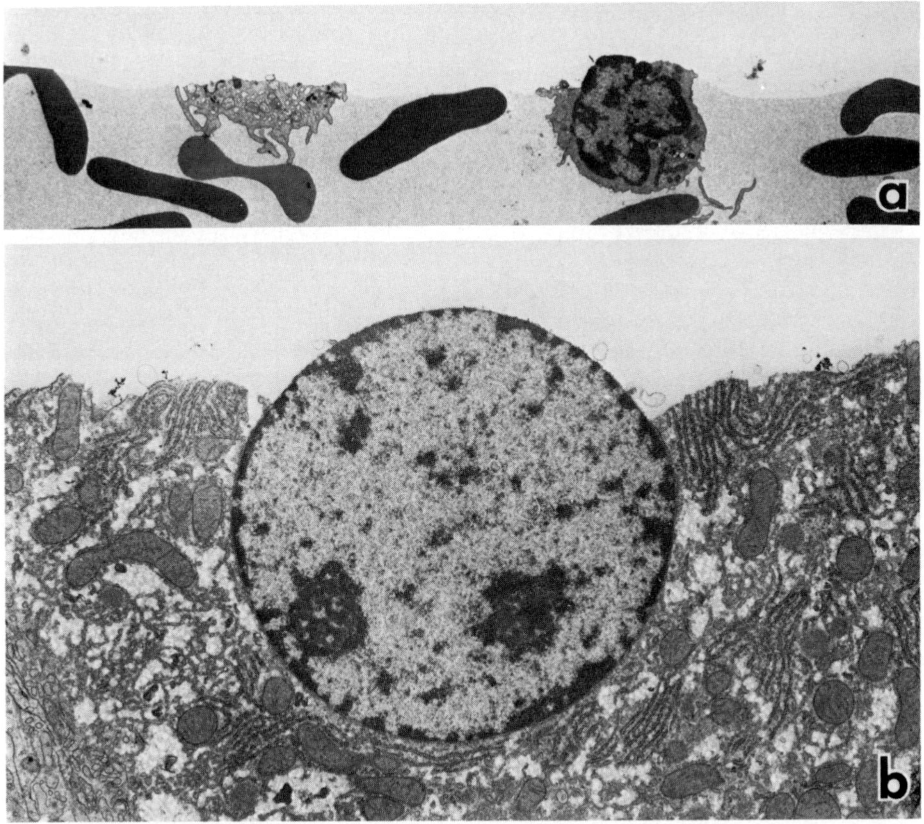

Fig. 1. Thin-section views of freeze-fractured cells. (a) Isolated cells (human erythrocytes and leukocytes, magnification = ×3,000); (b) rat hepatocyte, magnification = ×9,000.

2.1. Thin section fracture-label

In this first method, freeze-fracture is performed by repeated crushing of frozen tissues (or of cell suspensions co-cross-linked by glutaraldehyde within a matrix of bovine serum albumin (BSA)) within a vessel* filled with liquid nitrogen [11, 12]. Thawing is accomplished in a solution of glycerol/glutaraldehyde and de-glycerination in water/glycyl-glycine (to quench free aldehyde groups). The specimens (a suspension of fine fragments) are then labelled and processed for thin sectioning. Thick sections are stained and examined with the light microscope and the blocks are trimmed to include the edges of fragments that appear to contain the greatest proportion of membrane fracture faces (see Fig. 1; reference [12]). With the electron microscope, the fractured edges are inspected for favourable fractures and photographed (Figs. 1, 2c and 4a and c).

* The vessel can be of any design, but care must be taken to avoid porcelain mortars or vessels that might give off ceramic or glass 'chips' that could later destroy a diamond knife.

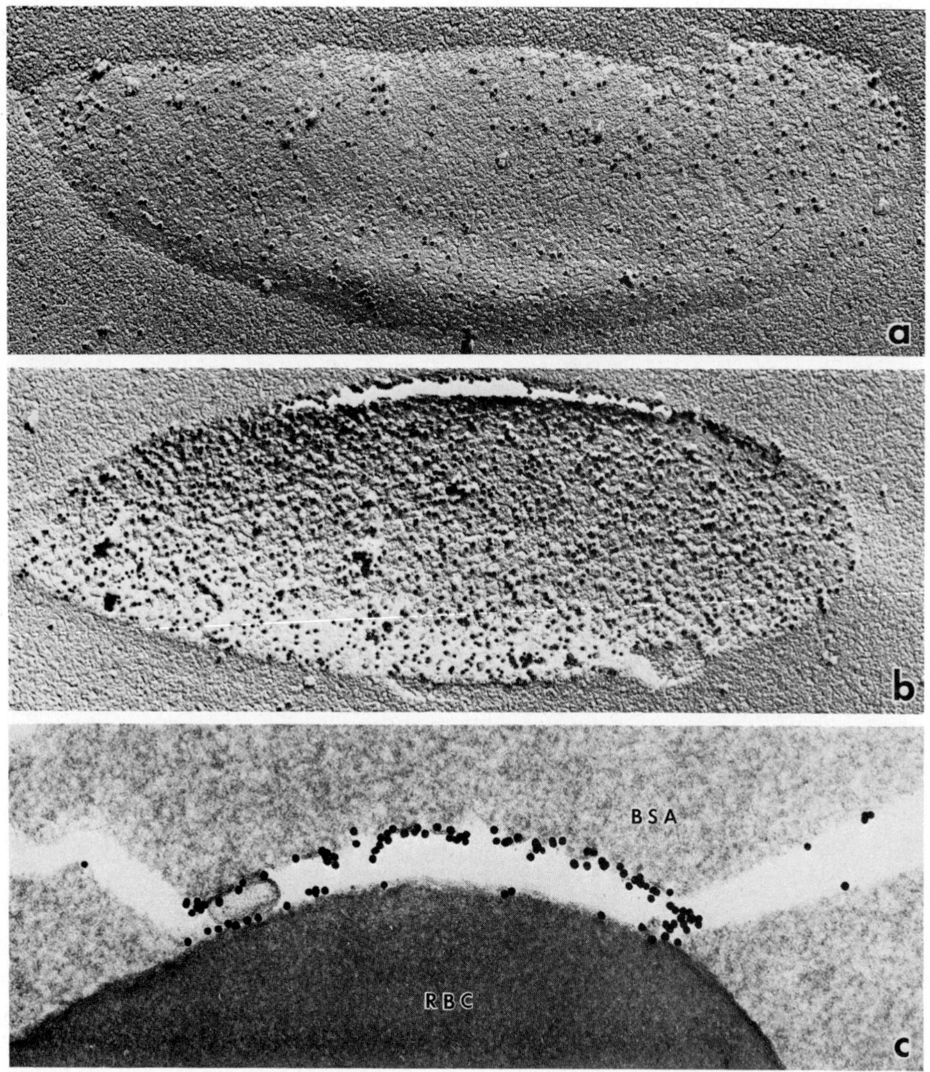

Fig. 2. Partition of wheat germ agglutinin-binding sites between P and E faces of freeze-fractured erythrocytes treated with wheat germ agglutinin and labelled with ovomucoid-coated colloidal gold. (a, b) Critical point-dried fracture-label; (c) thin-section fracture-label. (a, b) magnification = ×36,000; (c) magnification = ×72,000.

2.2. *Critical point-drying fracture-label*

Critical point-drying fracture-label involves the production of platinum/carbon casts of fracture-labelled preparations [13–15]. These are also easy to obtain. This procedure does not involve fine crushing of frozen tissues or of cell-BSA gels;

instead, fairly large (~1–2 mm) fragments are obtained by splitting specimens with a scalpel under the surface of liquid nitrogen. The specimens are then thawed, de-glycerinated, and labelled. After labelling, the fragments are critical-point dried, placed on a stage with an adhesive backing, replicated in a platinum/carbon evaporator and observed with an electron microscope. At very low magnifications these preparations appear similar to conventionally freeze-fractured specimens; at higher magnification they are different and, also, less attractive. Significantly, intramembranous particles are not observed as the texture of membrane fracture faces suffers considerable alterations (Figs. 2a, b and 4b).

3. Interpretation

Technically, fracture-label is easy; however, interpretation is much more complex. As with any new technique, methods to interpret the new results must accompany technical development; the reader is referred to the original work for detailed interpretation of both the morphology and labelling of fracture-labelled membranes [10–18]. The following paragraphs contain a brief discussion on the labelling of protoplasmic and exoplasmic membrane halves.

At first, it was intriguing that 'split' bilayered membranes should be labelled since the plane of fracture followed and exposed the hydrophobic matrix of the membrane. We have shown that, upon thawing, exposure of the apolar regions of the membrane to a hydrophilic environment causes the split monolayers to

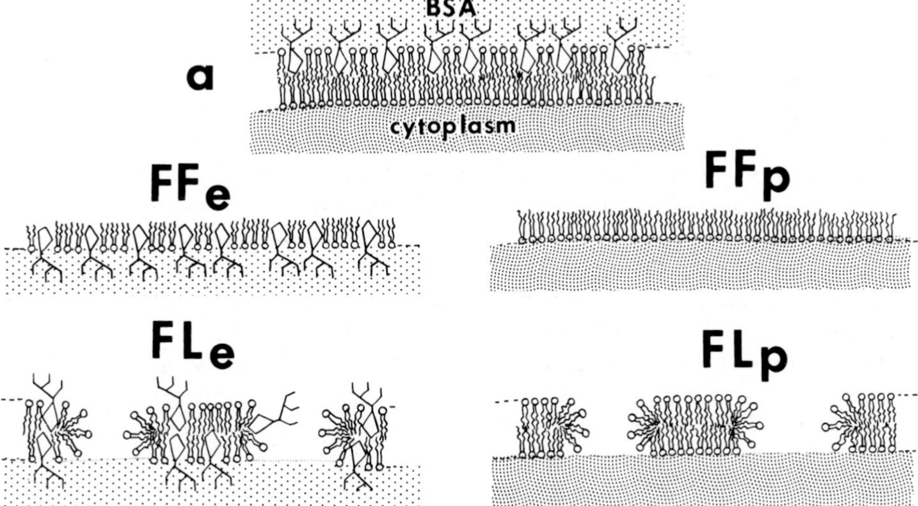

Fig. 3. Freeze-fracture splits the membrane in two complementary halves (FF_e, endoplasmic; FF_p, protoplasmic). Post-fracture exposure of freeze-fractured membranes to an aqueous environment causes reorientation of membrane components with reconstitution of bilayered structures (FL_c, FL_p). Only phospholipids and glycolipids are depicted.

reorganise into bilayered structures [11, 12, 15, 17] (Fig. 3). Lectins (Concanavalin A and wheat germ agglutinin) generally label strongly the exoplasmic halves of plasma membranes. The protoplasmic halves of biological membranes may also be labelled by lectins [11, 15, 16, 19]. As expected, and as we have confirmed in experiments using lysed cells [16, 17], no lectin-binding sites exist at the inner surface of the membrane. Therefore, lectin-binding sites on protoplasmic faces must represent transmembrane glycoproteins that partition preferentially with the inner (i.e. 'protoplasmic') membrane half (see section 4, below). The course of fracture is deviated from the centre of the membrane to the outer surface, as transmembrane proteins are dragged, with their lectin-binding heterosaccharides, across the outer ('exoplasmic') half of the membrane.

4. What fracture-label can do

With fracture-label, all membranes—plasma and intracellular—in isolated cells or in tissues can be labelled. Cytoplasmic components exposed by cross-fracture can also be labelled. Fracture-label uses established, time-tried fixation, impregnation, and cytochemical labelling procedures. Cross-contamination of membrane halves is apparently non-existent. This is shown by common instances where only exoplasmic membrane halves are densely labelled (see reference [17]), in particular in 'cracks' (Fig. 2c) in the preparation where P and E membrane halves are separated by a narrow gap [12, 15, 18]. So far, the only instance of contamination that we have been able to detect [17] was non-specific (i.e. of membranes, cytoplasm and BSA matrix) by glycogen granules effused from cross-fractured glutaraldehyde-fixed *Amoeba* cells (in intact cells the granules, trapped but not cross-linked, are released upon fracture).

In principle, therefore, fracture-label can detect any membrane component, lipid or protein, that is capable of cytochemical labelling; it determines their position in any membrane; in many instances, it can detect transmembrane proteins, in particular those preferentially associated with the inner half of the membrane. The absence of recognisable intramembrane particles does not constitute as severe a limitation as it might look at first. Only seldom are the particles organised (or organisable) into the distinct arrays that must exist in order for their relation to labelling patterns to be possible; comparison of fracture-labelled and freeze-fractured specimens can also circumvent this limitation [19].

In the initial studies of fracture-label [11, 12, 15], we used human erythrocytes since the main features of the organisation of their membranes are established. Fracture-label with Concanavalin A and wheat germ agglutinin-ferritin or gold conjugates (Fig. 2) showed that on freeze-fracture two principal transmembrane proteins in human erythrocyte membranes, Band III and glycophorin, have symmetric destinies. Band III partitions preferentially with the inner half of the membrane, with approximately three quarters of the label (Concanavalin A) observed over protoplasmic membrane halves, whilst glycophorin partitions preferentially with the outer membrane half, as a similar proportion of receptors (wheat germ agglutinin) is observed over the exoplasmic half of the membrane (Fig. 2).

These initial results led to our current hypothesis (detailed below) on the behaviour of transmembrane proteins during fracture. Proteins heavily expressed at the outer surface and not anchored to other components at the inner surface of the membrane may tend to partition with the outer half of the membrane; conversely, those—like Band III—whose relative expression at the outer surface is poor and which are bound to molecules of the membrane skeleton or the cytoskeleton will preferentially partition with the inner half of the membrane. This process must involve dragging across the outer membrane half, with retention of the receptors and antigens on protoplasmic fracture faces. The process is stochastic, as a minority of Concanavalin A or of wheat germ agglutinin-binding sites exposed at the surface of the erythrocyte membrane are observed over E and P faces, respectively.

Fracture-labelling of plasma and intracellular membranes [12–14, 19] with Concanavalin A and wheat germ agglutinin easily reveals the general pathways followed during the glycosylation of membrane proteins. Concanavalin A, a label for mannose-rich glyco-components, labels intensely the exoplasmic halves of membranes of the endoplasmic reticulum (Fig. 4a) and nuclear envelope, as well as the plasma membrane. Wheat germ agglutinin, a label for sialic acid and N-acetyl glucosamine, fails to label either endoplasmic reticulum (Fig. 4b) or nuclear envelope membranes but labels strongly plasma membranes (Fig. 4b), lysosomes, phagocytic vacuoles, and, in endocrine pancreas, secretory vesicle membranes (Fig. 4c). We have shown that labelling of Golgi membranes by wheat germ agglutinin can be surprisingly low (rat hepatocyte, salivary gland, human lymphocytes (see reference [18]), a finding that accords with the view of the Golgi as an 'assembly line' where few fully assembled glycoproteins may be found at any given time.

In many instances lectin labelling of fractured membranes cannot define the molecular species (glycolipid; glycoprotein). If, as discussed, label is found over the P face, then (assuming or proving that no lectin-binding sites exist at the inner surface) it can be ascribed to a protein, viz. a transmembrane glycoprotein. In other cases, identification can be made because glycoproteins are not present, as in the case of *Acanthamoeba castellani*. Here, with fracture-label, we have been able to provide the first demonstration of the restriction of glycolipid molecules to the exoplasmic half of plasma [17] and intracellular membranes (Barbosa and Pinto da Silva, in preparation).

As fracture-label can detect the presence of transmembrane glycoproteins (P face labelling), it can also be used to determine heterogeneity of their expression in a population of cells. It can also define their regionalisation within specific areas of the membrane. With human T lymphocytes [16] we showed that wheat germ agglutinin-binding transmembrane proteins are heterogeneously expressed, as defined by labelling of P plasma membrane faces. Finally, in the highly polarised sperm cell [19], it was concluded that transmembrane sialoglycoproteins are accumulated preferentially within the region of the plasma membrane that overlays the acrosome.

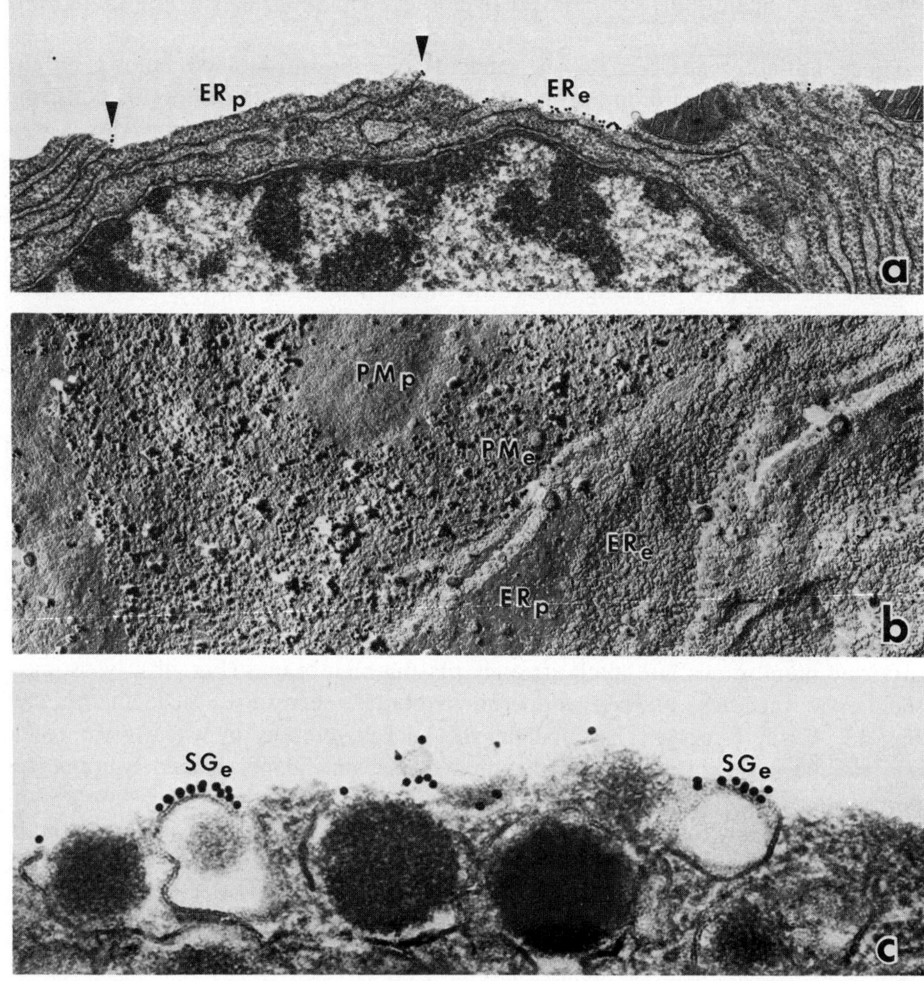

Fig. 4. Fracture labelling of pancreatic cells. (a) Concanavalin A binding sites (colloidal gold/peroxidase method) strongly label the E face of freeze-fractured endoplasmic reticulum but fail to label P face (arrow: label of cross-fractured membrane; note absence of labelling of mitochondrion; (b) wheat germ agglutinin (colloidal gold/ovomucoid method) labelled E face of plasma membrane (P face is not labelled) demonstrating the absence of terminally glycosylated glyco-conjugates in endoplasmic reticulum membranes (ER_p and ER_c); (c) wheat germ agglutinin-binding sites are revealed on E faces of fractured secretory granules of endocrine pancreatic cells. (a) magnification = ×30,100; (b) magnification = ×21,500; (c) magnification = ×73,100.

5. Outlook

Fracture-label is in its infancy. Labelling has been mostly confined to lectins/colloidal gold or ferritin conjugates, although we know that immuno-cytochemical techniques can be used [13]. As exposure of components is performed by fracture and not by 'loosening' of the cytoplasm, glutaraldehyde, rather than formaldehyde, can be used with obvious advantages, although fixation protocols that do not result in obliteration of the antigen must be followed.

The use of random fracture through the cell interior is also being explored in order to expose the cytoplasm to permeation of electron-dense markers (for example, native ferritin) or of radioactive macromolecules. Here, fracture is used to explore the nature and distribution of spaces within the cytoplasm and nucleoplasm to test theories of cytoplasmic organisation, topology, and dynamics.

One of the unique advantages of fracture-label is that the fracture process exposes any cell membrane without disruption of the surrounding regions, as will happen during slicing of fresh or of fixed tissues. Fracture-label can approach many problems whose solution currently involves the isolation and purification of membrane fractions. Biochemical isolation, i.e. fractionation techniques, have inherent problems of cross-contamination. These led, for instance, to the proposal of a back flow of fully glycosylated glycoproteins, a finding denied by the observation of fracture-labelled preparations, where both nuclear envelope and endoplasmic reticulum membranes remain unlabelled by wheat germ agglutinin [14] (Fig. 4b). Generalised access to plasma membranes or intracellular membranes opens the possibility of labelling membranes such as myelin, photoreceptor cells, or chloroplast membranes not normally accessible to cytochemical labels. However, fracture-label is no cytochemical panacea and, being in its developmental stages, particular care must be exercised in the interpretation of the results it so easily produces.

6. Acknowledgements

I thank Drs. M. Luisa F. Barbosa and Artur Aguas for critical reading of the manuscript and Ms. Mary Jo Elsasser and Mr. Cliff Parkison for expert editorial assistance.

7. References

1. Pinto da Silva, P. and Branton, D. (1970) Membrane splitting in freeze-etching. Covalently labelled ferritin as a membrane marker. J. Cell Biol. 45, 598–605.
2. Pinto da Silva, P., Douglas, S.D. and Branton, D. (1971) Localization of A antigen sites on human erythrocyte ghosts. Nature (London) 232, 194–196.
3. Tillack, T.W., Scott, R.E. and Marchesi, V.T. (1972) The structure of erythrocyte membranes studied by freeze-etching. II. Localization of receptors by phytohemagglutinin and influenza virus to the intramembranous particles. J. Exp. Med. 135, 1209–1227.
4. Pinto da Silva, P. and Nicolson, G. (1974) Freeze-etch localization of Concanavalin A receptors to the membrane intercalated particles in human erythrocyte membranes. Biochim. Biophys. Acta 363, 311–319.
5. Pinto da Silva, P., Moss, P.S. and Fudenberg, H.H. (1973) Anionic sites on the membrane intercalated particles of human erythrocyte ghosts. Freeze-etch localization. Exp. Cell Res. 81, 127–138.

6. Pinto da Silva, P. (1972) Translational mobility of membrane intercalated particles of human erythrocyte ghosts: pH dependent, reversible aggregation. J. Cell. Biol. 53, 777–787.
7. Van Harreveld, A., Trubatch, J. and Seiner, J. (1974) Rapid freezing and electron microscopy for the arrest of physiological processes. J. Microsc. 100, 189–198.
8. Heuser, J.E., Reese, T.S. and Landis, D.M.D. (1979) Organization of acetylcholine receptors in quick-frozen, deep-etched, and rotatory-replicated *Torpedo* postsynaptic membrane. J. Cell Biol. 82, 150–173.
9. Fisher, K.A. (1976) Analysis of membrane halves: cholesterol. Proc. Natl. Acad. Sci. U.S.A. 73, 173–177.
10. Edwards, H.H., Mueller, T.J. and Morrison, M. (1979) Distribution of transmembrane peptides in freeze-fracture. Science 1343–1345.
11. Pinto da Silva, P., Parkison, C. and Dwyer, N. (1981) Fracture-label: cytochemistry of freeze-fracture faces in the erythrocyte membrane. Proc. Natl. Acad. Sci. U.S.A. 78, 343–347.
12. Pinto da Silva, P., Parkison, C. and Dwyer, N. (1981) Freeze-fracture cytochemistry: thin sections of cells and tissues after labelling of fracture faces. J. Histochem. Cytochem. 29, 917–928.
13. Pinto da Silva, P., Kachar, B. Torrisi, M.R., Brown, C. and Parkison, C. (1981) Freeze-fracture cytochemistry: replicas of critical point dried cells and tissues after fracture-label. Science 213, 230–233.
14. Pinto da Silva, P., Torrisi, M.R. and Kachar, B. (1981) Freeze-fracture cytochemistry: Localization of wheat-germ agglutinin and Concanavalin A binding sites on freeze-fractured pancreatic cells, J. Cell Biol. 91, 361–372.
15. Pinto da Silva, P. and Torrisi, M.R. (1982) Freeze-fracture cytochemistry: Partition of glycophorin in freeze-fractured erythrocyte membranes. J. Cell Biol. 93, 463–469.
16. Torrisi, M.R. and Pinto da Silva, P. (1982) T lymphocyte heterogeneity: WGA labelling of transmembrane glycoproteins. Proc. Natl. Acad. Sci. U.S.A. 79, 5671–5674.
17. Barbosa, M.L.F. and Pinto da Silva, P. (1983) Asymmetric topology of glycolipids in membranes: Concanavalin A labelling of membrane halves in *Acanthamoeba castellani*. Cell 33, 959–966.
18. Torrisi, M.R. and Pinto da Silva, P. (1984) Compartmentalization of intracellular membrane glycocomponents is revealed by fracture-label, J. Cell Biol. 98, 29–34.
19. Aguas, A. and Pinto da Silva, P. (1983) Regionalization of transmembrane glycoproteins in the plasma membrane of boar sperm head is revealed by fracture-label. J. Cell Biol. 97, 1356–1364.

Immunolabelling for Electron Microscopy (Polak/Varndell, eds)

CHAPTER 15

Scanning electron microscope immunocytochemistry in practice

Gisèle M. Hodges[1], Marie A. Smolira[1] and D.C. Livingston[2 †]

[1] *Tissue Interaction Laboratory, and* [2] *Chemistry Laboratory, Imperial Cancer Research Fund, Lincoln's Inn Fields, London WC2, UK*

CONTENTS

† Deceased.

1. Introduction

More than a decade has now elapsed since the first publications on the application of scanning electron microscopy (SEM) to histochemical and immunocytochemical studies. There remains, however, a putative appreciation of the potential use of the SEM in investigations on the molecular organisation of supramolecular structures. Three reasons may be given to account for this. First, because of the relatively low resolving capability of early instruments it was not generally realised that the SEM could possess sufficient resolution to yield information on the macromolecular elements of biological structures. Secondly, there was the belief that it would be difficult, if not impossible, to obtain by scanning electron microscopy topographical information other than from the 'natural face' of biological specimens. And thirdly, were the relative inherent limitations of early marker systems used for SEM.

However, SEM immunocytochemistry is theoretically a very sensitive method and, other than as a complement to the detection and identification of specific cell residues achieved by transmission electron microscopy (TEM) or light microscopy techniques, offers intrinsic advantages for the study of supramolecular structures. First, numerous cells can be viewed such as to allow for a ready three-dimensional interpretation and statistical evaluation of molecular patterns in relation to structure. Secondly, new investigative techniques now provide for effective study of the tissue and cell interior at continuously variable levels of resolution. Thirdly, SEM instrumental modalities such as back-scattered electron imaging (BEI), cathodoluminescence, and X-ray microanalysis have significant roles to play in the molecular mapping of biological structures. Thus SEM in bridging the gap between light microscopy and transmission electron microscopy can add a new and flexible dimension of resolution to immunocytochemical investigations. Several recent articles give further examples of the general considerations briefly outlined above [1–10].

The purpose of this chapter is to consider some procedures central to SEM immunocytochemistry, to point out sources of artefacts, to catalogue some representative contributions and, to establish those areas where SEM would be potentially most useful in the future.

2. Marker systems for scanning electron microscopy

A variety of marker systems have been developed for the visual localisation, topographical analysis and quantification of target molecules in supramolecular structures. These target molecules have been identified, in general, by immunochemical methods or by lectin reactions in combination with suitable tracers or markers; such tracers have been chosen in relation to the mode(s) of microscopy and to the instrumental modalities selected and, according to the information required of a particular experimental design.

In the selection of suitable markers for SEM immunocytochemistry there are several prerequisites to be met which are summarised here and detailed in recent reviews and publications [4, 8, 9, 11]. Firstly, the marker must be of a size and shape that can be readily visualised against the possibly complex topographical background of a supramolecular structure, yet be small enough to allow good target site localisation. The choice of marker size and shape may be further dictated by the type of conductive coating used in sample preparation for the SEM [12–14] (see section 4.1). Secondly, the marker should be chemically stable such that during storage, immunocytochemical labelling, and sample preparation for SEM there is both minimal marker degradation or aggregation and marker size or shape change. Thirdly, the marker should be interactive, of preference, across a range of ligand molecules. This should be with retention of ligand activity and, with strong interaction between marker and ligand molecules, either through high-affinity specific binding, non-specific adsorption, or covalent bonding, such as to minimise ligand dissociation during labelling and the consequent competition by both free ligands and marker-ligand conjugates for the same target molecules which could reduce the density of cell labelling that may be visualised by SEM. Fourthly, the marker should demonstrate an absence of or, at the most, a minimal natural binding affinity for biological surfaces; this imposes the need for the level of non-specific marker binding to be determined prior to specific labelling studies (see section 4.1).

Marker systems can be categorised into three main groups—particulate, enzymatic and emissive—and markers from all three groups have been explored in the SEM as visualisation tags for target molecules. These markers are listed in Table 1 and are reviewed in several recent publications together with discussions concerning the advantages and disadvantages of these various particulate, enzymatic and emissive marker systems [4, 8, 9, 11, 15–18]. Of the three types of marker systems attention has focussed on particulate systems and more especially on colloidal gold markers (see sections 3 and 5). This is a consequence of certain intrinsic and technical limitations shown at present by most enzymatic and emissive marker systems which restrict their general use for SEM; such limitations can be of minimal consequence in certain experimental designs. Inevitably, the choice of marker system and marker size will be dictated by the kind of information required of a particular investigation. Marker systems can be further categorised, therefore, according to their applicability to low or high resolution studies and to analysis by given modes of microscopy (transmitted light or fluorescence microscopy, transmis-

TABLE 1
Marker systems for SEM immunocytochemistry

Marker	Size (nm)	Shape	Reference
Particulate markers			
Haemocyanin	35 × 50	Cylindrical	4, 57, 58, 63, 87–95
Ferritin	12, total diameter; 5, central iron containing micella	Spherical	31, 96–101
Tobacco mosaic virus (TMV)	15 × 300	Rod-shaped	18, 102–106
Bushy stunt virus (BSV)	30	Spherical	18, 76, 104, 106
Southern bean mosaic virus (SBMV)	25	Spherical	107, 108
SV40	40–45		98, 109
Turnip mosaic virus Alfalfa mosaic virus			110
Bacteriophage T4	200, head-tail; 100, head	Hexagonal head-tail assembly	103, 111–113
E. coli f2 phage			107
Polystyrene latex spheres	50–300	Spherical	11, 67, 114–126
Polymethacrylate latex spheres	500–1,000	Spherical	127
Co-polymer microspheres	30–340	Spherical	8, 9, 64, 128–130
Defined silica spheres	7–25	Spherical	131, 132
Iron co-polymer microspheres	30–50	Spherical	9, 133, 134
Iron-dextran complexes	7 × 11–12 × 12	Rod-shaped	135–137
Mercury-ligand complexes	8–10	Spherical	138, 139
Colloidal gold spheroids	5–150	Spherical to oblate	16, 19, 20, 21, 22, 41, 43, 82
Silver grains autoradiography			140
Enzymatic markers			
Horseradish peroxidase: reaction product	30–50	Particles or crystals	141–145
Emissive markers			
Fluorescence			146
Cathodoluminescence			6, 166
X-Ray			6, 31, 147

sion or scanning electron microscopy) and instrumental modality. Thus large markers, selected for detection at relatively low magnifications by SEM, can allow rapid surveys of large sample areas with, for example, positive identification of cell subpopulations and statistical calculations of the ratio of labelled and unlabelled cells without relation to particular cell surface features which are masked by these larger markers. Among markers appropriate for such low resolution SEM studies may be included the larger diameter particulate populations of polystyrene latex spheres, polymethacrylate latex spheres, co-polymer microspheres, silver grains, colloidal gold and various emissive markers. By contrast, the topographical detection and mapping of receptor sites will demand application of small, clearly discernible, markers such that a close correlation between target molecule and marker is attained. As a consequence, for such studies preference has been given both to the smaller diameter synthetic particulate markers, and in particular to those showing strong electron emissivity such as colloidal gold and, to small biological particulate markers such as haemocyanin, tobacco mosaic virus (TMV) and T4 bacteriophage. Ultrastructural enzymatic markers such as horseradish peroxidase are limited by a relative diffusion of the reaction product such that the actual ligand-receptor site can be difficult to determine.

3. Colloidal gold as a marker for scanning electron microscopy

Of the particulate marker systems, colloidal gold has generated increasing interest in recent years as an electron-dense, non-cytotoxic and stable cytochemical marker which is readily prepared in a size range from 3 to 150 nm mean diameter, and can be observed by various modes of microscopy (reviewed by [16, 19–22]). Although investigated early on by electron microscopy, and used in 1962 as a particulate tracer in TEM studies [23], the modern era of colloidal gold as a marker system dates from its application first as a TEM specific marker by Faulk and Taylor [24] and then as a SEM marker by Horisberger and his colleagues [25]. The colloidal gold marker system has since been introduced for light [26–28] and fluorescence [28] microscopy; for freeze-etch and surface shadow replica electron microscopy [29, 30], for multiple labelling electron microscopical experiments using ferritin and gold [31, 32] or colloidal gold of different particle sizes [33, 34], for combined SEM and TEM studies [35], and for back-scattered electron imaging (BEI) [36] and X-ray mapping analysis [17].

Gold markers of reproducible and consistent size have been produced by the controlled reduction of tetrachloroauric acid using a wide range of reducing agents [16, 21, 22] including ultrasonics [37], while the preparation of radioactive colloidal gold has been recently reported [38]. Of the 50 or so tetrachloroauric acid reducing agents detailed in the literature, the most commonly used to date have been sodium citrate [39] for the preparation of gold particles in the size range of 15–150 nm mean diameter, and sodium ascorbate [40] or ethereal phosphorus [22] for smaller gold particles of 8–13 and of 3–12 nm mean diameters respectively.

Several comprehensive reviews have appeared on the colloidal gold marker system which provide both methodologies on the preparation and characterisation of gold markers [16, 19–22] and various discussions including that on the chemistry and physics of metallic colloids [16]; on the properties of colloidal gold [16, 19, 21, 22]; and on factors influencing the stability of colloidal gold such as particle size, ionic composition of the solute, dialysis, and cleanliness of glassware and solutions to avoid spurious aggregation centres [16, 19].

Gold markers produced by the method of Frens [39] have been most widely used for SEM immunocytochemical studies. However, these markers, which are in the form of oblate spheroids, demonstrate an increasing heterogeneity with increasing size [19, 41]. Such relative heterogeneity of gold marker preparations has underlined the importance of establishing the physical characteristics of gold markers by direct TEM measurement and monitoring of the absorption spectra [19, 41] (Appendix 7.1). Also, has been emphasised the necessity to select gold markers for multiple labelling experiments that ensure minimal overlapping (i.e. two standard deviations from the mean diameter of each marker population) if unequivocal identification of markers of different size is required [19, 41].

An important surface property of colloidal gold is the negative charge carried by the gold particles in water by virtue either of adsorbed hydroxyl ions or, by dissociation of possible aurocomplexes formed on the surface of particles (for further discussion see reference [22] and Chapter 2, pp. 17–26). Stability of the colloid in water is maintained, as a consequence, by electrostatic repulsion. Addition of increasing electrolyte concentrations will lead, however, to a progressive destabilisation of the colloidal gold that is related to particle size with large particles being less stable against electrolyte flocculation than small particles. Such destabilisation results in a colour change from red to blue. This arises through the electrostatic shielding of gold particles by electrolytes which can result in particles approaching to a critical distance where van der Waals-London attractive forces become dominant and flocculation of the colloidal gold ensues. Such flocculation can be prevented by the addition of solutions of various macromolecules whereby positively-charged macromolecules attracted into the van der Waals radius of the gold particles, become strongly adsorbed to the surface and thereby stabilise the colloidal gold against electrolyte-induced flocculation. Protein adsorption to gold particles is, however, a highly complex and incompletely understood phenomenon with a number of physico-chemical factors influencing the adsorption process [16, 19, 42, 43]. Of importance among these parameters is the pH value of the colloidal gold, there being clear evidence that adsorption of protein to gold is pH dependent and related to the isoelectric point (pI) of the protein with maximal binding starting at or just on the basic side of the pI [19, 27, 41–44]. Also, particle size may be of importance in that some low molecular weight substances have been found to stabilise only gold particles smaller than 40 nm [34, 43]. Various approaches have been used to determine the appropriate pH for optimal protein-gold conjugation, such as spectrophotometric determination of absorption isotherms [42], isoelectric focusing [27, 43], and radioassay [43]. The microtitration assay, a visual semi-quantitative estimation of gold flocculation by colour change

[43] (Appendix 7.2.3), offers a convenient and rapid approach for routine applications, yielding results closely comparable to those obtained by more sophisticated methods. In determining the optimal amount of macromolecule necessary to stabilise colloidal gold against flocculation, various studies have shown that this can be assessed from the resistance of colloidal gold to salt-induced flocculation as a function of protein concentration at the appropriate pH for a given protein-gold conjugation, with flocculation being monitored spectrophotometrically [42] or by microtitration assays [41, 43] (Appendix 7.2.3). However, for certain proteins, saturation of gold markers with radiolabelled compounds has not been achieved even at concentrations of protein far in excess of that required for stabilisation in the presence of electrolytes [41]. This suggests that with increasing protein concentrations, increasing amounts of protein may be adsorbed to gold as 'multilamellar' shells. As a consequence, it has been thought inappropriate to speak of 'saturating' concentrations of protein, but rather that the gold marker system should be defined in terms of 'stabilising' concentrations viz. stability to electrolytes [16, 19, 41].

In formulating a general protocol for the preparation of protein-gold conjugates (see Appendix 7.2.4) various basic information is required including data on the isoelectric point or isoelectric point range of the selected proteins; on the stability range of the selected gold marker with changes in pH (Appendix 7.2.1), on the optimal pH for protein-gold marker adsorption (Appendix 7.2.2), and on the protein-stabilising levels (Appendix 7.2.3), all of which may be determined by microtitration assays. A recommendation has been made that proteins should be purchased salt-free and solubilised in water or very low molarity sodium chloride in order to prevent salts interfering with protein adsorption and inducing colloid flocculation [16]. Also, the presence of phosphate and borate ions should be avoided as these bind to many types of surfaces and could prevent the binding of macromolecules to gold particles. Where protein preparations containing salts are purchased these should be dialysed against water wherever possible in that flocculation of colloidal gold has been seen to occur at concentrations of protein well in excess of the required stabilising concentrations and prior to the salt-induced flocculation invoked by 10% sodium chloride (see Appendix 7.2.3). This may be accounted for by both the purchase of high salt-containing protein preparations and their subsequent solubilisation in buffers rather than water resulting in final salt concentrations at levels such as to induce flocculation of the colloidal gold despite the high protein concentration. In all cases, it has been recommended either to centrifuge or membrane filter buffers and macromolecule solutions immediately prior to the preparation of the protein-gold conjugates in order to remove microaggregates which could serve as spurious gold nucleation sites. In general, protein solutions in a 10% excess of the optimal stabilising amount have been used and preference given for the addition of gold colloid to the macromolecule solution [16, 21, 22] rather than the reverse order of dropwise addition of protein to gold marker which has been, however, successfully followed by some investigators [19, 41–43]. Upon completion of the protein-gold adsorption reaction various reagents have been used as further stabilisation agents with, as an objective, the 'neutralisation' of possible free hydroxylated sites at the gold surface and maintenance of appropriate interparticular distances thereby minimising possible aggregation and enhancing probe stability [16, 19, 21, 22]. Polyethylene glycol has emerged as the stabilising agent of choice with the best protective effect being observed in the molecular weight range of 15,000–20,000 [16].

Following stabilisation, free, unconjugated macromolecules and unconjugated gold particles have been generally separated from the conjugated marker population by appropriate washing and centrifugation protocols; and the protein-gold conjugates, after suspension in an appropriate buffer containing a stabilising agent such as polyethylene glycol, sterile membrane-filtered prior to storage [16, 41, 43] (Appendix 7.2.4.). Fractionation of the protein-gold conjugates into homogeneous sub-fractions has also been done by glycerol or sucrose gradient centrifugation [16, 45], or by gel-column filtration. Various studies indicate clearly that, in general, proteins adsorbed on colloidal gold maintain similar characteristics to those of the non-adsorbed protein, and maintain their reactivity for many months when stored at 4°C. Such observations have been made for immunoglobulins (antisera, Ig fractions, monoclonal antibodies) [24, 26, 46]; protein A [21, 22]; horseradish peroxidase [42]; avidin [29]; toxins [47, 48]; and lectins [22, 49]. Accurate determinations of the amount of protein bound as a ratio of the bioactivity of the gold probe are of obvious importance in defining this marker system [16, 19, 41]. Effective procedures for the assessment of the activity, stability and behaviour of protein-gold conjugates on storage can be provided by direct and indirect radio-binding assays and by agglutination reactions, though these different assays may not give identical values for protein binding to gold [19, 24, 41, 43]. Whereas direct radioassays provide an estimate of total protein bound, only bioassays (e.g. agglutination or indirect radioassay) may evaluate effective protein-binding activity, in that the orientation of protein on gold is unknown, active sites may be sterically hindered, and protein multilamellation may have occurred. Of these approaches, agglutination procedures have been adapted and proven as reliable and convenient routine systems for the demonstration that adsorbed proteins are present on colloidal gold in a bioactive form [19, 41] (Appendix 7.2.5).

Concern has been expressed in the literature that the non-covalent electrostatic adsorption process binding protein to colloidal gold may be inherently unstable allowing possible exchange of originally bound protein with other competing proteins. However, the evidence available to date would tend to support the premise that proteins, in general, bind tightly to particles of colloidal gold with protein competition being moderated by the addition of polyethylene glycol [16, 19, 41]. Nevertheless, on the basis of the protein-gold marker multilamellation concept discussed earlier, desorption of some protein could occur though enough protein would be left bound for the probe to be fully bioactive. Release of macromolecules into the buffer solutions can occur, however, the extent of which will depend on the macromolecule adsorbed to colloidal gold. Appropriate washing and centrifugation procedures are essential, therefore, both after storage and prior to labelling applications in order to minimise reduction in labelling efficiency as a consequence of competing proteins in free solution [16, 41] (Appendix 7.3.2.).

4. Specimen preparation and labelling procedures for scanning electron microscopy

4.1. Some general considerations

A variety of labelling procedures, broadly grouped into direct and indirect or sandwich techniques, have been devised for the identification and visualisation of target molecules of supramolecular structures. Originally developed for fluorescence microscopy, such labelling procedures have been subsequently adapted for transmitted light microscopy using phase or differential interference contrast optics; to transmission electron microscopy; and more recently to scanning electron microscopy. Comprehensive discussions and practical descriptions of these labelling procedures together with a consideration of their advantages and disadvantages have been given in various recent publications [15, 50–53]. Several sources also provide extensive background data on the specific subject of labelling procedures for SEM immunocytochemistry, and the reader will find there many details and numerous references [4, 8, 9, 18, 31, 49, 54–58].

Of the two principal approaches employed in immunocytochemical microscopy, the indirect labelling procedure has been the approach most favoured in SEM immunocytochemistry. Here, specimens are first exposed to an unmodified primary specific ligand and the bound primary ligand then detected in one or more additional steps by a second ligand-marker conjugate directed against the first. Among arguments favouring the indirect approach has been that only a limited variety of secondary conjugates are required and these can be made and stored for general use against a wide range of primary ligands thereby removing the complication of preparing many different primary ligand-marker conjugates. Also, the indirect approach is more sensitive than the direct binding approach with the binding of the primary ligand unconstrained by attached marker; furthermore, signal amplification may be provided by attachment of several ligand-marker conjugates (estimated at between 4 and 10-fold) to each first ligand. In recent times there have been various developments in indirect labelling methodology and a summary of the main indirect labelling models used with SEM is schematically illustrated in Figure 1. Among useful developments of interest in SEM immunocytochemistry is the establishment of protein A-marker conjugates providing valuable visualiser tags based on the ability of protein A to bind in a specific manner with Fc portions of various mammalian IgG molecules [15, 21, 22, 29, 59]. Another useful modification employs the specific high affinity interaction between biotin and avidin where specific primary ligands derivatised with biotin are detected by interaction with markers derivatised with avidin [59, 60]. Localisation of lectin-reactive macromolecules may be also mediated via covalently bound carbohydrate residues which occur naturally on some glycoprotein electron-dense markers such as haemocyanin or horseradish peroxidase (HRP) [16, 53, 56, 61, 62].

Direct methods, in which primary ligand-marker conjugates are used to visualise the target molecules in a one-step procedure have been far less extensively used in SEM immunocytochemistry than indirect methods. Direct binding has the advan-

Direct (one-step) method

Indirect (two-step) method

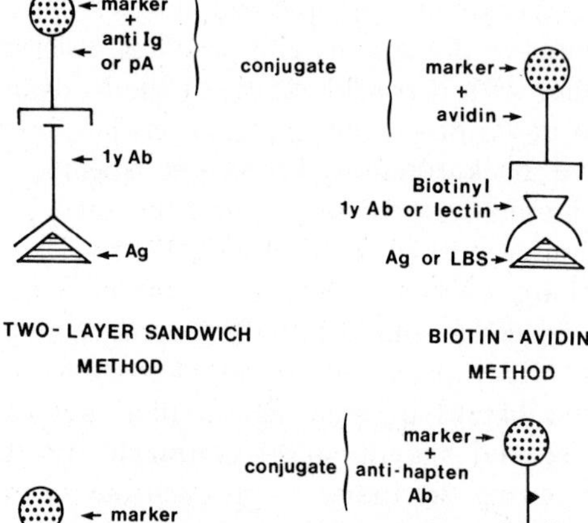

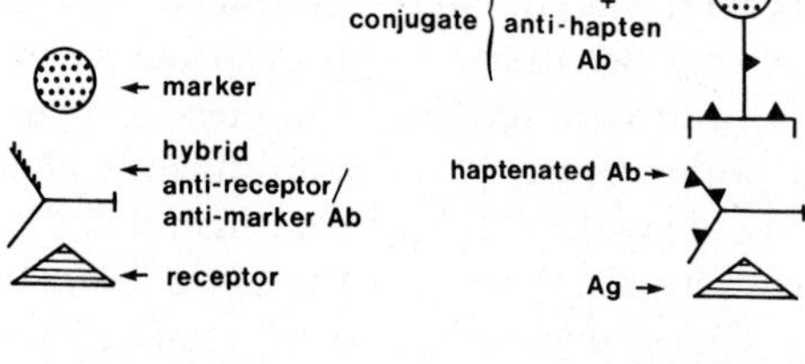

Fig. 1. Schematic representation of basic labelling protocols for SEM immunocytochemistry. The marker indicated is colloidal gold. Abbreviations: Ab = antibody; Ag = antigen; CBS = Concanavalin A binding site; HRP = horseradish peroxidase; Ig = immunoglobulin; LBS = lectin binding site; OVM = ovomucoid; pA = protein A; WBS = wheat germ agglutinin binding site. *Note:* (1) When IgG Ab is applied in excess, the individual Ab molecules compete with each other for their specific Ag

Indirect (multistep) bridge method

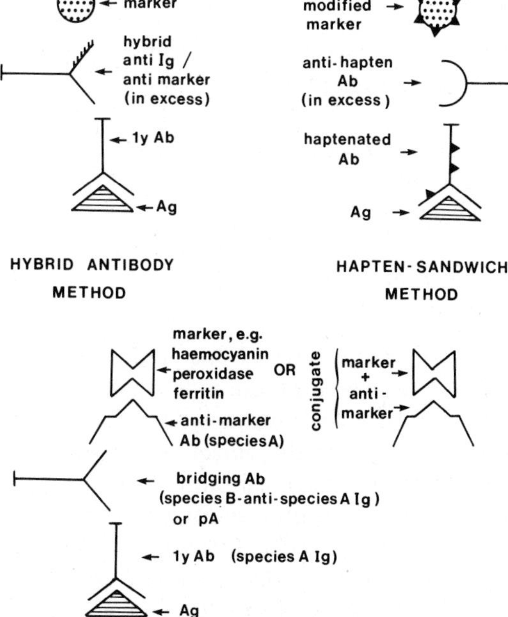

receptors such that only one Ag-binding site per IgG molecule binds to the Ag. The second Ag-binding site is thus free to bind specifically to a suitable reagent subsequently applied. The Ab applied in excess thus acts as a bridge between reagents. (2) Care should be taken when using Fab or $F(ab')_2$ Ig fragment to ensure that any Ab used to detect the fragments is directed against these portions of the Ig molecule, and not against the Fc portion. For references see text and also [161–165].

tage though, in giving a greater specificity of labelling and providing quantitative data on the number of binding sites of a target receptor. Disadvantages of the direct approach are, however, that relatively large amounts of primary ligand are required for the preparation of active ligand-marker conjugates (though use of monoclonal antibodies and more efficient ligand-marker procedures may minimise this problem); and, that each specific ligand must be separately conjugated with the marker. Direct labelling has been mostly used with SEM to document the distribution of lectin-binding sites as demonstrated by specific lectin-marker conjugates prepared from a range of lectins and markers; and, in a few studies to visualise antigenic sites on various cell systems.

Labelling procedures are influenced by a variety of, generally, well-documented factors [9, 15, 18, 50, 51, 53] and many of the constraints are of consequence irrespective of labelling or of microscopical system. Among the various constraints encountered those that should be considered for SEM immunocytochemistry include the influence of:

a) specimen preparation procedures both for labelling and for subsequent SEM;
b) label-induced rearrangements of target moieties;
c) type of sample;
d) marker or ligand-marker conjugate size;
e) steric hindrance of the marker;
f) valence of the visualising moiety;
g) stoichiometry of macromolecule binding to marker;
h) free unconjugated macromolecule and,
i) non-specific adsorption.

Specimen preparation is probably one of the major problems encountered in immunocytochemistry and, at present, there is no single method of choice that can simultaneously provide for an excellent preservation of specimen ultrastructure and adequately immobilise and maintain the bioactive integrity of target moieties. In essence, accurate identification and localisation of target molecules in supra-molecular structures will depend on the conditions of specimen preparation; and the choice of such conditions will be a function of the objectives of a given investigation. In unfixed specimens, while reactivity of target molecules can be expected to be maximal, maintenance of structural integrity can be poor and lateral artefactual label-induced rearrangements of target moieties can occur [4, 15]. Such redistribution of target sites can be minimised by conditions of lowered temperature (labelling at 4°C) or presence of metabolic inhibitors (e.g. 0.02% sodium azide). These conditions may not be adequate, however, in preventing small-scale clustering of labelled target sites [4, 11, 15]. Furthermore, the morphological integrity of unfixed cells can be significantly altered both by changes in physiological environment during labelling and by low temperature conditions [63]. Also, sodium azide, even at low concentrations, may induce morphological changes in some cell types [64].

As a consequence, specimens are ordinarily exposed to fixation procedures in order to preserve structure, and to restrict both lateral mobility (i.e. redistribution) of target molecules and loss of those moieties or of the ligand-marker conjugate through shedding or endocytosis during the labelling process. Chemical fixation can result, however, in a significant alteration and diminution in target molecule reactivity, and certain molecular structures can prove unresponsive to conventional fixatives. Furthermore, given target molecules may have both different tolerances to different fixatives, and different fixation requirements in different tissues; while, different target molecules in the same cell can have different tolerances to given fixatives. Therefore, conditions of pH, osmolarity, temperature and time, may need to be established for particular target molecule and specimen circumstance and level of microscopical analysis. Further comprehensive discussions of these points can be found in several recent reviews [15, 17, 21, 22, 52, 53, 65, 66].

Among chemical fixatives used in electron microscopy, aldehydes (viz. glutaraldehydes and paraformaldehyde) have been generally chosen for immunocytochemistry studies. But, glutaraldehyde, while generally regarded as the fixative of choice for adequate preservation of cellular fine structure can impose major losses of target site reactivity and cross-link cytoplasmic components sufficiently to make penetration of immunoreagents difficult. Paraformaldehyde, on the other hand, can achieve minimal loss of immunoreactivity but with less than adequate maintenance of ultrastructural integrity. Prefixation of specimens with aldehydes, in particular glutaraldehyde, can also cause non-specific attachment of ligand-marker conjugates by the reactivity of free aldehyde groups with ligand ε-amino groups. Therefore, the precaution should always be taken of blocking any cell-bound free aldehydes with reactive amino groups prior to labelling: tissue culture medium, glycine, lysine, ammonium chloride or ammonium carbonate buffer, and sodium borohydride are effective blocking agents [4, 8, 15, 18, 52, 53, 67, 68]. Nevertheless, blockage of such free reactive aldehyde groups may not always be complete and these various treatments could lead to alterations in antigenicity (cf. reference [69]). Inclusion of 7% sucrose in post-fixation washing solutions has been suggested as a possible aid in the preservation of antigen reactivity [70]. Occasionally traces of antigens and especially polysaccharides may leak from glutaraldehyde-fixed cells into the medium and cause agglutination of the ligand-marker conjugate [71, 72]. For some immunocytochemical applications, low concentrations of glutaraldehyde either alone or in combination with paraformaldehyde or acrolein have been successfully used [53, 65, 66, 73]. Other fixatives, such as solvent fixatives (e.g. alcohols, acetones), the carbodiimides and diimidoesters and a miscellany of routine histological fixatives, tentatively assessed for ultrastructural immunocytochemistry [65, 66] remain to be evaluated for SEM immunocytochemistry. Reference should be made to the literature for a detailed description of different fixatives, of their action on various tissue components and target site reactivity and, the consequent artifacts they may impose [15, 53, 65, 69]. Factors influencing the exposure of hidden antigens and proteolytic enzyme pretreatment for the unmasking of antigens concealed during fixation [65, 69, 74] are among other considerations yet to be critically studied in SEM immunocytochemistry.

Another major problem encountered in SEM immunocytochemistry is the difficulty in establishing a good stoichiometry between marker and target site. The resolution of labelling by SEM is limited both by present-day instrumental capability (3–10 nm) and a requirement for some conductive specimen coating (5–20 nm) [12–14]. As a consequence, methods devised for SEM immunocytochemistry generally require markers of a relatively large size and/or distinctive shape such that they are readily recognised against the often irregular topographical contours of supramolecular structures. The precision with which most target molecules can be localised is severely limited, therefore, and attempts to adopt the high-resolving capabilities of electron-dense markers used in immuno-TEM to labelling in SEM have yet to be realised.

Various factors, including the above problem of steric hindrance of the marker, influence both relative or absolute quantification and the statistical determination of binding sites on supramolecular structures by electron microscopy (for details see, for example, [4, 9, 15, 16, 53, 75]). In brief:

a) The larger size markers used in SEM will in general sterically preclude a one-to-one correspondence of marker to target molecule dependent on the density and pattern of distribution of the binding sites. Also, some binding sites may be inaccessible to large markers where masked by other molecular components which inhibit marker penetration. Furthermore, binding sites close to the supramolecular structure may be less accessible to large markers than those target molecules extending out from the surface and positive or high labelling density may be dependent on the choice of markers below a certain size limit [9, 16]. As a consequence, in multiple labelling studies the size and order of addition of ligand-marker conjugates may need to be carefully established such as to avoid marker steric hindrance and allow marker accessibility to binding sites [16, 20, 34];
b) Marker steric hindrance may be also a problem in direct labelling procedures in that absence of labelling may not necessarily indicate absence of target molecules on the basis that accessibility of markers to binding sites is often dependent on marker size [16, 20]. By contrast, indirect labelling procedures usually achieve a more intense marking of binding sites by step-wise amplification of the initial target-ligand interaction. Such procedures, however, reduce the precision of receptor localisation by the distance interposed between target site and marker;
c) Labelling efficiency and the precision of quantification of binding sites can be significantly influenced by the ratio of primary ligand (or primary ligand-marker conjugate) to target site and of secondary ligand-marker conjugate to primary ligand. The valence of both ligand and of target molecule is important here in determining the number of attachment sites [4, 15, 16]. Labelling efficiency is also affected by the amount of specific ligand conjugated to the marker. This is seen as influencing the density of labelling and, probably determining the strength of ligand-marker conjugate binding to target sites and therefore the stability of this binding in the course of labelling and SEM preparation procedures. There is evidence that specifically bound conjugates may be lost through these various preparative treatments [9, 16, 19].

Among other experimental factors integral to the design of meaningful SEM immunocytochemical studies must be considered:

a) those conditions which are specifically appropriate for the preparation and handling of specimens for SEM;
b) those strategies appropriate for labelling (e.g. concentration and reactivity of ligand and conjugate; incubation time; buffer system); and
c) the validation of immunocytochemical procedure and verification of SEM preparation by adequate experimental controls.

SEM labelling studies have been carried out mostly on cells (either attached to solid supports or in suspension) or on tissues such as to provide macromolecular information of natural surfaces. Alternative sources of molecular information available from within tissues and cells remain still to be adequately exploited by SEM immunocytochemistry. Labelling of cells in suspension has several disadvantages. Both prefixation and incubation in multivalent ligand or with ligand-marker conjugate can cause clumping of cells with the possible introduction of artefacts by inaccessibility of surfaces to ligands, or by trapping of ligand-marker conjugates. Also, cells subjected to repeated centrifugation and resuspension may show alteration in cell surface morphology, while the shear forces may induce loss of cell-bound ligand-marker complexes. Cells in suspension can be collected onto various surfaces (e.g. membrane filters; glass coverslips coated with positively-charged polymers) the choice of which may need to be carefully monitored with respect to cell type in order to avoid the introduction of morphological artefacts [4, 8, 15, 17, 61, 76–78].

As far as possible, specimens for SEM immunocytochemical studies should be carefully manipulated to minimise trauma-induced morphological alterations. Also, they should be carefully washed prior to labelling or prefixation to remove dead or dying cells (which can give non-specific labelling) and extraneous surface materials (which may obscure specific binding sites, or produce non-specific labelling) using buffer systems of physiological pH and osmolarity and at temperatures established as appropriate for a given biological system. In that residual surface materials can be left in the course of specimen preparation and labelling which may be mistaken for a marker or which can obliterate the marker when viewed by SEM, it is advised:

a) to use the same buffer system throughout to avoid formation of precipitates (as a consequence of mixing different buffers or imposing osmolarity changes);
b) to standardise on an appropriate number of buffer washes; and
c) to use serum-free buffers or tissue culture media.

Bovine serum albumin or ovalbumin (0.1–0.2% up to 1–2%) is, however, frequently added to the buffer or tissue culture medium to reduce non-specific binding of certain ligands or conjugates and for better maintenance of specimens during the labelling process [8, 15, 17, 20, 61]. An objection to the use of bovine serum albumin has been that it is known to bind, albeit weakly and reversibly, to various cell-associated molecules [15, 19] but, at low concentrations it is thought not to interfere with specific labelling [15]. Non-specific adsorption on prefixed specimens may be limited also by treatment with glycine or lysine. In multistep labelling procedures, washes must separate each labelling step to ensure the specificity of labelling; but caution must be used in that these washes can also remove the ligand attached to the target molecule under study.

The concentration of ligand and of ligand-marker conjugate to be used for labelling will depend on the particular experimental system; variation in the concentration of the labelling reagents can result in different patterns of target site distribution [15, 53]. Appropriate concentrations sufficient to label all sterically accessible target molecules may need therefore to be established empirically, for example, by serial dilution of the labelling reagent and determination of the lowest concentrations which will saturate specific sites as visualised by SEM or analysed by some other quantitative method. Various estimates of appropriate concentrations may be provided from the literature [15, 53]. For practical reasons, viz. to avoid dilution or waste of labelling reagents, specimens are usually processed in as small a volume as possible relative to cell number and potential target sites and such as to avoid any drying out of the specimen during labelling or washing procedures.

Incubation times for labelling may also need to be empirically established with respect to the experimental system under study. If the ligand-marker conjugate is not too dilute then labelling within 5 min or, more generally, between 10 and 30 min can be readily demonstrated at ambient temperature or at 37°C; longer incubation times (30–60 min to several hours) may be needed at lower temperatures of 0–4°C [15, 53].

After labelling and appropriate washings, including a final wash in a protein-free physiological buffer or tissue culture medium, specimens are prepared for SEM by standard procedures as discussed in the literature [9, 12–14, 79–81]. Briefly, post-fixation (generally with glutaraldehyde) must be adequate to preserve certain types of markers (e.g. haemocyanin) [4] and to maintain both the morphology of the specimen and the spatial localisation of the ligand-marker conjugate. An additional osmium-thiocarbohydrazide step is sometimes used in order to minimise problems of specimen 'charging' in the SEM [13, 14]; this may impose, however, a certain granularity to the specimen surface impeding adequate visualisation of the marker. Specimens, following dehydration (through graded alcohols or acetone) and drying (using either critical point-drying or freeze-drying) are generally carbon- or metal-coated. It should be noted that loss of bound conjugates may occur through the dehydration or drying steps. For example, a loss of 10–15% of bound ligand-gold conjugates has been reported during the dehydration step in a graded series of ethanols [19]. Careful monitoring and homogeneity of the metal coating and refinements in metal coating procedures [12] are also essential to minimise the masking of very small markers.

4.2. Experimental controls

In the application of SEM immunocytochemistry, appropriately selected controls which establish method and ligand specificity, and run in parallel with experimental specimens, are essential for meaningful evaluation of the labelling data. Such controls include those commonly performed in the field of immunocytochemistry: comprehensive details and discussion of these can be found in various recent publications [15, 50, 51, 53, 68, 69]. Sources of non-specific labelling may originate from the non-specific adsorption of one or more of the labelling reagents; the binding of 'natural' antibodies directed against unrecognised target moieties other than those under study; hydrophobic or ionic interactions; fixation artefacts; and, endogenous peroxidase. Conversely, negative or low densities of labelling may not automatically imply the absence or a sparse distribution of the target sites under investigation. Sources of error may include the modification or loss of target molecules during fixation; inaccessibility of ligand or ligand-marker complex to binding site; steric hindrance of marker; modification or inactivation of labelling reagents.

In indirect labelling the controls employed to establish the validity of the immunocytochemical procedure may include:

a) replacement of the specific primary antiserum by either pre-immune IgG (or equivalent), some other 'irrelevant' primary antiserum, or buffer solution;
b) omission of the specific first-layer primary ligand incubation step; and, incubation with only the second-layer ligand-marker conjugate;
c) incubation with the specific first-layer primary ligand, then with non-conjugated second-layer ligand, and last with the second-layer ligand-marker complex;
d) blocking of the primary ligand with a specific inhibitor prior to labelling (e.g. presorption of specific antibody with its corresponding antigen; addition of a specific inhibitory sugar to a lectin);
e) change in sequence of double labelling to check that large markers do not exchange with smaller ones or fill unsaturated sites.

Other labelling controls appropriate for the validation of direct immunocytochemical methods include:

a) the inhibition of specific labelling both by (i) preincubation of the specimen with excess amount of unconjugated primary ligand followed by incubation with the primary ligand-marker conjugate; and (ii) blocking specific binding sites of the primary ligand-marker conjugate by preincubation of the complex with a specific inhibitor (e.g. addition of a specific inhibitory sugar or glycopeptide to a lectin-marker conjugate will block the sugar-binding sites of the lectin, [61]);
b) incubation of the specimen with a marker conjugate of pre-immune IgG (or equivalent) instead of specific antibody.

Furthermore, in SEM immunocytochemistry additional verification of labelling specificity by correlative TEM or light microscopical studies may be necessary in that precipitates or small blebs might be mistaken for the marker. This emphasises the importance of those markers with distinctive size and shape and with physico-chemical characteristics allowing their analysis by such SEM instrumental modalities as backscattered electron imaging, X-ray microanalysis, or cathodoluminescence.

5. Some applications of scanning electron microscope immunocytochemistry

From the discussion in the preceding sections it appears clear that a large number of SEM markers are available for cell labelling studies and can provide flexibility in selecting the optimal marker for a given study. Furthermore, it is evident that labelling methods for SEM are essentially similar to those developed for light microscopical immunohistochemistry and for TEM immunocytochemistry.

Of the particulate labelling systems currently available for SEM, colloidal gold has generated increasing interest in that it offers several special advantages as an immunocytochemical marker. In brief, as previously reviewed [16, 19, 22, 41, 43, 55, 82]:

a) monodispersed gold particles can be rapidly and inexpensively prepared in a size range of 3–150 nm mean diameter and stored at 4°C for many months;
b) gold probes may be prepared with a wide range of macromolecules including antibodies, lectins, hormones and toxins;
c) correlative TEM and SEM studies are possible because of the strong electron density and emissivity of the gold particles;
d) different target sites can be separately labelled with probes of different size;
e) correlative light microscope visualisation is also possible where sufficient labelling occurs, the gold particles appearing as an orange-, red- or purple-surface coating depending on their mean diameter; this can provide a convenient alternative to the fluorescent technique with the added advantage that the preparations are stable, also;
f) the high electron back-scattering coefficient of gold suggests that the enhanced contrast of back-scatter electron imaging could provide a superior alternative for visual or computer-aided quantitative analysis of receptor sites; while

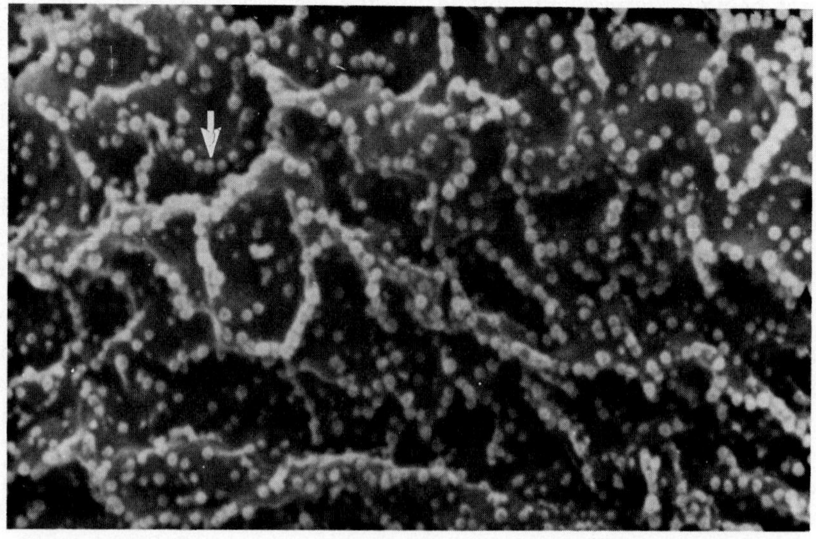

Fig. 2

g) the characteristic X-ray signals emitted by gold could be used to image and quantify cell-bound gold markers by application of X-ray microanalytical techniques.

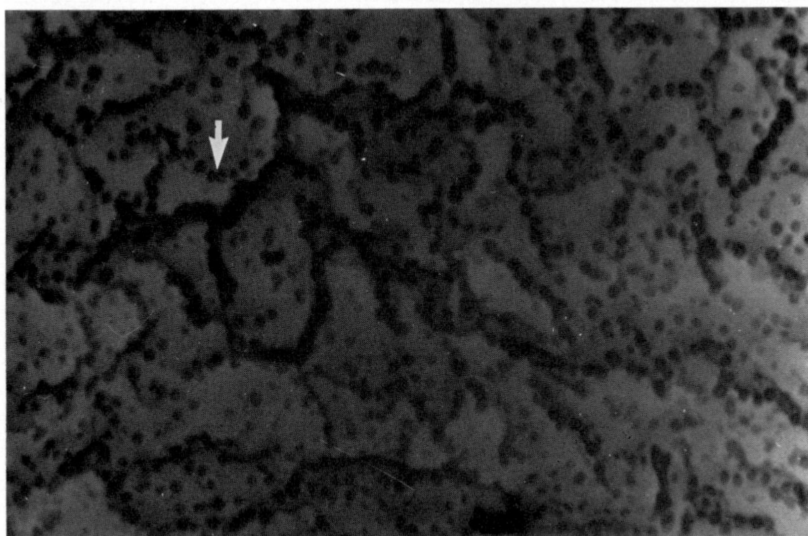

Fig. 3

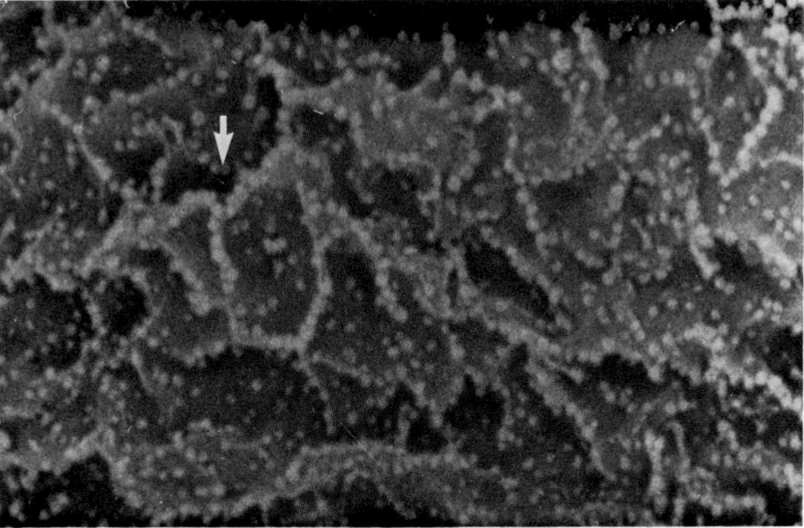

Fig. 4

Figs. 2–4. SEM of luminal surface of superficial cell of rat bladder urothelium labelled for Con A receptor sites using an indirect (two-step) procedure (see Fig. 1) viz. treatment with Con A followed by 45 nm gold-HRP conjugate. Specimens platinum-coated and viewed by SEI (Fig. 2); BEI normal polarity (Fig. 3); and BEI reverse polarity (Fig. 4). Gold particles (↓) show a uniform distribution over the cell surface (see reference [85]). Magnification = ×20,000.

Table 2

Selected list of antigen and lectin receptors detected by SEM immunocytochemical methods

Receptor	Cell, tissue or organ	Reference
Cell surface antigens		
Ig; membrane	B and T lymphocytes, human (normal and malignant)	99, 121
IgG, IgM, IgA, IgD, IgGMAD, IgAGM	B and T lymphocytes, human (normal and malignant)	67, 109, 111, 114, 118, 148
Ig:membrane	B and T lymphocytes, mouse	64, 105, 106, 119, 120, 129
IgG, IgM, IgA	B and T lymphocytes, mouse	76
Ig:membrane	B and T lymphocytes, rabbit	119, 120
Thyl 1; H2	B and T lymphocytes, mouse	67, 102
Rhodopsin	Retina	91
Fibronectin	Chick fibroblasts; Human, hamster and mouse fibroblasts	36, 149, 150
Polyamine; Forssman	Chick embryo cells	125
Teratoma-defined antigens; H2	Mouse fibroblasts (L cells)	63
Cell surface antigen	Chick embryo erythroid cells	57, 92, 93
Influenza	Kidney (MDCK and GMK) cells; erythrocytes.	103, 104, 112, 151, 152
	HeLa cells	131, 132
Respiratory syncytial virus	Vero cells	127
Murine leukaemia virus	Murine cells	54, 90
Mouse mammary tumour virus	Mouse mammary tumour cells	18, 58, 96, 106
Anti-*Candida* antibodies	*Candida albicans*	100
Anti-Saliva antibodies	Microbial cells	126
Mannan	*Candida utilis; Schizosaccharomyces pombe*	33, 55, 71, 72, 153
Cell surface lectins		
Concanavalin A	Rat lymphocytes	95
	Mouse thymocytes	113
	Murine leukaemic lymphocytes	113
	Mouse fibroblasts (L cells)	89, 95
	Mouse neuroblastoma cells	155
	Mouse mammary tumour cells	18, 58
	Rat spermatozoa	135
	Rat bladder urothelium	85
	Hepatocytes	156–158

Receptor	Cell, tissue or organ	Reference
	Retina; rod outer segments	123
	Erythrocytes	49, 55, 95, 159
	Milk fat globule (MFG) membrane	35, 55
	Higher plant protoplasts	154
	Gametes-*Ulva mutabilis*	88
	Dictyostelium discoideum	160
	Candida utilis	49, 55
	Schizosaccharomyces pombe	153
Wheat germ agglutinin	Mouse neuroblastoma cells	155
	Hepatocytes	49, 55, 158
	Erythrocytes	49, 55, 159
	Milk fat globule (MFG) membrane	35, 55
	Dictyostelium discoideum	160
	Candida utilis	49, 156
Peanut agglutinin	Rat bladder urothelium	41
	Erythrocytes	85
Ricinus communis agglutinin	Mouse neuroblastoma cells	155
	Mouse fibroblasts (L cells)	89
	Hepatocytes	49, 55, 158
Soybean agglutinin	Hepatocytes	49, 55, 158
	Erythrocytes	49, 55, 159
	Milk fat globule (MFG) membrane	35, 55

Correct interpretation of marker distribution patterns is not always straightforward and may sometimes present considerable difficulties in SEM immunocytochemistry through inappropriate choice either of marker, for a given type of specimen or investigation, and/or of conductive coating. The interpretation of labelling patterns must also take into account the tendency of some marker systems to aggregate, though these difficulties may be overcome by use of appropriate methodology. For example, the tendency of gold particles to aggregate, particularly if stored for more than a few days after gold-probe preparation, imposes the need for the probes to be lightly centrifuged and the supernatant membrane-filtered immediately prior to use (Appendix 7.3.2). Particle aggregation may also occur in labelling experiments as a consequence of inadequate washing of the specimen and/or leaking of specific macromolecules from the specimen surface: such labelled preparations may show an optical colour change from red to blue, and the aggregates may occur either detached from the specimen surface or adsorbed onto it (e.g. Fig. 7) [16].

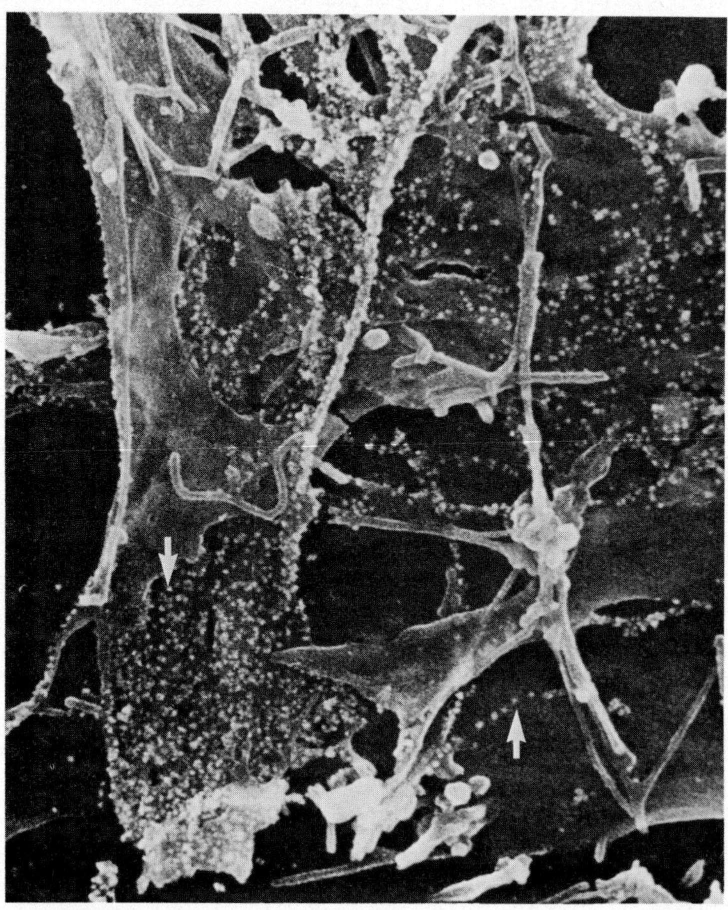

Fig. 5. SEM (by SEI) of HE117 fibroblasts treated with anti-human fibronectin antibodies followed by 45 nm gold-Ig conjugate using an indirect (two-step) procedure (see Fig. 1). The gold particles (↓) are located and aligned along the fibronectin strands and over occasional dense mats of fibronectin (see reference [36]). Magnification = ×12,850.

In order to illustrate the wide range of the actual or potential application of SEM immunocytochemical methods some selected examples are given in Table 2 and Figs. 2–7. This is not meant, however, to represent a comprehensive review of the studies carried out in the various fields which is obviously outside the province of this chapter.

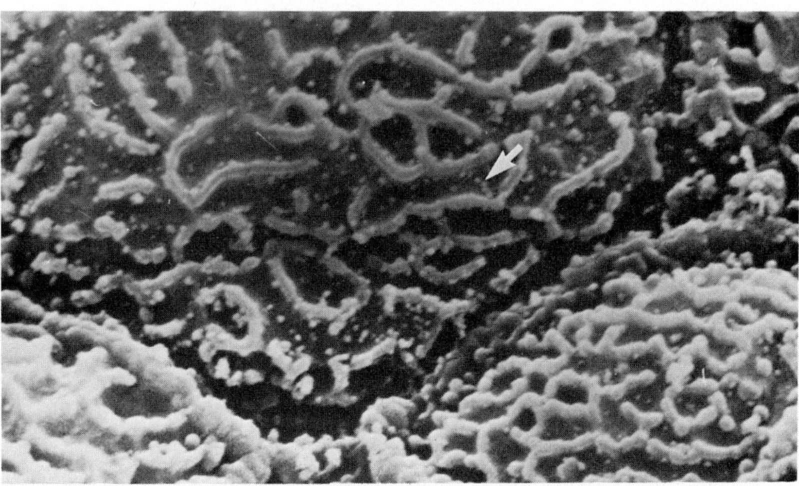

Fig. 6. SEM (by SEI) of luminal surface of late maturing intermediate cell of rat bladder urothelium labelled for a urothelial-specific differentiation antigen (UMA) using an indirect (two-step) procedure (see Fig. 1) viz. treatment with anti-UMA followed by 35 nm gold-Ig conjugate. Gold particles (↓) are located mostly at the base of the micropliacae (see reference [85]). Magnification = ×15,300.

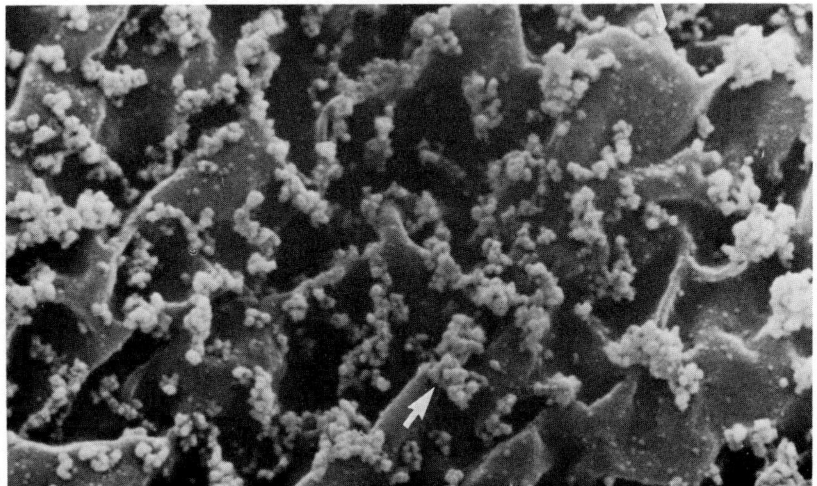

Fig. 7. SEM (by SEI) of superficial cell of rat bladder urothelium labelled for Con A receptor sites using an indirect (three-step) procedure (see Fig. 1) viz. treatment with Con A, then rabbit anti-Con A, followed by 38 nm gold anti-rabbit Ig conjugate. This illustrates non-specific aggregation of gold particles (↓) (see Section 5). Magnification = ×20,000.

6. Concluding remarks

From the studies carried out to date it is evident that the SEM can both offer a unique system for positive identification, quantification and localisation of specific membrane components on cell surfaces, and provide, in combination with fluorescence and transmission electron microscopy a powerful approach to the study of cell membrane properties. Yet, at the time of writing still only limited attention is being given to the localisation of molecular elements in supramolecular structures using SEM immunocytochemical procedures.

Since the introduction of SEM immunocytochemistry to biological research within the last decade, it has been used mostly for surface observations of cells, tissues and organs with little attention to intracellular structures. Recent developments in both specimen preparation technique and in SEM instrumentation can now give much more high resolution information about sub-surface features of biological specimens. As such, and by virtue of its continuously variable levels of resolution, the generation of a variety of different signals, the large specimen size and surface area available for analysis, the SEM can be seen to combine functions similar to those of light microscopy and TEM. The groundwork is therefore laid for the use of SEM as an attractive alternative to light microscopy and TEM providing a means of amplifying the evidence obtainable from immunocytochemical methodology with its role formulated in terms of providing a further spatial dimension to the analysis of the molecular topography of supramolecular structures.

7. Appendix

7.1. *Preparation of gold markers*

7.1.1. *Glassware*

To avoid spurious nucleation of the colloid from contaminants on vessel walls, scrupulous cleanliness is essential. Glassware is soaked overnight at 60°C in 0.25% Decon 90 detergent (Decon Laboratories, Hove, Sussex) prepared in double glass distilled water (DGDW); rinsed twice in DGDW; 4–5 times in tap water; 2–3 times in DGDW; 2–3 times in triple glass distilled water (TGDW); stored in TGDW; and sterilised by autoclaving.

7.1.2. *Solutions*

Solutions for gold marker production are prepared from TGDW and membrane-filtered (0.22 μm pore size) to avoid spurious aggregation of the colloidal gold from extraneous dust particles, organic substances, etc.

7.1.3. *Colloidal gold*

Colloidal gold in the range of 20–150 nm mean diameter is prepared by the method of Frens [39]. Tetrachloroauric acid (50 mg) is added to TGDW (500 ml) and the solution heated to boiling; various quantities (2–20 ml) of freshly-prepared aqueous 1% trisodium citrate are added rapidly; the solution is refluxed for a time period (5–30 min) as determined by the solution colour which progresses from yellow–blue–burgundy–red following gold chloride reduction and completion of gold nucleation. The colloidal gold is sterile membrane-filtered (0.22 μm pore size) into sterile glass containers, stored at 4°C, and checked routinely at monthly intervals for colour change and presence of gross aggregates. The volume of citrate added will determine the particle size of the monodisperse gold colloid (Fig. 8).

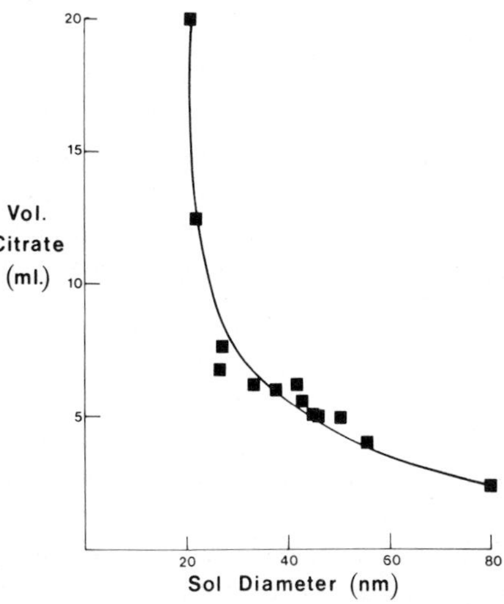

Fig. 8. Mean diameter of gold particles produced by the method of Frens [39] as a function of the volume of 1% sodium citrate added to 500 ml 0.01% chloroauric acid [41].

Colloidal gold in the range of 8–13 nm mean diameter is prepared by the method of Stathis and Fabrikanos [40]. Tetrachloroauric acid (100 mg) is added to TGDW (100 ml) and an aliquot of this mixed at 4°C with potassium carbonate (0.2 M) and TGDW in a ratio of 1 part tetrachloroauric acid solution (0.1%): 1.5 parts 0.2 M K_2CO_3: 25 parts TGDW. One part aqueous ascorbic acid (0.7%) is then added with stirring to this solution which changes immediately to a purple-red colour. The solution is made up to 100 ml with TGDW and heated until the solution colour progresses to red. The colloidal gold is then sterile membrane-filtered and stored as above.

7.1.4. Physical characterisation of gold markers

7.1.4.1. Transmission electron microscopy. 5 μl samples of colloidal gold are air-dried onto carbon-formvar coated 400-mesh copper grids, examined by TEM using catalase crystals to calibrate magnification. Approximately 100 particles are photographed (in a magnification range of ×50–60,000), and printed to give a particle size of at least 0.75 cm diameter. Mean diameters and eccentricities (ratio of major:minor axes) are determined by digitising both axes of 70–100 particles.

7.1.4.2. Absorption spectra. Samples are spectrophotometrically scanned from 450–700 nm and absorbance curves obtained. Unflocculated red solutions of colloidal gold display a single peak of absorption in the visible spectrum ranging between 520 and 550 nm. The position and shape of the peak is influenced by the mean size and shape of the gold marker population. With increasing mean particle diameter the peak moves to a longer wavelength; with increasing heterogeneity in particle shape there is a broadening of the peak.

7.1.5. Membrane filtration

To avoid possible retention of colloidal gold on membrane filter (0.22 μm pore size), cellulose acetate (Sartorius Ltd., Belmont, Surrey, U.K.) or polycarbonate (Bio-Rad Laboratories Ltd., Watford, Herts, U.K.) membrane filters are used in preference to cellulose nitrate (Millipore (U.K.) Ltd., Harrow, Middlesex, U.K.). Also, as proteins can bind to nitrocellulose filters, where these filters are used they should be presaturated with buffer containing bovine serum albumin (1–5 mg/ml) [16, 43].

7.2. Routine protocol for ligand-gold conjugation

7.2.1. Microtitration assay for the determination of colloidal gold stability with pH change

1) Membrane-filter (0.22 μm) the selected colloidal gold preparation.
2) Prepare a range of 100 mM buffers (citrate-phosphate, pH 4–6.5; phosphate, pH 7–7.6; Tris-HCl, pH 8–9; carbonate, 9.5–11) (see Table 3).
3) Place 7 μl of the 100 mM buffers (to cover a pH range of 4–11) into each well of duplicate rows of a 96-well conical bottom microtitre tray.
4) Add to these wells 93 μl of the colloidal gold. This gives a final buffer concentration of 7 mM. Mix by tapping tray. Leave for 5 min at ambient temperature.
5) Colloidal gold stability is visually assessed from its change in colour with pH value. Stability of the colloidal gold (i.e. absence of flocculation) at a given pH is demonstrated by its red colour. This colour changes to violet and finally pale blue with destabilisation of the colloidal gold (i.e. increasing flocculation). The pH range over which colloidal gold remains stable is determined therefore from the range of wells showing maintenance of red colour.

7.2.2. Microtitration assay for the determination of the optimal pH for ligand-gold marker adsorption

1) Prepare a 5–10 mg/ml stock protein solution in DGDW (low molarity buffers, e.g. up to 5 mM NaCl, may be used if essential for a given protein; see Section 3).
2) Dilute an appropriate aliquot of the protein stock solution to give a working protein concentration of 50–500 or 1,000 μg/ml in DGDW depending on the protein (i.e. such that the protein concentration is in excess of the stabilising level required for the gold marker system, see section 3).
3) Using the colloidal gold stability microtitration assay (Appendix 7.2.1), add 10 μl of the working protein dilution to the pH-buffered colloidal gold in the

TABLE 3

Selected range of 100 mM pH buffers

0.1 M citric acid (ml)	*0.2 M Na_2HPO_4 (ml)*	*DGDW (ml)*	*pH*
11.09	6.96	6.95	4.0
9.28	7.86	7.86	4.5
8.00	8.50	8.50	5.0
6.88	9.06	9.06	5.5
5.65	9.68	9.67	6.0
4.48	10.26	10.26	6.5
0.2 M NaH_2PO_4 (ml)	*0.2 M Na_2HPO_4 (ml)*	*DGDW (ml)*	*pH*
4.87	7.63	12.505	7.0
3.50	9.00	12.50	7.2
2.00	10.50	12.50	7.5
1.80	10.70	12.50	7.6
0.2 M HCl (ml)	*0.2 M Tris (ml)*	*DGDW (ml)*	*pH*
4.36	8.14	12.50	8.0
2.77	9.73	12.50	8.5
1.14	11.37	12.49	9.0
0.2 M Na_2CO_3 (ml)	*0.2 M $NaHCO_3$ (ml)*	*DGDW (ml)*	*pH*
2.50	10.00	12.50	9.5
6.00	6.50	12.50	10.0
9.20	3.30	12.50	10.5
11.40	1.10	12.50	11.0

range of those wells showing a red colour (i.e. showing colloidal gold stability). Mix by tapping tray. Leave for 10–15 min at ambient temperature.

4) Stability of the protein-gold mixture with pH change is scored visually and given as stable (red, unflocculated) to unstable (blue, flocculated) on a numerical scale.

In general, the protein-gold mixtures show stability over a broad pH range (i.e. 2–7 pH units) that includes the pI value for the protein. In such cases, the optimal pH for ligand-gold conjugation is set on the basis of the protein pI value (previously established from published or experimental isoelectric focusing data) and chosen such as to be within 0.5 units basic of the mean protein pI.

7.2.3. Microtitration assay for the determination of the optimal protein concentration required to stabilise colloidal gold

1) Place 20 μl DGDW in each of the wells of one series of duplicate rows of a 96-well conical bottom microtitre tray.
2) Add 20 μl of a 500–1,000 μg/ml protein-working dilution (prepared in DGDW from a 5–10 mg/ml stock-protein solution) to the first wells of each row of the above. Then prepare 2-fold serial protein dilutions leaving a total volume of 40 μl in the final wells.

3) Prepare an aliquot of colloidal gold buffered at the optimal pH for ligand-gold marker adsorption based on the ratio of 0.93 ml colloidal gold: 0.07 ml buffer (of suitable pH as determined in Appendix 7.2.2) such as to give a final buffer concentration of 7 mM. Note that 2.5 ml buffered colloidal gold is sufficient to test 24 wells.
4) Add 100 μl of the buffered colloidal gold to the serially diluted protein in each but the final wells. Mix by tapping tray. Leave for 15 min at ambient temperature.
5) To assess resistance of the buffered colloidal gold-protein mixtures to salt-induced flocculation add 20 μl of 10% NaCl (prepared in DGDW) to all but the final wells of the duplicate rows. Leave for 5 min at ambient temperature.
6) Stability of the protein-gold mixture with protein concentration is scored visually. At saturating levels of protein adsorption, stabilisation of the gold marker is achieved with absence of flocculation. Therefore, the last well in the sequence of serial-protein dilutions to show maintenance of red colour represents the dilution endpoint at which the protein is able to stabilise 100 μl of buffered colloidal gold.
7) Calculate the protein stabilising concentration per ml of buffered colloidal gold required to stabilise the gold marker. For example, an endpoint in well 6 indicates that 20 μl of protein diluted to $1/2^6$ (1/64) stabilises 100 μl of buffered gold sol. For 1 ml of buffered gold sol the protein stabilising concentration will be $20/64 \times 10$ μl, i.e. 3.21 μl of the working protein dilution. Scale up these values to appropriate amounts (e.g. a 15 ml total) for the routine preparation of ligand-gold conjugates.

7.2.4. Preparation of gold probes (i.e. gold-ligand conjugates)
Approximately 1.5–2 ml of gold probe may be prepared from 15 ml of a buffered colloidal gold [19, 41].

1) First establish a) the stability of colloidal gold over a range of pH values (see Appendix 7.2.1); b) the optimal pH for ligand-gold marker adsorption (see Appendix 7.2.2); c) the optimal protein concentration for stabilisation of the colloidal gold (see Appendix 7.2.3).
2) Prepare stock solutions of a) 2.5% polyethylene glycol (PEG; 20,000 daltons) in DGDW; and b) 0.4 mg PEG (20,000 daltons) per millilitre isotonic phosphate-buffered saline (isoPBSa) (i.e. 8 parts PBSa + 2 parts DGDW) (PBSa composition: NaCl 8 g, KCl 0.2 g, Na_2HPO_4 1.14 g, KH_2PO_4 0.2 g, 1,000 ml DGDW; pH 7.0; if required adjust to pH 7.3–7.4 with N NaOH); microfuge (see reference [41]) suitable quantities of PEG solutions before use at maximum speed for 10–15 min.
3) Place 13.95 ml colloidal gold in a sterile 50 ml glass beaker with magnetic flea, and stir gently on a magnetic stirrer.
4) Add 1.05 ml of a 100 mM buffer of suitable pH (Appendix 7.2.1, 7.2.2). This gives a final buffer concentration of 7 mM.
5) Gradually add the appropriate volume of diluted protein working solution

required to stabilise the colloidal gold as calculated above (Appendix 7.2.3). Leave to stir for 15 min at ambient temperature. Note that colloidal gold may be added, of preference, to the protein solution (see section 3).

6) Quench further protein-gold adsorption by the addition of 240 μl of a 2.5% PEG solution. This gives a final PEG concentration of 0.4 mg/ml established as adequate for the stabilisation of the gold-protein complex (see reference [19, 41]). Leave to stir for 5 min at ambient temperature.
7) Transfer and distribute the gold probe into 12 microfuge tubes (1.5 ml capacity); or, for small diameter probes (3–12 nm) into appropriate ultracentrifuge tubes containing approximately 0.5 ml glycerol or 35 M sucrose (see section 3).
8) Centrifuge at 4°C for time periods and at speeds appropriate for the concentration of the gold probe to an easily mobile (red) pool at the bottom of the tube. For 30–40 nm diameter gold particles use 12,000 *g* for 2–3 min; 12 nm diameter gold particles use 40,000 *g* for 60 min; 5 nm diameter gold particles use 60,000 *g* for 60 min.
9) Remove the supernatant retaining a 1 ml aliquot to assay for protein activity. This is of relevance should the gold probe prove to be inactive (see Appendix 7.2.5). Centrifuge the supernatant two or three times until colourless or very light pink, then discard, retaining from the last centrifugation a further 1 ml aliquot to assay for protein bioactivity.
10) Resuspend the mobile pools of gold probe by vortex action; wash twice in 0.4 mg/ml PEG-PBS solution; and sterile membrane filter (0.22 μm pore size) the collected gold probe into sterile microfuge tubes.

7.2.5. Passive haemagglutination assay of gold probe bioactivity

7.2.5.1. Preparation of indicator cell systems.

1) Fix 1 volume of fresh erythrocytes (e.g. human or rat) washed 3 times in isoPBSa (see Appendix 7.2.4) in 4–5 volumes of modified Karnovsky's fixative [83] (1 volume 4% paraformaldehyde in 0.1 M Sorensen's phosphate buffer pH 7.4 + 2 volumes 6% glutaraldehyde in Sorensen's buffer + 1 volume Sorensen's buffer). Rotate for 5 min at ambient temperature.
2) Add 5 volumes of 2 M glycine (aq.) to block free reactive aldehyde groups. Rotate for 20 min at ambient temperature.
3) Centrifuge to sediment the cells; discard the supernatant; add 10 ml of 2 M glycine; resuspend the cells by vortex action; rotate for 20 min at ambient temperature; wash three times by centrifugation and suspension in isoPBSa.
4) Resuspend fixed erythrocytes in assay diluent (isoPBSa + 5 mg/ml bovine serum albumin + 0.05% sodium azide) to a 10% haematocrit; aliquot; and store at 4°C.

These indicator cells can be used to assay:

a) Concanavalin A-gold conjugates: Concanavalin A binds directly to normal erythrocyte surface sugars.
b) PNA-gold conjugates: the erythrocytes must be first neuraminadase-treated to

remove terminal sialic acid residues and expose D-galactose to which PNA binds.

Following step 4 above, prepare a 10% haematocrit in isoPBSa (see 7.2.4, step 2). Add 6 μl of stock neuraminidase (1 U/ml) per ml of cell suspension. Rotate cells for 20 min at ambient temperature. Wash cells three times in isoPBSa; once in assay diluent (see step 4); resuspend as a 10% haematocrit in the assay diluent; and store at 4°C.

c) antibody-gold conjugates: the erythrocytes are coated with an immunoglobulin (Ig) that is recognised by the antibody.

Following step 4 above, wash erythrocytes three times in isoPBSa. Solubilise 1 mg of immunoglobulin (usually rabbit, mouse, or goat IgG) in 0.5 ml isoPBSa. Add protein to packed erythrocytes such as to give 1 mg immunoglobulin per ml of packed cells. Rotate cells for 2 min at ambient temperature to distribute protein. Wash cells three times in isoPBSa; resuspend in assay diluent (see step 4); rotate for 5 min; and wash once in assay diluent. Test for the presence of antibody-labelled erythrocytes by agglutination in the presence of specific antibody (see below).

d) avidin-gold conjugates: the erythrocytes are first biotinylated.

Wash unfixed erythrocytes twice in isoPBSa; prepare a 10% haematocrit in isoPBSa; place on ice; add sodium metaperiodate to give 1 mM in 10 ml suspended erythrocytes; mix and leave on ice for 15 min. Wash erythrocytes three times in ice cold isoPBSa. To the periodised cells add a freshly prepared solution of 2.5 mg biotin hydrazide in 1 ml isoPBSa; leave for 30 min at ambient temperature; and wash cells three times in isoPBSa. Fix and glycine-treat erythrocytes as above (steps 1–4).

7.2.5.2. Assessment of indicator cells. This assay checks for a) autoagglutinability of indicator cells; b) working dilutions of indicator cells which give a minimum visible pellet.

1) Wash indicator cells in assay diluent (see Appendix 7.2.5.1, step 4); resuspend pelleted cells by vortex action; prepare a 1/10 dilution of a 10% haematocrit stock suspension.
2) Add 45, 40, 30, 20 and 10 μl of assay diluent to wells in duplicate rows of a 96-well conical bottom microtitre tray.
3) Add 5, 10, 20, 30, 40 and 50 μl of test cell suspension to the wells such that the following dilutions of test cell are obtained.

Well pair	1	2	3	4	5	6
Vol. diluent (μl)	0	10	20	30	40	45
Vol. erythrocytes (μl) (1% haematocrit)	50	40	30	20	10	5
i.e. Dilution of stock Erythrocytes (10% haematocrit)	1/10	8/100	6/100	4/100	2/100	1/100

4) Leave for 1–2 h or until the cells sediment to form defined pellets in the bottom of the conical wells.
5) Select the 1% haematocrit dilution that gives a clearly defined pellet for use in

the assay of the protein and protein-gold conjugates. Where autoagglutination occurs, i.e. formation of a diffuse precipitate, discard the cells and prepare a fresh stock.

7.2.5.3. Gold Probe (i.e. gold-ligand conjugate) Bioactivity Assay. The assay uses indicator cells (e.g. erythrocytes, bacteria) previously sensitised (Appendix 7.2.5.1) with, for example, a protein, which will interact specifically with that used in the preparation of gold probe. The presence of reactive ligands on gold probes is determined, therefore, by the addition of appropriate indicator cells followed by the scoring of agglutination patterns (e.g. reactive anti-rabbit IgG on a gold probe is determined by indicator cells treated with rabbit IgG). By the comparison of the agglutination end-points between known concentrations of unconjugated ligand and the ligand-gold conjugates, the amount of ligand bound to a gold probe can be calculated. But in that the amount of gold particles in a given solution is a function of its absorbance at peak absorption wavelength (A 1 cm: λ max) then effective ligand binding values can be more appropriately expressed in terms of μg/ml/Aλmax; this allows a more meaningful comparison between different batches of gold probes though these protein values could be but a fraction of the total protein adsorbed [19, 41, 43].

1) Place 20 μl DGDW into each well of four groups of duplicate rows in a 96-well conical bottom microtitre tray.
2) Into the first wells of the first group of duplicate rows (rows 1 and 2) add 20 μl of a known concentration of the ligand standard used in the preparation of the gold probe. Leave rows 3 and 4 available for further dilution of the ligand standard.
3) Into the first wells of the second group of duplicate rows (5, 6) add 20 μl of the gold probe under test.
4) Into the first wells of the third group of duplicate rows (7, 8) add 20 μl of the supernatant retained from the first centrifugation of the gold probe (see Appendix 7.2.4, step 9). This estimates ligand loss from the ligand-gold conjugate (*control* I).
5) Add 20 μl of DGDW to two wells. This serves to assess the sedimentation characteristics and working concentration of the indicator cells (*control* II).
6) Serially dilute the ligand standard, gold probe and supernatant (using 2-fold serial dilutions; 20 μl transfer volumes) leaving total volumes of 40 μl of each in the final wells available for further dilution as necessary.
7) Add 50 μl of appropriate indicator cells, diluted to a working concentration (see Appendix 7.2.5.2), to all the wells except those containing 40 μl volumes.
8) Into the first and second wells of the fourth group of duplicate rows (rows 9, 10) add 20 μl of gold probe at dilutions of 1:2 and 1:4, respectively. Mix 60 μl of the ligand standard with 60 μl of appropriate indicator cells (e.g. erythrocyte-rat IgG + anti-rat IgG): this serves to block the binding sites to which the gold probe is directed. Add 50 μl of this erythrocyte mix to the probe-containing wells. This serves to check the specificity of the reaction between probe and indicator cell (*control* III).

9) Mix by tapping tray. Leave for 1–2 h at ambient temperature.
10) Visually assess the wells and calculate the bioactivity of the gold probe. The endpoints for the gold probe, ligand standard, or supernatant serial dilutions are represented by the last wells to show partial or complex agglutination. Erythrocytes should sediment to form defined pellets in the control wells. Agglutination in control I will give an estimate of free ligand in the gold probe preparation.

7.2.5.4. Horseradish peroxidase (HRP)-gold probe bioactivity assay. This assay provides a rapid method for determining HRP-gold probe bioactivity by obviating the need for indicator cells. The assay may be used also for assaying HRP activity prior to probe conjugation or immunoperoxidase labelling.

1) Prepare 2-fold serial dilutions of the HRP-gold probe, supernatant and HRP as described above (Appendix 7.2.5.3, steps 1–6).
2) Prepare 0.5 mg/ml 3,3′diaminobenzidine tetrahydrochloride (DAB) in iso PBSa. This solution can be stored for several weeks at 4°C. Immediately before use add concentrated (i.e. 30%) hydrogen peroxide (H_2O_2) to give a final concentration of 0.03% H_2O_2. Filter solution through Whatman No. 1 filter paper before use.
3) Add 50 μl of freshly H_2O_2-activated DAB solution to all of the wells (but except those containing 40 μl volumes).
4) Mix by tapping tray.
5) After a time period of precisely 5 min visually assess the endpoint. This is determined by a colour change (from a very faint brown to a definite brown colouration) which represents the reaction product formed between HRP and its substrate (H_2O_2-activated DAB). Note that since the enzyme-substrate reaction will continue until all the substrate is expended, then the 5 min time period must be strictly adhered to in order to maintain a consistency between assays.

7.3. Representative protocol for gold labelling of specimens for scanning electron microscopy

In this section gold labelling for scanning electron microscopy (SEM) is outlined in basic terms since detailed procedures will change from application to application. The instructions summarise details from the literature (see section 4; Fig. 2; Table 2).

7.3.1. Choice of specimen preparation method

7.3.1.1. Handling of specimens.

a) Cells in suspension: suspend cells in 10 ml washing buffer solution (WBS) (*see Comment 1*) containing bovine serum albumin (*see Comment 2*) (1 mg/ml; WBS-BSA). Centrifuge at 250 g for 5 min. Discard the supernatant. Repeat the suspension and centrifugation. Resuspend the final cell pellet as a single cell suspension in 1 ml WBS-BSA, or in a volume to give approximately 10^5–10^6 cells/ml. Dispense 50 μl of the cell suspension into each of a series of 1–2 ml capacity polystyrene round-bottomed tubes.

b) Attachment of cells to solid support: cells in suspension, after washing with WBS-BSA, may be attached to various solid supports as follows: i) place poly-L-lysine (100 μg/ml in DGDW) on acetone- or ethanol-cleaned coverglasses; leave for 30 min at 4°C; drain; wash well with DGDW; and air dry. Pipette cells onto coverglass and allow to settle; ii) gently cytocentrifuge cells onto coverglass; iii) layer gently an appropriately diluted suspension of cells onto a fine pore membrane filter (*see Comment 3*) (e.g. 13 mm diameter; 0.22 or 0.45 μm pore size) using a 1 ml syringe attached to a membrane filter assembly; do not allow specimens to dry out. Note that specimen labelling and preparation for SEM can be done by passage of reagents through the filter assembly; the membrane filter is removed prior to critical point drying.

Cells and tissues may be attached by growth on coverglasses (e.g. 13 mm diameter) or other appropriate solid supports. A procedure economical of time and reagents and allowing a variety of simultaneous procedures (e.g. checkerboard titrations, controls etc.) is the use of multitest slides (e.g. Multitest slides, Flow Laboratories Ltd., Irvine, Scotland).

c) Solid tissues: dissect tissues in a serum-free tissue culture medium or balanced salt solution (e.g. Hanks saline) to size dictated by experimental design and mode(s) of microscopy using appropriate microsurgical instrumentation to minimise trauma-induced damage to tissue; wash tissues in 3 changes of WBS-BSA to remove extraneous protein and other contaminating materials—note that certain mucosal surfaces can present special problems of cleaning (cf. reference [84]); where necessary pin out tissues (using entomological pins), for example, on dental wax (cf. reference [85]).

d) Tissue sections: i) Paraffin wax sections (5 μm thick), prepared from appropriately fixed specimens, are dewaxed and rehydrated through 3 changes of xylene, 2 changes of absolute alcohol and 1 change of 70% ethanol; washed in 1 change of DGDW and 3 changes (5 min each) of WBS-BSA. Do not allow specimens to dry out; ii) Tissue slices (20–200 μm thick) are prepared from unfixed or from appropriately fixed specimens using either the Smith-Farquhar tissue sectioner or the Oxford Vibratome®. The tissue slices are placed into droplets of WBS-BSA of sufficient volume to preclude any drying out of the specimens.

Comment 1: An extensive range of buffer systems is available for immunolabelling methodology and reference should be made to wide literature regarding their properties. For immunocytochemistry the washing buffer solution (WBS) should be critically controlled with respect to pH, osmolarity, as well as composition. This is particularly pertinent to studies using unfixed specimens where the use of serum-free tissue culture medium is recommended. In those cases where ligands require ionic additives (e.g. certain lectins, see Table 4), care should be taken to avoid deleterious interactions between these and WBS components (e.g. precipitation of phosphates) (see also Section 4.1): in such cases the use of Tris buffers may be advised rather than phosphate buffers. Representative examples of buffer systems used in labelling protocols include: 0.1 M phosphate-buffered saline (pH 7.2–7.4); 50 mM Tris-buffered saline (pH 7.6); 20 mM HEPES-MEM (pH 7.2–7.4) or 20 mM HEPES-RPMI 1640 (pH 7.2–7.4).

TABLE 4

Ionic requirements and specific inhibitors for a selected series of lectins

Lectin (abbreviation)	*Major sugar specificities*	*Ionic requirements (final conc. 1 mM)*	*Binding inhibitor*[a]	*Blood group specificity*
Concanavalin A (Con A)	α-D-Glucose α-D-Mannose	Mn^{2+} Ca^{2+}	α-D-Methylmannose	non-specific
Dolichos biflorus agglutinin (DBA)	*N*-Acetylgalactosamine	—	*N*-Acetylgalactosamine	$A_1 \gg A_2$
Peanut agglutinin (PNA)	Gal-β-(1–3)-GalNAc[b]	—	α-D-Galactose	Neuraminidase-digested A, B, O or T antigen
Ricin communis agglutinin I (RCA I)	β-D-Galactose	—	Lactose	non-specific
Soybean agglutinin (SBA)	*N*-Acetylgalactosamine α-D-Galactose	Mn^{2+} Ca^{2+}	*N*-Acetylgalactosamine	A > O > B
Ulex europalus agglutinin I (UEA I)	α-L-Fucose	—	α-L-Fucose	$O \gg A_2$
Wheat germ agglutinin (WGA)	$(\beta\text{-}(1\text{–}4)\text{-D-GlcNAc})_2$[b]	Mn^{2+} Ca^{2+} Zn^{2+}	*N*,*N*-Diacetyl-chitobiose, *N*-Acetylglucosamine	non-specific

[a]Inhibitor concentrations required for total inhibition of lectin binding can vary widely (0.2 mM–0.5 M). 0.2 M concentrations are commonly used and increased if total inhibition not obtained.
[b]Gal: Galactose; GalNAc: *N*-Acetylgalactosamine; GlcNAc: *N*-Acetylglucosamine.

Comment 2: Addition of 0.1–0.5% bovine serum albumin or ovalbumin to the washing buffer solution (WBS) provides a more physiological environment for living cells and, 'conditions' specimens such as to avoid non-specific binding of marker conjugates (see Section 4.1). Note that the carbohydrate nature of the albumins precludes their use in labelling with lectins.

Comment 3: Polycarbonate filters may be preferred in that they provide a flat background and the pores give a useful reference guide to size.

7.3.1.2. Pre-fixation.

a) Prior to pre-fixation, rinse specimens in three changes of WBS-BSA (5 min each).
b) Fix specimens (*see Comment 4*) in an appropriate fixative. Choice, concentration and fixation time (*see Comment 5*) should be based on prior performance tests to assess suitability for adequate preservation of structural integrity and ligand reactivity (see section 4.1).

   Two examples of commonly used fixatives are paraformaldehyde (1–4%) and glutaraldehyde (0.1–0.5%, 2.5%): e.g. i) Paraformaldehyde (4%) in 0.1 M Sorensen's phosphate buffer pH 7.2–7.4. Dissolve 4 g paraformaldehyde in 95 ml 0.1 M Sorensen's phosphate buffer, heat on a warm plate with stirring to approximately 70–80°C; if paraformaldehyde incompletely dissolved add, while stirring, a few drops of 1 N NaOH until solution is clear. Adjust pH to 7.2–7.4, if necessary, with 1 N HCl or 1 N NaOH; then adjust volume of paraformaldehyde solution to 100 ml with Sorensen's buffer. Membrane filter (0.22 μm pore size) before use. Paraformaldehyde should be freshly prepared before use. For convenience, the solution can be prepared in the afternoon, stored overnight at 4°C, and used the following morning, ii) Glutaraldehyde (2.5%) in 0.1 M Sorensen's buffer pH 7.2–7.4. Add 10 ml of a 25% glutaraldehyde stock solution to 85 ml 0.1 M Sorensen's buffer; adjust pH to 7.2–7.4 as above i); then adjust volume of glutaraldehyde solution to 100 ml with Sorensen's buffer. iii) 0.1 M Sorensen's phosphate buffer, composition per litre: 405 ml of 0.2 M $Na_2\ HPO_4$ stock, 95 ml of 0.2 M $NaH_2PO_4 \cdot 2H_2O$ stock, and 500 ml DGDW. Adjust pH to 7.2–7.4 with 1 N HCl or 1 N NaOH.
c) Wash fixed specimens with three changes (5–15 min each) of WBS-BSA at ambient temperature.
d) Inactivate residual reactive aldehyde groups on aldehyde-fixed specimens by washing specimens at ambient temperature with either i) 0.1–0.5 M ammonium chloride in PBS for 30–60 min; ii) 0.05–0.2 M glycine or lysine-HCl in PBS for 30–60 min; or 0.5 mg/ml^{-1} sodium borohydride in PBS for 10 min. (Note that sodium borohydride is made just before use; see section 4.1.)
e) Wash specimens with three changes (5–15 min each) of WBS-BSA at ambient temperature.

Comment 4: Cells in suspension are best fixed by dropwise addition under constant gentle stirring (over a period of 1–2 min) of an equal volume of double strength fixative; and then handled as described above (Appendix 7.3.1.1). Packed cells (2–10 μl) may be fixed by resuspension in 5–10 ml of normal strength fixative.

Comment 5: Prefixation times of between 10 to 60 min at ambient temperature (18–22°C) and between 30–120 min at 4°C are commonly used. Note that prolonged periods of storage in fixative (up to 1 month) prior to labelling have been reported (cf. reference [39]).

7.3.1.3. Special specimen treatments.

a) Neuraminidase treatment of specimens is employed for the removal of terminal sialic acid residues and exposure of β-galactose residues, for example, in peanut agglutinin labelling protocols. Prepare a 1% neuraminidase solution in WBS from a 1 U/ml commercial stock; pre-warm to 37°C; incubate specimens (pre-warmed to 37°C) in an appropriate volume of neuraminidase solutions for 15 min at 37°C. Terminate enzyme reaction by washing with 3 changes (10 min each) of WBS-BSA.
b) Peroxidase and peroxidase-like enzymes are present in many tissues and a number of methods are available for inhibiting endogenous peroxidase and avoiding false-positives due to its presence. Endogenous peroxidase can be blocked by the sequential incubation of specimens with: i) 7.5% hydrogen peroxide in WBS (5 min); 2.28% sodium metaperiodate in WBS (30 min); 0.03% sodium borohydride in WBS (2 min) [86]. Use freshly prepared solutions; incubate at ambient temperature; terminate treatment with 3 changes (10 min each) of WBS-BSA.
c) If necessary, antigens may be unmasked by treatment of specimens with, for example, either: i) 0.1% trypsin in 0.1% calcium chloride in DGDW (pH 7.8) (20–30 min at 37°C); ii) 0.1% protease in PBS; iii) 0.4% pepsin in 0.01 N HCl (pH 2.0) (20–30 min at 37°C); iv) 0.06% pronase in 0.5 M Tris buffer (pH 7.5) (10 min at 4°C). Pre-warm (or cool) solutions and specimens (contained in WBS) for 10 min prior to enzyme treatment. Terminate enzyme reactions with several changes (3–5 min each) of WBS-BSA. For further methodological details see Reference [74].

7.3.2. Outline of SEM gold-labelling procedure

1) Prepare the specimens according to the aim of the study and experimental design: for choice of specimen preparation method viz. handling, prefixation (*see Comment 6*) and special treatment regimens see above (Appendix 7.3.1). Prepare the ligand solutions as appropriate to their specific requirements: for lectins see Table 4 regarding ionic requirements and specific inhibitors.
2) Wash specimens with three changes (5 min each) of WBS-BSA (*see Comment 7*).
3) [OPTIONAL] (*see Comment 7*). Non-specific labelling can be reduced by exposure of specimens to normal serum (e.g. at a 1:5–1:25 dilution in WBS) from the species in which the antibodies were raised, for 5–10 min at ambient temperature.
4) a) In the one-step procedure (see Fig. 1): react specimens with an adequate volume (*see Comment 8*) of optimally diluted (*see Comment 9*) gold-primary ligand conjugate for time periods and incubation temperatures (*see Comment 10*) as appropriate.

b) In the two or multiple step procedures (see Fig. 1): i) react specimens with an adequate volume (*see Comment 8*) of optimally diluted (*see Comment 9*) unconjugated ligand for time periods and incubation periods (*see Comment 10*) as appropriate; ii) wash specimens with three changes (5 min each) of WBS-BSA (*see Comment 11*); iii) [OPTIONAL] treat specimens with normal serum or IgG (at a 1:5–1:25 dilution in WBS) of species in which the primary antibody is raised: this will block free antigen sites on the middle stage antibody labelling step; iv) react specimens with gold-secondary ligand conjugate using volume (*see Comment 8*), dilution (*see Comment 9*), time period, and incubation temperature (*see Comment 10*) as appropriate (two-step procedure); or react specimens with intermediary ligands and final gold-ligand conjugate with appropriate washes between steps (multiple-step procedure).

5) Wash specimens to remove excess gold-ligand conjugate with three changes (5 min each) of WBS-BSA (*see Comment 11*).
6) Assess specificity of the gold labelling by a series of controls (see section 4.2). These should give negative results or reasonably low background labelling.
7) Immerse specimens in 2.5% glutaraldehyde in 0.1 M Sorensen's buffer for times appropriate to specimen. Then, as required, post-fix in 1% aqueous osmium tetroxide for 1–2 h; or, process through a thiosemicarbazide-osmium tetroxide schedule [13, 14, 81]. Dehydrate specimens by passing through increasing concentrations of ethanol, methanol or acetone to replace water within the specimen; and, critical point-dry the specimens from liquid CO_2 [81]. Mount the specimens [80]; deposit a conductive coating as necessary [12–14, 81] and examine by SEM using appropriate instrument modalities [6, 81].

Comment 6: Incubation of unfixed specimens at 4°C or in the presence of 0.02–0.1% sodium azide will almost completely prevent lateral movement of surface targets; whereas, incubation at 37°C will promote lateral redistribution into patches and allow for cap formation as well as internalisation (see section 4.1). Aldehyde-fixed specimens must be reacted with blocking agents (e.g. 0.1 M glycine) to bind free aldehydes (see section 4.1; Appendix 7.3.1) and thereby reduce non-specific labelling.

Comment 7: The same buffer is generally used for the various washing steps (see section 4.1; Appendix 7.3.1.1; Comment 1). Background labelling may be reduced by incubating specimens with 1% BSA or ovalbumin, or 1:5 normal non-immune serum in buffer, for 5 min prior to labelling (see section 4.1). Use of bovine serum albumin or ovalbumin is preferred as a blocking treatment of non-specific binding sites (but see section 4.1). Use of diluted non-immune serum may not be always recommended, for example, protein A-gold complexes can bind 'non-specifically' to immunoglobulins adsorbed from the normal serum and give false-positive results. Background labelling can sometimes be reduced by including 0.05–0.1% Triton X-100 in all washes of fixed specimens.

Comment 8: Labelling reagent volumes will vary in relation to specimen size; for example, as little as 5–10 μl is needed for 12 mm diameter coverglasses; 50–100 μl volumes are adequate for most purposes; while 200–400 μl may be needed for tissue biopsies. Specimen support surfaces should be dried with absorbent clean filter paper before applying labelling reagent to specimens. But note that specimens must always be kept adequately moist. Supports should be gently rocked to ensure full coverage of specimen with labelling reagent. Drying of specimens must be avoided; therefore, labelling reactions should be done in a closed humid chamber.

Comment 9: a) The optimal dilution of immunocytochemical reagent which gives intense specific labelling and low background has to be determined empirically for each labelling reagent and for each specimen preparation using the labelling procedure itself. Both the primary ligand and the gold-secondary ligand conjugate should be tested in a dilution range generally up to 1/1000 and used at the highest possible dilution in order to minimise background labelling (use of highly concentrated antibody solutions, for example, can result in high background labelling) [53]; b) 100–200 μg/ml lectin solutions are usually prepared for labelling of unfixed specimens; higher concentrations ($\geqslant$200 μg/ml) may be necessary for labelling of aldehyde-fixed specimens [61]; c) Dilution of antibodies may be done in 1% BSA or ovalbumin to minimise background labelling; (but see section 4.1; Appendix 7.3.2; Comment 7); d) Ligands and gold conjugates may need to be lightly centrifuged (10–12,000 *g* for 30 sec) and the supernatant membrane-filtered (0.22 μm pore size) in order to remove possible aggregates (see sections 3; 4.1).

Comment 10: Incubation of specimens with labelling reagents for 15–60 min is generally sufficient to saturate accessible target sites at ambient temperature (18–22°C) or 37°C. Longer incubation times may be needed if labelling is carried out at 4°C; lower background labelling can be obtained by incubating at 4°C overnight. By extending the time of labelling with each labelling reagent to 24–48 h at 4°C there can be considerable improvement in the uniformity of labelling at high antibody dilutions. Some antigens will yield significantly better labelling when a detergent (e.g. Triton X-100) is added to the antiserum at low concentrations (0.05–0.5%).

Comment 11: Each labelling step should be followed by extensive washing of the specimen. An important washing stage in the unlabelled antibody technique follows the application of the primary reagent: failure to remove non-specific antibody at this stage will lead to excessive background labelling.

8. References

1. Albrecht, R.M. and Wetzel, B. (1979) Ancillary methods for biological scanning electron microscopy. SEM Symposium, 12/III, 203–222.
2. Becker, R.P. and Sogard, M. (1979) Visualization of subsurface structures in cells and tissues by backscattered electron imaging. SEM Symposium, 12/II, 835–870.
3. Becker, R.P. and Geoffroy, J.S. (1981) Backscattered electron imaging for the life sciences: introduction and index to applications—1961 to 1980. SEM Symposium, 14/IV, 195–206.
4. Brown, S.S. and Revel, J.P. (1978) Cell Surface Labelling for the Scanning Electron Microscope. In: J.K. Koehler (Ed.) Advanced Techniques in Biological Electron Microscopy, Vol 2. Springer-Verlag, Berlin–Heidelberg–New York, pp. 65–88.
5. Carr, K.E., McLay, A.L.C., Toner, P.G., Chung, P. and Wong, A. (1980) SEM in Service Pathology: a review of its potential role. SEM Symposium, 13/III, 121–138.
6. Echlin, P. (1981) The analysis of organic surfaces. SEM Symposium, 14/I, 1–20.
7. Haggis, G.H. (1982) Contribution of scanning electron microscopy to viewing internal cell structure. SEM Symposium, 15/II, 751–763.
8. Molday, R.S. (1977) Cell surface labelling techniques for SEM. SEM/IITRI, 10/II, 59–74.
9. Molday, R.S. and Maher, P. (1980) A review of cell surface markers and labelling techniques for scanning electron microscopy. Histochem. J. 12, 273–315.
10. Tanaka, K. (1980) Scanning electron microscopy of intracellular structures. Int. Rev. Cytol. 68, 97–125.
11. Polliack, A. and Gamliel, H. (1983) Surface Immunocytochemistry of lymphocytes using Scanning Electron Microscopy. In: G.M. Hodges and K.E. Carr (Eds.) Biomedical Research Applications of Scanning Electron Microscopy, Vol. 3. Academic Press, London and New York, pp. 31–77.
12. Echlin, P., Broers, A.N. and Gee, W. (1980) Improved resolution of sputter-coated metal films. SEM Symposium 13/I, 163–170.

13. Murphy, J.A. (1978) Non-coating techniques to render biological specimens conductive. SEM Symposium, 12/II, 175–194.
14. Murphy, J.A. (1980) Non-coating techniques to render biological specimens conductive/1980 update. SEM Symposium, 13/I, 209–220.
15. De Petris, S. (1978) Immunoelectron microscopy and immunofluorescence in membrane biology. In: E.D. Korn (Ed.) Methods in Membrane Biology, Vol. 9. Plenum Press, New York and London, pp. 1–201.
16. Horisberger, M. (1981) Colloidal gold: a cytochemical marker for light and fluorescent microscopy and for transmission and scanning electron microscopy. SEM Symposium, 14/II, 9–31.
17. Hoyer, L.C., Lee, J.C. and Bucana, C. (1979) Scanning immunoelectron microscopy for the identification and mapping of two or more antigens on cell surfaces. SEM Symposium, 12/III, 629–636.
18. Nemanic, M. (1975) On cell surface labelling for the SEM. SEM/IITRI, 8, 342–349.
19. Goodman, S.L., Hodges, G.M. and Livingston, D.C. (1980) A review of the colloidal gold marker system. SEM Symposium, 13/II, 133–146.
20. Horisberger, M. (1979) Evaluation of colloidal gold as a cytochemical marker for transmission and scanning electron microscopy. Biol. Cellul. 36, 253–258.
21. Roth, J. (1982) The protein-A gold (pAg) technique—a qualitative and quantitative approach for antigen localization on thin sections. In: G.R. Bullock and P. Petrusz (Eds.) Techniques in Immunocytochemistry, Vol. 1. Academic Press, London and New York, pp. 108–133.
22. Roth, J. (1983) The colloidal gold marker system for light and electron microscopic cytochemistry. In: G.R. Bullock and P. Petrusz (Eds.) Techniques in Immunocytochemistry, Vol. 2, Academic Press, London and New York, pp. 217–284.
23. Feldherr, C.M. and Marshall, J.M. (1962) The use of colloidal gold for studies of intracellular exchange in amoeba. *Chaos chaos*. J. Cell Biol. 12, 640–645.
24. Faulk, W.P. and Taylor, G.M. (1971) An immunocolloid method for the electron microscope. Immunochemistry 8, 1081–1083.
25. Horisberger, M., Rosset, J. and Bauer, H. (1975) Colloidal gold granules as markers for cell surface receptors in scanning electron microscopy. Experientia 31, 1147–1149.
26. De Mey, J., Moeremans, M., Geuens, G., Nuydens, R. and De Brabander, M. (1981) High resolution light and electron microscopic localization of tubulin with the IGS (immunogold staining) method. Cell Biol. Int. Reports 5, 889–899.
27. Geoghegan, W.D., Scillian, J.J. and Ackerman, G.A. (1978) The detection of human B lymphocytes by both light and electron microscopy utilizing colloidal gold labelled anti-immunoglobulin. Immunol. Commun. 7, 1–12.
28. Horisberger, M. and Vonlanthen, M. (1979) Fluorescent colloidal gold: a cytochemical marker for fluorescent and electron microscopy. Histochemistry 64, 115–118.
29. Tolson, N.D., Boothroyde, B. and Hopkins, C.R. (1981) Cell surface labelling with gold colloid particulates: the use of avidin and staphylococcal protein A-coated gold in conjunction with biotin and fc-bearing ligands. J. Microsc. 123, 215–226.
30. Wagner, M., Roth, J. and Wagner, B. (1976) Gold labelled protectin from Helix pomatia for the localization of blood group A antigen of human erythrocytes by immunofreeze-etching. Exp. Pathol. 12, 277–281.
31. Hoyer, L.C. and Bucana, C. (1982) Principles of Immunoelectron Microscopy. In: J.J. Marchalonis and G.W. Warr (Eds.) Antibody as a Tool. J. Wiley and Sons Ltd., pp. 233–271.
32. Wagner, M. and Wagner, B. (1977) Electron microscopic detection of cryptantigen A42/HP (Friedenreich antigen) on human erythrocytes by means of gold-labelled agglutinin from Helix pomatia. Z. Immun. Forsch. 153, 450–456.
33. Horisberger, M. and Vonlanthen, M. (1979) Multiple marking of cell surface receptors by gold granules; simultaneous localization of three lectin receptors on human erythrocytes. J. Microsc. 115, 97–102.
34. Roth, J. and Binder, M. (1978) Colloidal gold, ferritin and peroxidase as markers for electron microscopic double labelling lectin techniques. J. Histochem. Cytochem. 26, 163–169.

35. Horisberger, M., Rosset, J. and Vonlanthen, M. (1977) Location of glycoproteins on milk fat globule membranes by scanning and transmission electron microscopy using lectin-labelled gold granules. Exp. Cell Res. 109, 361–369.
36. Trejdosiewicz, L.K., Smolira, M.A., Hodges, G.M., Goodman, S.L. and Livingston, D.C. (1981) Cell surface distribution of fibronectin in cultures of fibroblasts and bladder derived epithelium; SEM-immunogold localization compared to immunoperoxidase and immunofluorescence. J. Microsc. 123, 227–236.
37. Baigent, C.L. and Muller, G. (1980) A colloidal gold prepared with ultrasonics. Experientia 36, 472–473.
38. Kent, S.P. and Allen, F.B. (1981) Antibody coated gold particles containing radioactive gold in the demonstration of cell surface markers. Histochemistry 72, 83–90.
39. Frens, G. (1973) Controlled nucleation for the regulation of particle size in monodisperse gold suspensions. Nature Phys. Sci. 241, 20–22.
40. Stathis, E.C. and Fabrikanos, A. (1958) Preparation of colloidal gold. Chem. Ind. (London) 27, 860–861.
41. Goodman, S.L., Hodges, G.M., Trejdosiewicz, L.K. and Livingston, D.C. (1981) Colloidal gold markers and probes for routine application in microscopy. J. Microsc. 123, 201–213.
42. Geoghegan, W.D. and Ackerman, G.A. (1977) Adsorption of horseradish peroxidase, ovomucoid and anti-immunoglobulin to colloidal gold for the indirect detection of Concanavalin A, wheat germ agglutinin and goat anti-human immunoglobulin G on cell surfaces at the electron microscopic level; a new method, theory and application. J. Histochem. Cytochem. 25, 1187–1200.
43. Goodman, S.L., Hodges, G.M., Trejdosiewicz, L.K. and Livingston, D.C. (1979) Colloidal gold probes. A further evaluation. SEM Symposium, 12/III, 619–628.
44. Geoghegan, W.D., Ambegaonkar, S. and Calvanico, W.J. (1980) Passive gold agglutination. An alternative to passive haemagglutination. J. Immunol. Methods 34, 11–21.
45. Slot, J.W. and Geuze, H.J. (1981) Sizing of protein A-colloidal gold probes for immunoelectron microscopy. J. Cell Biol. 90, 533–536.
46. Romano, E.L., Stolinski, C. and Hughes-Jones, N.C. (1974) An antiglobulin reagent labelled with colloidal gold for use in electron microscopy. Immunochemistry 11, 521–522.
47. Montesano, R., Roth, J., Robert, A. and Orci, L. (1982) Non-coated membrane invaginations are involved in binding and internalization of cholera and tetanus toxin. Nature (London) 296, 651–653.
48. Schwab, M.E. and Thoenen, H. (1978) Selective binding, uptake and retrograde transport of tetanus toxin by nerve terminals in the rat iris. J. Cell Biol. 77, 1–13.
49. Horisberger, M. and Rosset, J. (1977) Colloidal gold, a useful marker for transmission and scanning electron microscopy. J. Histochem. Cytochem. 25, 295–305.
50. Bullock, G.R. and Petrusz, P. (Eds.) (1982) Techniques in Immunocytochemistry, Vol. 1. Academic Press, London and New York.
51. Bullock, G.R. and Petrusz, P. (Eds.) (1983) Techniques in Immunocytochemistry, Vol. 2. Academic Press, London and New York.
52. Osborn, M. and Weber, K. (1982) Immunofluorescence and immunocytochemical procedures with affinity purified antibodies: tubulin-coating structures. Methods Cell Biol. 24, 98–132.
53. Sternberger, L. (1979) Immunocytochemistry, 2nd Edn. J. Wiley, New York.
54. Gonda, M.A., Gilder, R.V. and Hsu, K.C. (1979) Immunologic techniques for the identification of virion and cell surface antigens by correlative fluorescence, transmission electron and scanning electron microscopy. SEM Symposium, 12/III, 583–594.
55. Horisberger, M. and Rosset, J. (1977) Gold granules a useful marker for SEM. SEM/IITRI, 10/II, 75–82.
56. Lotan, R. (1979) Qualitative and quantitative aspects of labelling cell surface carbohydrates using lectins as probes. SEM Symposium, 12/III, 549–564.
57. Miller, M.M., Strader, C.D. and Revel, J.P. (1980) Hemocyanin-Protein A, as immunochemical reagent for scanning and transmission electron microscopy. SEM Symposium, 13/II, 125–131.

58. NEMANIC, M.K. (1979) Prospective on cell surface labelling using the hapten-sandwich method: an integrated approach. SEM Symposium, 12/III, 537–547.
59. ROMANO, E.L. and ROMANO, M. (1977) Staphylococcal protein A bound to colloidal gold: A useful reagent to label antigen-antibody sites in electron microscopy. Immunochemistry 14, 711–715.
60. HOPKINS, C.R., BOOTHROYD, B. and GREGORY, H. (1979) Identification of receptors for epidermal growth factors in the electron microscope using an epidermal growth factor—biotin/avidin-gold procedure. Biochem. Soc. Trans. 7, 956–957.
61. NICOLSON, G.L. (1978) Ultrastructural localization of lectin receptors. In: J.K. Koehler (Ed.) Advanced Techniques in Biological Electron Microscopy, Vol. II, Springer-Verlag, Berlin–Heidelberg–New York, pp. 1–38.
62. ROTH, J. (1978) The Lectins: molecular probes in cell biology and membrane research. Exp. Pathol. Suppl. 3, 5–186.
63. OSTRAND-ROSENBERG, S., EDIDIN, M. and WETZEL, B. (1979) Differential localization of H-2 and teratoma-defined antigens on mouse L cell fibroblasts. SEM Symposium, 12/III, 595–600.
64. MOLDAY, R.S., DREYER, W.J., REMBAUM, A. and YEN, S.P.S. (1975) New immunolatex spheres: visual markers of antigens on lymphocytes for scanning electron microscopy. J. Cell Biol. 64, 75–88.
65. BRANDTZAEG, P. (1982) Tissue preparation methods for immunohistochemistry. In: G.R. Bullock and P. Petrusz (Eds.) Techniques in Immunocytochemistry, Vol. 1. Academic Press, London and New York, pp. 1–75.
66. GRZANNA, R. (1982) Light microscopic immunocytochemistry with fixed, unembedded tissues. In: G.R. Bullock and P. Petrusz (Eds.) Techniques in Immunocytochemistry, Vol. 1. Academic Press, London and New York, pp. 183–204.
67. GAMLIEL, H., LEIZEROWITZ, R., GURFEL, D. and POLLIACK, A. (1981) Scanning immunoelectron microscopy of human leukaemia and lymphoma cells: a comparative study of techniques using immunolatex spheres as markers. J. Microsc. 123, 189–199.
68. VAN LEEUWEN, F. (1982) Specific immunocytochemical localization of neuropeptides: a utopian goal? In: G.R. Bullock and P. Petrusz (Eds.) Techniques in Immunocytochemistry, Vol. 1. Academic Press, London and New York, pp. 283–299.
69. PEARSE, A.G.E. (1980) Histochemistry, Theoretical and Applied, 4th Edn., Vol. 1: Preparative and Optical Technology. Churchill Livingstone, London.
70. TAKAMIYA, H., BATSFORD, S. and VOGT, A. (1980) An approach to postembedding staining of protein (immunoglobulin) antigen embedded in plastic: prerequisites and limitations. J. Histochem. Cytochem. 28, 1041–1049.
71. HORISBERGER, M. and VONLANTHEN, M. (1977) Location of mannan and chitin on thin section of budding yeasts with gold markers. Arch. Microbiol. 115, 1–7.
72. HORISBERGER, M., ROSSET, J. and BAUER, H. (1976) Localisation of mannan at the surface of yeast protoplasts by scanning electron microscopy. Arch. Microbiol. 109, 9–14.
73. KRAEHENBUHL, J.P., RACINE, L. and GRIFFITHS, G.W. (1980) Attempts to quantitate immunocytochemistry at the electron microscope level. Histochem. J. 12, 317–332.
74. FINLAY, J.C.W. and PETRUSZ, P. (1982) The use of proteolytic enzymes for improved localisation of tissue antigens with immunocytochemistry. In: G.R. Bullock and P. Petrusz (Eds.) Techniques in Immunocytochemistry, Vol. 1. Academic Press, London and New York, pp. 239–249.
75. PETRUSZ, P., ORDRONNEAU, P. and FINLEY, J.C.W. (1980) Criteria of reliability for light microscopic immunocytochemical staining. Histochem. J. 12, 333–348.
76. CARTER, D.P. and WOFSY, L. (1976) Immunospecific labelling of mouse lymphocytes in the scanning electron microscope. J. Supramol. Struct. 5, 139–153.
77. NEWELL, D.G. (1980) The white cell system. In: G.M. Hodges and R.C. Hallowes (Eds.) Biomedical Research Applications of Scanning Electron Microscopy, Vol 2. Academic Press, London and New York, pp. 219–303.
78. WETZEL, B., CANNON, G.B., ALEXANDER, E.L., ERIKSON, JR., B.W. and WESTBROOK, E.W. (1974) A critical approach to the scanning electron microscopy of cells in suspension. SEM/IITRI 7, 581–588.
79. BELL, P.B. and REVEL, J.P. (1980) Scanning Electron Microscope Application to cells and tissues

in culture. In: G.M. Hodges and R.C. Hallowes (Eds.) Biomedical Research Applications of Scanning Electron Microscopy, Vol. 2. Academic Press, London and New York, pp. 1–63.

80. MURPHY, J.A. (1982) Considerations, materials and procedures for specimen mounting prior to scanning electron microscopic examination. SEM Symposium, 15/II, 657–696.
81. WATERMAN, R.E. (1980) Preparation of embryonic tissues for SEM. SEM Symposium, 13/II, 21–44.
82. MARCHOL, J.B., BRELINSKA, R. and HERBERT, D.C. (1982) Analysis of colloidal gold methods for labelling proteins. Histochemistry 76, 565–575.
83. KARNOVSKY, M.J. (1965) A formaldehyde-glutaraldehyde fixative of high osmolarity for use in electron microscopy. J. Cell Biol. 27, 137A–138A.
84. TONER, P.G. and CARR, K.E. (1979) The Digestive System. In: G.M. Hodges and R.C. Hallowes (Eds.) Biomedical Research Applications of Scanning Electron Microscopy, Vol. 1, Academic Press, London and New York, pp. 203–272.
85. HODGES, G.M., SMOLIRA, M.A. and TREJDOSIEWICZ, L.K. (1982) Urothelium-specific antibody and lectin surface mapping of bladder urothelium. Histochem. J. 14, 755–766.
86. HEYDERMAN, E. (1979) Immunoperoxidase technique in histopathology: applications, methods and controls. J. Clin. Pathol. 32, 971–978.
87. BEN-SHAUL, Y., OPHIR, I., COHEN, E. and MOSCONA, A. (1977) SEM study of dissociated embryonic cell surface activity. SEM/IITRI, 10/II, 29–35.
88. BRATEN, T. (1976) Concanavalin A-binding sites on the surface of gametes and zygotes of green alga Ulva mutabilis studied with the scanning electron microscope. Sixth Eur. Cong. Elec. Micros. Jer., 192–194.
89. BROWN, S.S. and REVEL, J.P. (1976) Reversibility of cell surface label rearrangement. J. Cell Biol. 68, 629–641.
90. GONDA, M.A., GILDER, R.V. and HSU, K.C. (1979) An unlabelled antibody macromolecule technique using hemocyanine for the identification of type B and type C retrovirus envelope and cell surface antigens by correlative fluorescence, transmission electron, and scanning electron microscopy. J. Histochem. Cytochem. 27, 1445–1454.
91. JAN, L.Y. and REVEL, J.P. (1975) Hemocyanin-antibody labelling of rhodopsin in mouse retina for a scanning electron microscope study. J. Supramol. Struct. 3, 61–66.
92. MILLER, M.M. and TEPLITZ, R.L. (1978) Detection of embryonic surface antigens using the SEM. SEM Symposium, 11/II, 893–898.
93. MILLER, M.M., STRADER, C.D., RAFTERY, M.A. and REVEL, J.P. (1981) Hemocyanin linked to protein A as an immunochemical labelling reagent for electron microscopy. J. Histochem. Cytochem. 29, 1322–1327.
94. SMITH, S.B. and REVEL, J.P. (1972) Mapping of Concanavalin binding sites on the surface of several cell types. Dev. Biol. 27, 434–441.
95. WELLER, N.K. (1974) Visualization of Concanavalin A-binding sites with scanning electron microscopy. J. Cell Biol. 63, 699–707.
96. GONDA, M.A., ARTHUR, L.O., ZERE, V.H., FINE, D.L. and NAGASHIMA, K. (1976) Surface localization of virus production on a glucocorticoid-stimulated oncornavirus-producing mouse mammary tumour cell line by scanning electron microscopy. Cancer Res. 36, 1084–1093.
97. HÄMMERLING, U., STACKPOLE, C.W. and KOO, G. (1973) Hybrid antibodies for labelling cell surface antigen. Methods Cancer Res. 9, 255–282.
98. KAY, M.M. (1978) Multiple labelling technique for immuno-scanning electron microscopy. In: M.A. Hayat (Ed.) Principles and Techniques of Scanning Electron Microscopy, Vol. 6. Van Nostrand Reinhold, New York, pp. 338–357.
99. MARCHALONIS, J.J., BUCANA, C. and HOYER, L. (1978) Visualization of a guinea pig T lymphocyte surface component cross-reactive with immunoglobulin. Science 199, 433–435.
100. TOKUNAGA, J., FUJITA, T., HATTORI, A. and MULLER, J. (1976) Scanning electron microscopic observations of immunoreactions on the cell surface: analysis of Candida albicans cell wall antigens by the immunoferritin method. SEM/IITRI, 9/I, 301–310.
101. UMEDA, A. and AMAKO, K. (1977) Evaluation of ferritin, phi X 174 and pyocin as the morphological

markers for the immunoelectron microscopy by the high resolution scanning electron microscope. J. Electron Microsc. 26, 87–93.

102. Hämmerling, U., Polliack, A., Lampen, N., Sabety, M. and de Harven, E. (1975) Scanning electron microscopy of tobacco mosaic virus labelled lymphocyte surface antigens. J. Exp. Med. 141, 518–523.

103. Kumon, H. (1976) Morphologically recognizable markers for scanning immuno-electron microscopy. II. An indirect method using T_4 and TMV. Virology 74, 93–103.

104. Kumon, H., Uno, F. and Tawara, J. (1976) Morphological studies on viruses by SEM and an approach to labelling. SEM/IITRI, 9/II, 85–92.

105. Lipscomb, M.F., Holmes, K.V., Vitetta, E.S., Hämmerling, U. and Uhr, J.W. (1975) Cell surface immunoglobulin. XII. Localization of immunoglobulin on murine lymphocytes by scanning immunoelectronmicroscopy. Eur. J. Immunol. 5, 255–259.

106. Nemanic, M.K., Carter, D.P., Pitelka, D.R. and Wofsy, L. (1975) Hapten-sandwich labelling. II. Immunospecific attachment of cell surface markers suitable for scanning electron microscopy. J. Cell Biol. 64, 311–321.

107. Hämmerling, U., Aoki, T., Wood, H.A., Old, L.J., Boyse, E.A., and De Harven, E. (1969) New visual marker of antibody for electromicroscopy. Nature (London) 223, 1158–1159.

108. Matter, A., Lisowska-Bernstein, B., Ryser, J.E., Lameline, J.P. and Vassalli, P. (1972) Mouse thymus-independent and thymus-derived lymphoid cells. II. Ultrastructural studies. J. Exp. Med. 66, 198–200.

109. Kay, M.M. (1975) Multiple labelling technique used for kinetic studies of activated human B lymphocytes. Nature (London) 254, 424–426.

110. Van Ewijk, W. and De Vries, E. (1977) Cell surface labelling of mononuclear cells with antisera associated to turnip yellow mosaic virus or alfalfa mosaic virus particles. A freeze-etch study. Histochem. J. 9, 329–340.

111. De Harven, E., Pla, D. and Lampen, N. (1979) Labelling of lymphocyte surface immunoglobulins with T_4 as a marker for scanning electron microscopy. SEM Symposium, 12/III, 611–618.

112. Kumon, H., Uno, F. and Tawara, J. (1976) A morphologically recognizable marker for scanning immunoelectron microscopy. I. T_4-bacteriophage. Virology 70, 554–557.

113. Tsutsui, K., Ichikawa, H., Kumon, H., Uno, F. and Tawara, J. (1978) Methods for visualization of Concanavalin A receptors on lymphoid cells by scanning electron microscopy. J. Electron Microsc. 27, 321–323.

114. Ben-Bassat, H., Mitrani-Rosenbaum, S., Gamliel, H., Naparstek, E., Leizerowitz, R., Korkesh, A., Sagi, M., Voss, R., Kohn, G. and Polliack, A. (1980) Establishment in continuous culture of a T-lymphoid cell line (HD-Mar) from a patient with Hodgkin's lymphoma. Int. J. Cancer 25, 583–590.

115. Ben-Shaul, Y., Hausman, R.E. and Moscona, A.A. (1979) Visualization of a cell surface glycoprotein, the retina cognin, on embryonic cells by immuno latex labelling and scanning electron microscopy. Dev. Biol. 72, 89–101.

116. Ben-Shaul, Y., Hausman, R.E. and Moscona, A.A. (1980) Age-dependent differences in cognin regeneration on embryonic retina cells: immunolabelling and SEM studies. Dev. Neurosci. 3, 66–74.

117. Gamliel, H. and Polliack, A. (1979) Scanning immunoelectron microscopic markers. Israel J. Med. Sci. 15, 639–646.

118. Gamliel, H. and Polliack, A. (1981) Positive identification of human leukaemic cells with scanning immuno-electron microscopy, using antibody coated polystyrene (latex) beads as markers. Scand. J. Haematol. 26, 297–305.

119. Linthicum, D.S., Sell, S., Wagner, R.M. and Trefts, P. (1974) Scanning immunoelectron microscopy of mouse B and T lymphocytes. Nature (London) 252, 1973–1976.

120. Linthicum, D.S. and Sell, S. (1975) Topography of lymphocyte surface immunoglobulin using scanning immunoelectron microscopy. J. Ultrastruct. Res. 51, 55–68.

121. LoBuglio, A.F., Rinehart, J.J. and Balcerzak, S.P. (1972) A new immunologic marker for scanning electron microscopy. SEM/IITRI, 5, 313–320.

122. Mannweiler, K. and Rutter, G. (1973) Comparative SEM and TEM studies on normal and

virus-infected culture cells (attempts to label virus-specific surface antigens) SEM/IITRI, 6, 513–520.
123. Molday, R.S. (1976) A scanning electron microscope study of Concanavalin A receptors on retinal rods labelled with latex microspheres. J. Supramol. Struct. 4, 549–557.
124. Molday, R.S., Dreyer, W.J., Rembaum, A. and Yen, S.P.S. (1974) Latex spheres as markers for studies of cell surface receptors by scanning electron microscopy. Nature (London) 249, 81–83.
125. Quash, G.A., Niveleau, S., Aupoix, M. and Greenland, T. (1976) Immunolatex visualization of cell surface Forssman and polyamine antigens. Exp. Cell Res. 98, 253–261.
126. Riviere, G.R., Cotton, W.R. and Derkowski, J.L. (1976) Demonstration of human salivary antibodies by latex spheres as immunologic markers. SEM/IITRI, 9/II, 67–74.
127. Fuchs, H. and Bächi, T. (1975) Scanning electron microscopical demonstration of respiratory syncytial virus antigens by immunological markers. J. Ultrastruct. Res. 52, 114–119.
128. Margel, S., Zisblatt, S. and Rembaum, A. (1979) Polyglutaraldehyde: A new reagent for coupling proteins to microspheres and for labelling cell-surface receptors. II. Simplified method by means of nonmagnetic and magnetic polyglutaraldehyde microspheres. J. Immunol. Methods 28, 341–353.
129. Molday, R.S. (1976) Immunolatex spheres as cell surface markers for scanning electron microscopy. In: M.A. Hayat (Ed.) Principles and Techniques of Scanning Electron Microscopy, Vol. 5. Van Nostrand Reinhold, New York, pp. 53–77.
130. Rembaum, A. and Dreyer, W.J. (1980) Immunomicrospheres: reagents for cell labelling and separation. Science 208, 364–368.
131. Peters, K.R., Gschwender, H.H., Haller, W. and Rutter, G. (1976) Utilization of high resolution spherical marker for labelling of virus antigens at the cell membrane in conventional scanning electron microscopy. SEM/IITRI, 9/II, 75–83.
132. Peters, K.R., Rutter, G., Gschwender, H.H. and Haller, W. (1978) Derivatized silica spheres as immunospecific markers for high resolution labelling in electron microscopy. J. Cell Biol. 78, 309.
133. Kronick, P., Campbell, G. and Joseph, K. (1978) Magnetic microspheres prepared by redox polymerization used in a cell separation based on gangliosides. Science 200, 1074–1076.
134. Molday, R.S., Yen, S.P.S. and Renbaum, A. (1977) Application of magnetic microspheres in labelling and separation of cells. Nature (London) 268, 437–438.
135. Baccetti, T.J. and Burrini, A.G. (1977) Detection of Concanavalin A receptors by affinity to peroxidase and iron dextran by scanning and transmission electron microscopy and X-ray microanalysis. J. Microsc. 109, 203–209.
136. Dutton, A.H., Tokuyasu, K.T. and Singer, S.J. (1979) An iron dextran-antibody conjugate for use in high resolution immunoelectron microscopy. J. Cell Biol. 83, 475a.
137. Marshall, P.R. and Rutherford, D. (1971) Physical investigation on colloidal iron-dextran complexes. J. Coll. Interface Sci. 37, 390–402.
138. Hodges, G.M. and Muir, M.D. (1974) X-ray spectroscopy in the scanning electron microscope study of cell and tissue culture material. In: T. Hall, P. Echlin and R. Kaufmann (Eds.) Microprobe Analysis as Applied to Cells and Tissues. Academic Press, London and New York, pp. 277–291.
139. Horisberger, M., Bauer, H. and Bush, D.A. (1971) Mercury-labelled Concanavalin A as a marker in electron microscopy—localization of mannan in yeast cell walls. FEBS Lett. 18, 311–314.
140. Hodges, G.M. and Muir, M.D. (1976) Scanning electron microscope autoradiography. In: M.A. Hayat (Ed.) Principles and Techniques of Scanning Electron Microscopy: Biological Applications, Vol. 5. Van Nostrand, Reinhold, New York, pp. 78–93.
141. Bretton, R., Clark, D.A. and Nathanson, L. (1973) The cytochemical detection of Concanavalin A-binding sites on cell surfaces by scanning electron microscopy. J. Microsc. 17, 93–96.
142. Hartman, A.L. and Nakane, P.K. (1981) Intracellular localization of antigens with backscatter mode of SEM using peroxidase-labelled antibodies. SEM Symposium, 14/II, 33–44.
143. McKeever, P.E. and Spicer, S.S. (1979) Demonstration of immune complex receptors on macrophage surfaces and erythrocytic endogenous peroxidase with correlated markers for light microscopy, scanning electron microscopy and transmission electron microscopy. SEM Symposium, 12/III, 601–610.

144. McKeever, P.E., Spicer, S.S., Brissie, N.T. and Garvin, A.J. (1977) Immune complex receptors on cell surfaces. III. Topography of macrophage receptors demonstrated by new scanning electron microscopic peroxidase marker. J. Histochem. Cytochem. 25, 1063–1068.
145. Nakane, P.K. and Hartman, A.L. (1980) Immunocytochemical localization of intracellular antigens with SEM. Histochem. J. 12, 435–447.
146. Springer, E.L., Riggs, J.L. and Hackett, A.J. (1974) Viral identification by scanning electron microscopy of preparations stained with fluorescein labelled antibody. J. Virol. 14, 1623–1626.
147. Abraham, J.L. (1979) Documentation of environmental particulate exposures in humans using SEM and EDAX. SEM Symposium, 12/II, 751–766.
148. Ito, S., Hattori, A., Ito, S., Ihzumi, T., Sanada, M. and Matsuoko, M. (1978) Surface immunoglobulins of human lymphocytes. Scand. J. Haematol. 20, 399–409.
149. Stenman, S., Wartiovara, J and Vaheri, A. (1977) Changes in the distribution of a major fibroblast protein, fibronectin, during mitosis and interphase. J. Cell Biol. 74, 453–467.
150. Wartiovara, J., Linder, E., Ruosilahti, E. and Vaheri, A. (1974) Distribution of fibroblast surface antigen. Association with fibrillar structures of normal cells and loss upon viral transformation. J. Exp. Med. 140, 1522.
151. Kumon, H., Uno, F., Saburi, Y. and Tawara, J. (1978) A morphologically recognizable marker for scanning immunoelectron microscopy. IV. Antibody induced redistribution of influenza virus antigens on the cell surface. J. Electron. Microsc. 27, 215–222.
152. Kumon, H., Uno, F., Uebo, O., Fujio, K., Uno, N. and Tawara, J. (1977) A morphologically recognizable marker for scanning immuno-electron microscopy. III. An application to virus research. J. Electron Microsc. 26, 349–350.
153. Horisberger, M., Vonlanthen, M. and Rosset, J. (1978) Localization of α-galactomannan and wheat germ agglutinin receptors in Schizosaccharomyces pombe. Arch. Microbiol. 119, 107–111.
154. Burgess, J. and Linstead, P.J. (1976) Ultrastructural studies of the binding of Concanavalin A to the plasmalemma of higher plant protoplasts. Planta (Berl.) 130, 73–79.
155. Maher, P. and Molday, R.S. (1979) Differences in the redistribution of Concanavalin A and wheat germ agglutinin binding sites on mouse neuroblastoma cells. J. Supramol. Struct. 10, 61–77.
156. Horisberger, M. and Rosset, J. (1976) Localization of wheat germ agglutinin receptor sites on yeast cells by scanning electron microscopy. Experientia 32, 998–1000.
157. Horisberger, M. and Vonlanthen, M. (1978) Simultaneous localization of hepatic binding protein specific for galactose and of galactose-containing receptors on rat hepatocytes. J. Histochem. Cytochem. 26, 960–966.
158. Horisberger, M., Rosset, J. and Vonlanthen, M. (1978) Location of lectin receptors on rat hepatocytes by transmission and scanning electron microscopy. Experientia 34, 274–276.
159. Horisberger, M. (1978) Agglutination of erythrocytes using lectin-labelled spacers. Experientia 34, 721–722.
160. Molday, R.S., Jaffee, R. and McMahon, D. (1976) Concanavalin A and wheat germ agglutinin receptors on Dictyostelium discoideum. Their visualization by scanning electron microscopy with microspheres. J. Cell Biol. 71, 314–322.
161. Molday, R.S. and Molday, L. (1979) Identification and characterization of multiple forms of rhodopsin and minor proteins in frog and bovine outer segment disc membranes. J. Biol. Chem. 254, 4653–4660.
162. Bayer, E.A., Wilchek, M. and Skutelsky, E. (1976). Affinity cytochemistry: the localization of lectin and antibody receptors on erythrocytes via avidin-biotin complex. FEBS Lett. 68, 240–244.
163. Heitzman, H. and Richards, F.M. (1974) Use of the avidin-biotin complex for specific staining of biological membranes in electron microscopy. Proc. Natl. Acad. Sci., U.S.A. 71, 3537–3541.
164. Wofsy, L. (1979) Hapten-antibody conjugates as probes of the lymphocyte surface. SEM Symposium, 12/III, 565–572.
165. Larsson, L.I. (1979) Simultaneous ultrastructural demonstration of multiple peptides in endocrine cells by a novel immunocytochemical method. Nature (London) 282, 743–746.
166. De Mets, M. (1974) Cathodoluminescence of organic chemicals In: M.A. Hayat (Ed.) Principles and Techniques of Scanning Electron Microscopy, Vol. 2. Van Nostrand, Reinhold, New York, pp. 1–20.

Immunolabelling for Electron Microscopy (Polak/Varndell, eds)

CHAPTER 16

Combined quick-freeze and freeze-drying techniques for improved electron immunocytochemistry

Ronald W. Dudek[1], Ian M. Varndell[2] and Julia M. Polak[2]

[1]Department of Anatomy, University of East Carolina, Greenville, NC, U.S.A. and [2]Department of Histochemistry, Royal Postgraduate Medical School, Hammersmith Hospital, Du Cane Road, London W12 0HS, U.K.

CONTENTS

1. INTRODUCTION

Currently, a considerable number of procedures are available for the immunocytochemical localisation of tissue-bound antigens at the electron microscopical level. With only minor modifications to the universally practised routine method of biological tissue preparation (fixation, dehydration, infiltration, embedding) a wide variety of antigens, notably proteins, can be identified in ultrathin resin sections using fairly simple post-embedding immunocytochemical techniques, the majority of which are described in detail in this book. One, and probably the most important, preparative modification is the omission of osmium tetroxide as a secondary fixative and contrasting agent from tissue destined for immunocytochemistry. Consequently, many published accounts, particularly those in which pre- and post-mortem pathological material has been investigated, demonstrate adequate antigenic preservation but largely at the expense of morphological appearance.

Several groups have attempted to improve morphological preservation whilst still retaining high levels of tissue antigenicity by introducing modifications to the basic technique. One such example is the unmasking of protein antigens in osmicated tissue using sodium metaperiodate by Bendayan and Zollinger [1]. On the whole, however, modifications to the basic techniques have not resulted in radical improvements except, perhaps, in selected model systems.

1.1. New approaches

The introduction of cryoultramicrotomy, its development into a reliable and reproducible technique, and the application of immunocytochemistry to frozen ultrathin sections has revolutionised antigen localisation studies at the electron microscopical level (see Chapter 6, pp. 71–82).

Similarly, the development of new low temperature embedding media, such as Lowicryl (see Chapters 3, pp. 29–36 and 9, pp. 113–121), has allowed structural integrity, particularly of membranes, to be retained in addition to protecting the viability of many antigens [2, 3] which were previously undetected in heat-polymerised resin sections.

One other approach, again utilising low temperature processing, is the combination of quick-freeze fixation and freeze-drying with electron immunocytochemistry which is described in some detail in the following sections using, as a model system, cultured rat pancreatic islets.

2. Quick-freeze fixation and freeze-drying

Briefly, this technique involves the ultrarapid cooling of tissue to liquid nitrogen temperature followed by dehydration in vacuo with slow warming to room temperature. The tissue is then exposed to osmium tetroxide vapour before infiltration and embedding in Araldite or Epon. Variations to this schedule include the replacement of osmium tetroxide with paraformaldehyde or parabenzoquinone vapour at 60°C followed by embedding in Lowicryl resin.

2.1. Why freezing techniques?

As previously mentioned in this chapter (section 1.1) several antigens have now been detected in tissue embedded in low temperature curing resins, notably

Lowicryl, when they were originally thought to be absent, presumably because they are heat-labile in the presence of mono- or oligomeric thermosetting resin components. This is one reason for resorting to freezing techniques, but there are several others:

a) Fixatives such as glutaraldehyde and formaldehyde cannot cross-link all biological macromolecules. Subsequent washing of the tissue and dehydration processes may re-distribute, or even elute, these diffusible substances. There is some evidence that up to 90% of immunoreactive luteinising hormone-releasing hormone (LHRH) [4], somatostatin [5] and vasopressin [6] can be extracted during ethanol dehydration.
b) The action of fixatives is not irreversible (see Chapter 6, pp. 71–82); immobilised macromolecules may be released from "fixed" tissue. Of particular importance in this case are degradative enzymes which may retain [7], or even regain, partial activity after fixation. It is also interesting that considerable differences in morphological appearance can be observed between serial ultrathin sections exposed to liquid media for varying lengths of time (for example, 6–24 h). In general, the longer the exposure period the higher the contrast, as cytosolic and organellar (for example, mitochondrial) components appear to lose the ability to stain with heavy metal salts.
c) Chemical fixation induces drastic, but unpredictable stresses in terms of effect and duration, on cellular and organellar membranes which may lead to changes in intraorganellar and/or intracellular volume. The efficacy of membranes as a barrier to the diffusion of fixatives is clearly differential, but also poorly understood. Thus apparent polarisations of cytosolic components may be a cellular response to the onset of fixative permeation.

Although liquid fixatives, particularly the aldehyde-based solutions, result in the creation of acceptable and generally reproducible architectural preservation, many electron microscopists will be able to add reasons to those given above for not using liquid fixatives in specific circumstances. Quick-freezing permits near instantaneous preservation of cellular structures without the artefacts induced by exposure to liquid fixatives.

2.2. Quick-freeze fixation

It is now 20 years since quick-freezing techniques were successfully applied to electron microscopical studies, although the general concept dates back to the turn of the century [8]. Indeed, rapid freezing for electron microscopical specimen preparation was introduced by Sjostrand [9, 10] four decades ago. It was not until 1964, however, when van Harreveld [11] introduced a quick-freezing device, that the significance of quick-freezing for electron microscopy was realised. Subsequent modifications [12, 13] to van Harreveld's device enable samples to be frozen instantaneously by rapid application against a copper block cooled to very low temperatures with, for example, liquid helium. One of these modifications [14] involves the freezing of a sample mounted on a thin cover glass.

In order to obtain the preservation of subcellular detail which reflects, as closely as possible, the appearance in the "living" condition by quick-freeze fixation, the tissue must be frozen so rapidly (10^4–10^6K s^{-1}) that most water molecules in the sample pass from the relatively amorphous liquid state (water) to the relatively amorphous solid state (vitreous ice) without forming detectable crystalline arrays (for review see Rash [15]). Slower freezing rates ($< 10^4$K s^{-1}) induce the water molecules to migrate to, and deposit in, ordered ice crystal lattices. Thus there must be no delay in the rate of heat transfer from sample to cryogen. Quick-freezing can be achieved by applying the sample to a highly polished, ultracold (liquid nitrogen—77 K or liquid helium—4 K), silver or copper bar (see section 2.3.1) either mechanically [11, 12] or manually [16]. However, to ensure reproducible high-quality ultrastructural preservation it is essential that the sample is applied to the cryosurface rapidly and without discontinuous contact ("bouncing") [17]. To prevent millisecond bounces of the tissue against the cryosurface a dampened mechanical device was designed (Fig. 1) which facilitates rapid delivery of sample to the freezing surface whilst ensuring low impact momentum of the tissue.

2.3. Preparation of the quick-freeze assembly

We recommend the "Gentleman Jim" quick-freezing assembly (Quick-Freezing Devices, P.O. Box 27038, South Station, Baltimore, MD 21230, U.S.A.; Fig. 1) because of its simplicity and dependability. The specific details of this instrument have been described previously [17, 18].

Briefly, the quick-freezing device is levelled on the laboratory bench. The digital electronic bounce monitor is engaged to check that the assembly is bounce-free. The bounce monitor should register 0 before the freezing of tissue is attempted. If the bounce monitor does not register zero:

a) change the level of glycerine;
b) adjust the damper, or
c) adjust the two screws on the striker near the probe tip (Fig. 1).

The stainless steel Dewar is filled with liquid nitrogen to the top. The polished copper bars (see section 2.3.1) are dropped into the liquid nitrogen and allowed to cool. At first, the rate of nitrogen evolution is high because of the steep temperature gradient between the copper bar and the liquid nitrogen. The process of nucleate boiling occurs in approximately 1 min. This can be observed by a sudden rush of nitrogen vapour from the top of the Dewar. The copper bars are left in the Dewar for 45 sec after nucleate boiling to ensure that they have reached the temperature of liquid nitrogen. A considerable amount of nitrogen boils away during the cooling of the copper bars, therefore it is important that the Dewar should be filled maximally at the start of the cooling process. The Dewar is aligned in the freezing assembly so that the probe tip and tissue will fall unobstructed onto the copper surface. Once the copper bars are at liquid nitrogen temperature and aligned in the freezing assembly the sample should be frozen without delay,

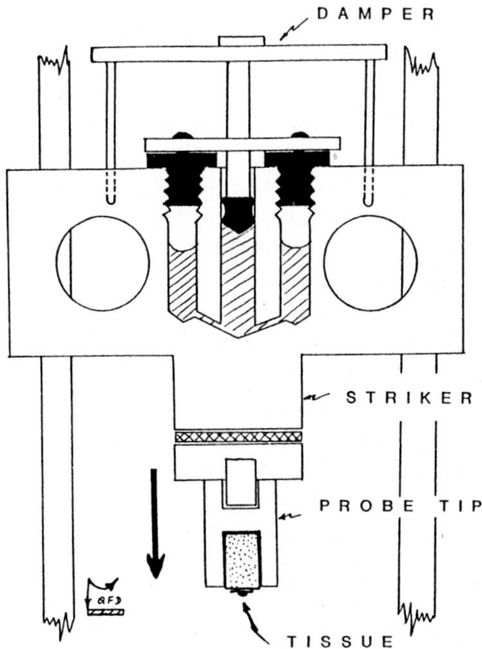

Fig. 1. An enlargement of the striker/damper mechanism shows that the damper consists of a weighted plunger inserted into a "W"-shaped, glycerine-filled reservoir in the striker. Both the striker and damper fall freely by gravity along stainless steel rods until the tissue and probe tip come into contact with the copper surface. The removable probe tip is attached to the striker/damper mechanism by a magnet. The slot in the probe tip is filled with P-990 foam. (Diagram by A. F. Boyne.)

otherwise one risks the chance of condensation forming on the surface of the metal which severely limits the rate of heat transfer.

2.3.1. Preparation of the copper bars

The surface of the copper bar is hand-polished with a 0.05 μm alumina suspension and Alpha A polishing cloth (both from Fisher, Pittsburgh, PA, U.S.A.). Jeweller's rouge, lens paper and acetone work well also. After polishing, the surface of the copper bar is rinsed with a stream of distilled water and dried carefully with filter paper. A rayon-velvet polishing cloth is used for the final polishing. If deep scratches or nicks occur on the freezing surface, more radical polishing methods with decreasing grades of Emery paper, or even machining, may be necessary.

If the bars are machine-polished the following protocol is recommended. Using an Ecomet I mechanical polisher (Buehler Ltd., Evanston, IL 60204, U.S.A.) the bars are polished with a diamond paste (Metadi II, 6 μm; also from Buehler) applied to a nylon polishing cloth together with AB Automet lapping oil. The final polish is obtained with Gamma Micropolish Alumina No. 3B (Buehler). This is followed by rinsing the bar in hot water which is then dried with SPI Polywipers

(SPI, P.O. Box 342, West Chester, PA, U.S.A.). A distinct improvement is evident when compared to hand polishing.

2.4. Handling and freezing of tissue

A limitation of applying quick-freezing to whole tissues which are covered by epithelia is that the interior needs to be sliced open (with a razor blade) so that the region which is applied to the copper surface includes the cells of interest. There is then the possibility that the mechanical trauma of slicing might induce artifacts which will be prominent in the well-frozen edge. A useful way of circumventing this problem is to use cultured or isolated cells.

2.4.1. Preparation of pancreatic islets

Pancreatic islets are isolated from rat, or other species, pancreata by treatment with Serva collagenase (Accurate Chemical and Scientific Corporation, Hicksville, NY, U.S.A.). The islets are collected and cultured for 7 days in RPMI 1640 culture medium containing 11.1 mM glucose, 10% heat-inactivated foetal bovine serum, 20 mM HEPES buffer, and antibiotics as described by Hellerstrom et al. [19]. The tissue culture step was instituted to permit cells at the islet periphery to undergo repair from any damage that may arise during the isolation procedure. It must be remembered that optimal quick-freeze fixation occurs only 15–20 μm deep to the surface of the islet. It is desirable not to waste this limited area by quick-freezing damaged cells. During or after the 7-day culture period the islets can be manipulated according to individual experimental design.

Using cultured islets of Langerhans we have obtained excellent ultrastructural morphology [20] (Fig. 2). In addition, exciting results have been obtained using monolayer cultures of cells grown on Thermanox coverslips or Cytodex micro-carriers (A.F. Boyne, personal communication).

2.4.2. Preparation of tissue holding probe tips

A piece of foam (No. P-990, 0.25 inch thick; Illbruk U.S.A., 3800 Washington Avenue North, Minneapolis, MN 55412, U.S.A.) is placed into a slotted probe tip (Fig. 1). The foam should extend 1–2 mm above the surface of the probe tip. A piece of aluminium foil is glued onto the surface of the foam.

2.4.3. Freezing of islets

We place 20–30 islets on a 3 × 5 mm piece of filter paper with the top right corner cut diagonally to mark the side the islets are on. The filter paper/islets is placed onto the aluminium foil and foam in the probe tip (see section 2.4.2) using forceps. The excess media or buffer is blotted off with filter paper and the probe tip is placed onto the striker (Fig. 1). The cap covering the copper metal surface is removed and the pin pulled to release the striker and damper mechanism (Fig. 1). After the tissue has been applied to the copper surface, we wait approximately 15 sec to chill the foil and foam. The probe tip is then held with large forceps whilst the striker and damper mechanism is re-pinned.

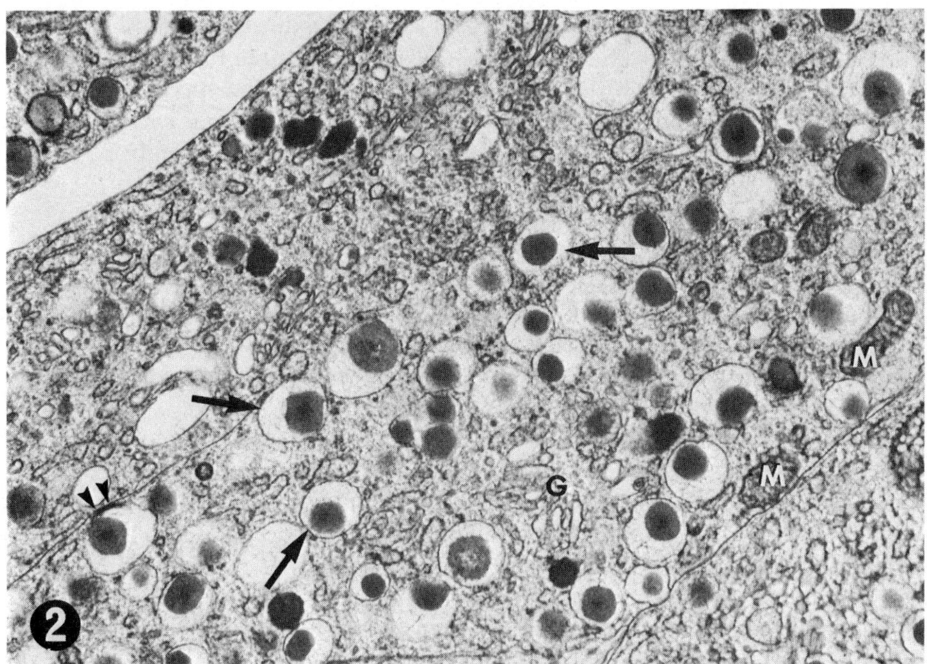

Fig. 2. Electron micrograph of quick-freeze fixation freeze-dried cultured rat pancreatic beta cell. Note the large dense-cored secretory granules (arrowed), mitochondria (M), Golgi (G) and cell junction (arrowheads). Osmium vapour fixation; uranyl acetate and lead citrate counterstains. Magnification = ×28,560.

The probe tip is then transferred to a container of liquid nitrogen using pre-cooled forceps. The frozen islet groups are then stored under liquid nitrogen in a polystyrene chest. Between each freeze the surface of the copper bar is gently wiped with a piece of P-990 foam to remove any condensed ice. The metal surface is used to freeze only two pieces of tissue. After this, the copper bar is removed from the Dewar and warmed with a heat-blow gun to room temperature. This warming takes about 6 min. The surface of the copper is then gently wiped with a rayon-velvet cloth. There should be no need to re-polish the bar. The copper bar is then re-equilibrated in liquid nitrogen.

After all the islet treatment groups have been frozen, a shallow polystyrene chest containing a "dissecting altar" is filled with liquid nitrogen. The "altar" is a cardboard box securely fastened to the bottom of the polystyrene chest with syringe needles. The level of liquid nitrogen is maintained just above the altar. The foam and tissue from the probe tip is transferred to the altar using pre-cooled forceps. The aluminium foil is carefully removed from the foam using a scalpel. It is imperative that the tissue is fully submerged below the level of the liquid nitrogen throughout this procedure. Using fine forceps with the points chilled to liquid nitrogen temperature, the aluminium foil with the filter paper and frozen islets is placed in a stainless steel "denture" and covered with a wire mesh cap. The "denture" is a semi-circle with five slots in which the frozen tissue is placed. The size of the denture is important because it must fit into the specimen trough of the Coulter-Terracio freeze drier. We routinely use two dentures each having five slots, therefore in one experiment we are able to handle ten pieces of frozen tissue.

2.5. Freeze-drying of islets

We recommend the Coulter-Terracio freeze-dry apparatus (Ladd, P.O. Box 1005, Burlington, VT 05402, U.S.A.) for freeze-drying pancreatic islets. Since the instruction manual amply describes the use of this apparatus and a discussion of its use has been published elsewhere [16, 21], we will not discuss its use in great detail. However, we will comment on some aspects indigenous to islets.

Firstly, as islets are small (100–500 μm), they cannot be manipulated with forceps. Consequently it is necessary to put 20–30 islets onto a small piece of filter paper using a finely-drawn pipette. Although the filter paper can be easily handled with forceps it has a tendency to float away in liquid nitrogen. In order to circumvent this, a slotted stainless steel denture with wire mesh caps was constructed as previously described. Secondly, the total time necessary to dehydrate and bring pancreatic islets to room temperature is 26–28 h (Fig. 3). Thirdly, freeze-dried islets may be exposed to osmium tetroxide vapour. We found that 10–15 min exposure to osmium vapour adequately imparts electron density to the islets and stabilises the membranes. In addition, islets may be exposed to paraformaldehyde or parabenzoquinone vapour for 1 h at 60°C. Fourthly, we use flat-embedding moulds and Epon 812/Araldite 502 to embed quick-freeze fixed and freeze-dried islets. It is important to orientate the islets correctly in the flat-embedding mould. The filter paper is orientated so that the islets are facing uppermost. This allows the islets to be sectioned perpendicular to the frozen surface so that every section contains a 10–15 μm thick area of well-fixed islet tissue.

3. Electron immunocytochemistry

In order to evaluate the use of quick-freeze fixation freeze-dried tissue for electron immunocytochemistry both the single [22, 23] and double [24] immunogold staining procedures were employed.

3.1. Single immunogold staining procedure [22, 23]

Ultrathin sections showing silver-grey interference colours (60–90 nm) of the Epon/Araldite-embedded freeze-dried islets were mounted onto cleaned, uncoated 300-mesh nickel grids and allowed to air-dry overnight. The modified immunogold staining procedure as described by Varndell and colleagues [23, 25] was applied. Ultrathin sections were also cut from glutaraldehyde-fixed (2.5%

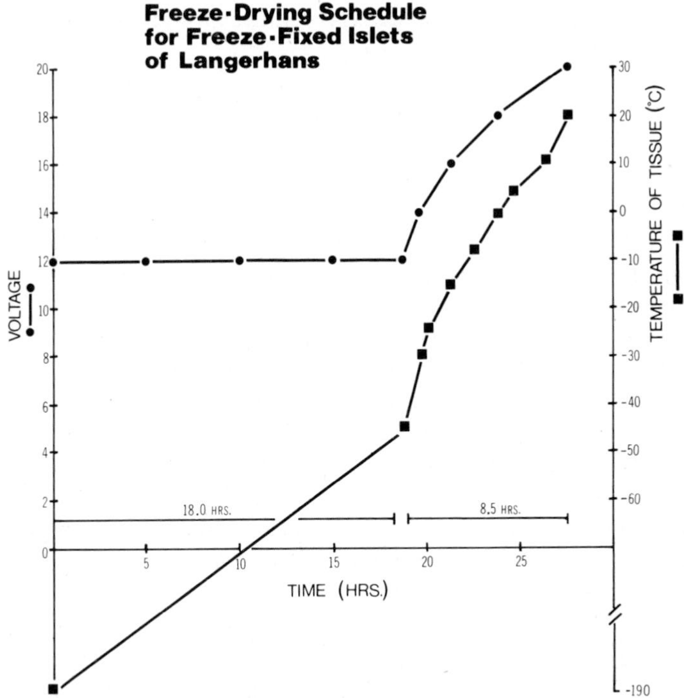

Fig. 3. This shows the freeze-drying schedule for pancreatic islets by plotting the relationship of time, temperature and voltage applied to the Vycor immersion heater.

glutaraldehyde in 0.1 M phosphate buffer, pH 7.2, 320 mOsm for 2 h), conventionally processed, Araldite-embedded blocks of human and rat pancreas, to serve as an immunostaining control.

The primary antisera were guinea pig anti-insulin (dilutions 1:1,600–1:80,000) and rabbit anti-rat C-peptide (dilutions 1:2,000–1:100,000). The C-peptide antiserum (reference R-901, kindly supplied by Professor N. Yanaihara, Shizuoka College of Pharmacy, Shizuoka, Japan) is known to cross-react extensively with pro-insulin. Incubation times varied from 1 h (room temperature) to 24 h (4°C). Antigen-antibody complexes were visualised using goat anti-guinea pig IgG or goat anti-rabbit igG, both adsorbed onto 20 nm gold particles, respectively. Incubation with the goat antisera was carried out for 1 h at room temperature. Following gold immunolabelling the sections were counterstained with uranyl acetate and lead citrate prior to viewing in the electron microscope.

3.2. Double immunogold staining procedure [24] (see also Chapter 13, pp. 155–177)

The ultrathin sections were incubated in a "cocktail" of both primary antisera for 45–60 min at room temperature. Following extensive washing the sections were incubated with a mixture of second layer antisera, diluted to optimal titre, for 30–40 min at room temperature.

4. Discussion

The morphological appearance of the quick-freeze fixation freeze-dried rat pancreatic islet material (Fig. 2) has been described previously [20]. Briefly, the inherent electron-density of the organelles is much greater than is normally encountered in conventionally processed tissue. Presumably, this is due to the fact that alcohol-soluble components are not eluted from the freeze-dried tissue. Osmium vapour fixation allows the clear definition of organellar and cellular membranes.

Electron immunocytochemistry revealed that the retention of antigenicity was many times greater than that observed in conventionally processed material. We were able to obtain acceptable immunostaining for insulin and C-peptide (pro-insulin) using dilutions 50 times greater than the optimal titres on glutaraldehyde-fixed, ethanol-dehydrated tissue sections. Figure 4 demonstrates the intensity of

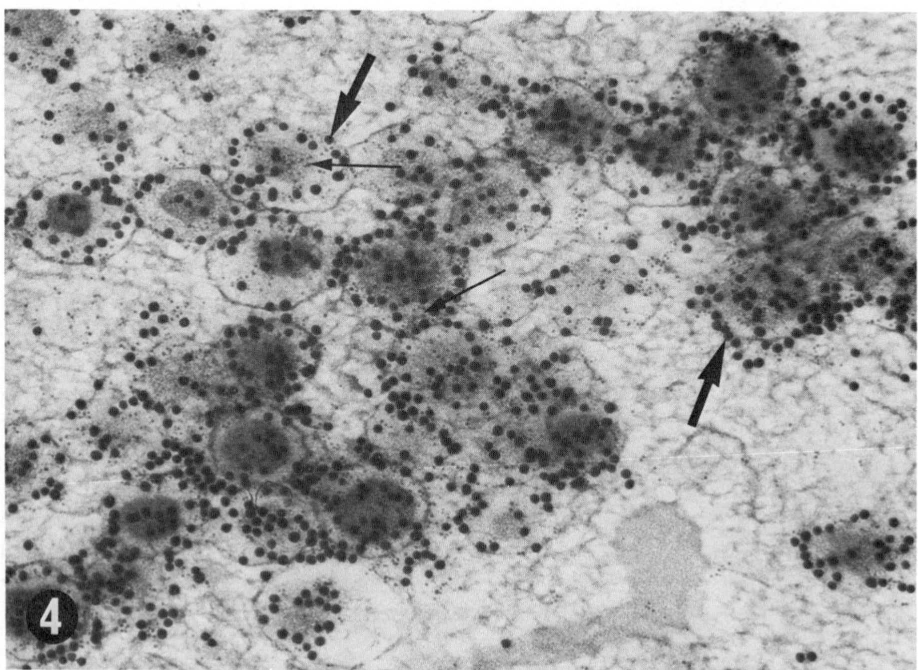

Fig. 4. Quick-freeze fixation freeze-dried rat pancreatic beta cell double immunostained with anti-insulin (dilution 1:32,000, visualised with goat anti-guinea pig IgG adsorbed to 20 nm gold particles; GAGP 20—large arrows) and anti-C-peptide (pro-insulin) (dilution 1:50,000, visualised with goat anti-rabbit IgG adsorbed onto 5 nm gold particles; GAR5—small arrows). Uranyl acetate counterstain. Magnification = × 73,000

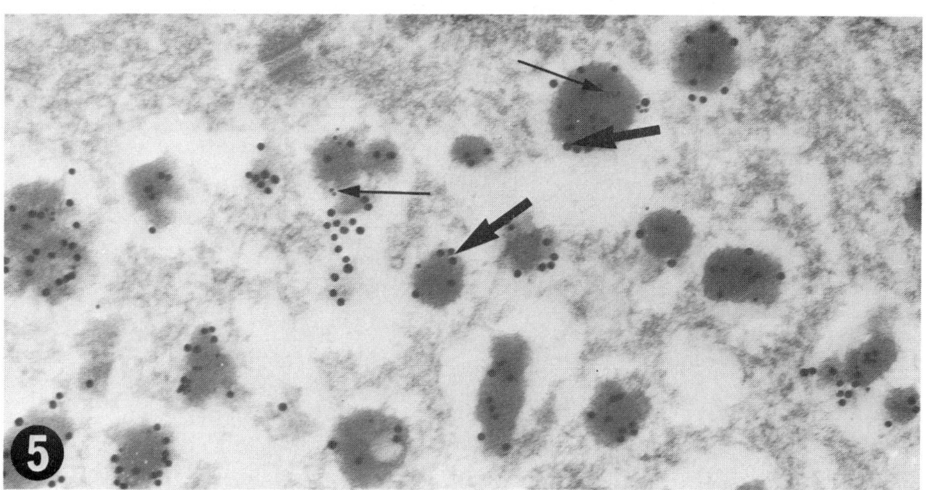

Fig. 5. Glutaraldehyde-fixed, ethanol dehydrated, ultrathin section of human pancreatic beta cell double immunostained to reveal insulin (dilution 1:1,600, GAGP20—large arrows) and C-peptide (pro-insulin) (dilution 1:2,000, GAR10—small arrows). Uranyl acetate and lead citrate counterstains. Magnification = ×65,000.

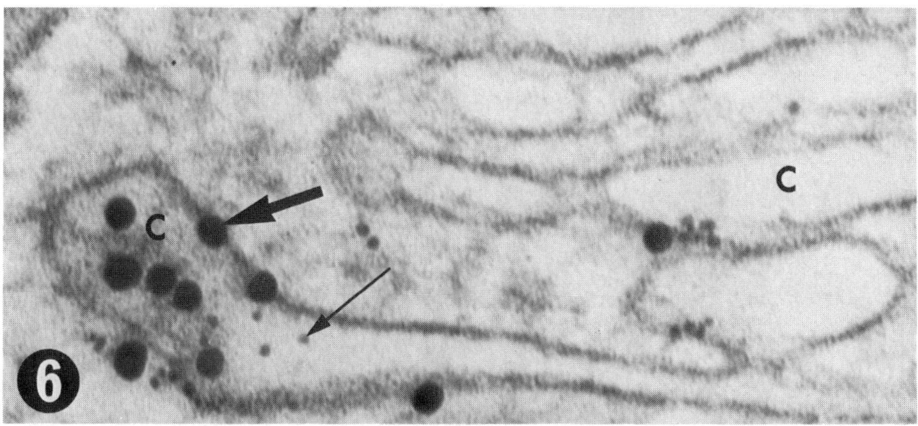

Fig. 6. Electron micrograph of freeze-dried rat pancreatic beta cell double immunostained for insulin (1:32,000, GAGP20—large arrows) and for C-peptide (pro-insulin) (1:50,000, GAR5—small arrows). The immunoreactivity is localised to an expanded cisterna (c) of the Golgi apparatus. Uranyl acetate counterstain. Magnification = ×20,000.

immunolabelling using anti-insulin and anti-C-peptide (pro-insulin) in a double immunogold staining procedure on freeze-dried tissue. For comparison, Fig. 5 shows the conventional (non-osmicated) result using both antisera 20 times more concentrated than in the preparations used for Fig. 4. It has also been possible to immunostain antigenic sites in elements of the Golgi apparatus (Fig. 6) using the quick-freeze fixation freeze-dried tissue, which has not been achieved in beta cell sections from conventionally processed pancreas.

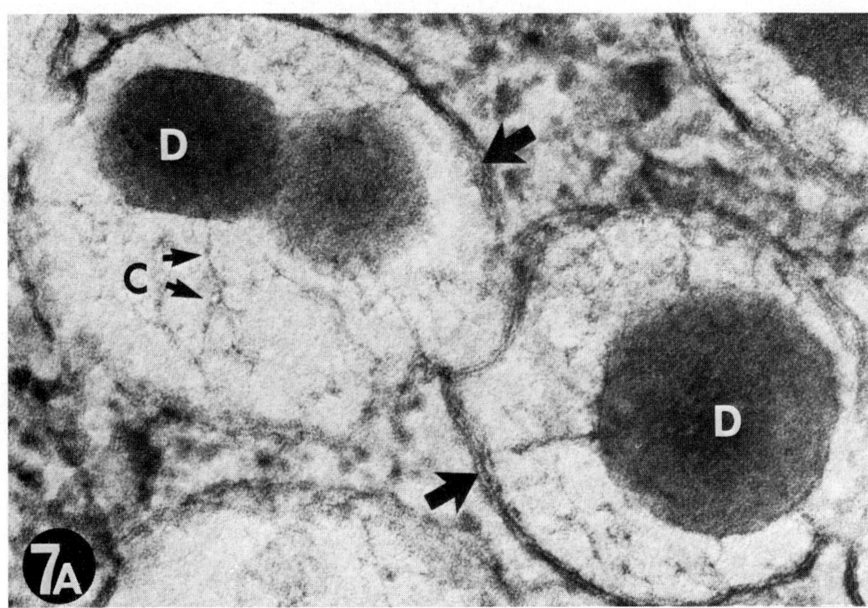

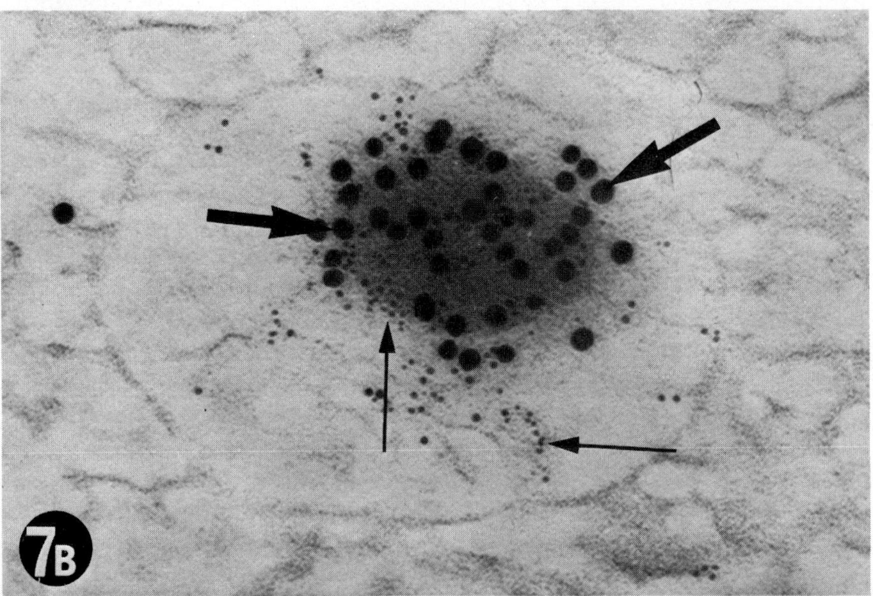

Fig. 7 (A). Quick-freeze fixation freeze-dried rat pancreatic beta cell secretory granules demonstrating the dense-core (D) and "cobweb" (C) of electron-dense material in the halo. Granule limiting membrane is arrowed. Osmium vapour fixation, counterstained with uranyl acetate and lead citrate. Magnification = ×137,500. (B) Similar preparation to (A). Double immunostained as in Fig. 6. Insulin immunoreactivity (large arrows) is largely restricted to the core whilst C-peptide (pro-insulin) (small arrows) is more widely distributed throughout the core and halo. Uranyl acetate counterstain. Magnification = ×137,500.

Using immunogold staining procedures we have been able to demonstrate quite clearly the improvement in antigenic retention obtained using quick-freeze fixation freeze-dried pancreatic islet tissue. The finding that the antisera used may be diluted many times more than the optimal titre for immunostaining conventionally processed tissue is of notable importance. Immunocytochemical specificity is generally improved by increasing dilution as there is less probability that low affinity or low avidity antibodies will be available to bind to antigens.

One interesting facet which has emerged from this work is the finding of some degree of regionalisation of insulin and C-peptide (pro-insulin) antigens within individual beta cell secretory granules. Figure 7 shows that insulin is restricted to the electron-dense core of the secretory granule, whereas C-peptide (pro-insulin) is found both in the core and also in the halo of the granule. As the C-peptide antiserum used in our study recognises antigenic determinants in both free and extended (pro-insulin) forms of C-peptide, it is difficult to interpret our findings accurately. However, it is tempting to speculate upon a possible analogy with the topographic segregation of glicentin and glucagon in pancreatic A cell granules [26]. In this study, Ravazzola and Orci were able to localise glicentin [27], the 69 amino acid component of pro-glucagon which encodes glucagon in its structure, to the halo of the alpha-type secretory granules whereas C-terminally reactive (pancreatic) glucagon was restricted to the electron-dense core. By using C-terminally reactive C-peptide antisera which do not cross-react with pro-insulin in combination with insulin antibodies we hope to be able to resolve the question of whether there is topographic segregation of peptides within beta cell secretory granules.

5. References

1. Bendayan, M. and Zollinger, M. (1983) Ultrastructural localization of antigenic sites on osmium-fixed tissues applying the protein A-gold technique. J. Histochem. Cytochem. 31, 101–109.
2. Roth, J., Bendayan, M., Carlemalm, E., Villiger, W. and Garavito, M. (1981) Enhancement of structural preservation and immunocytochemical staining in low temperature embedded pancreatic tissue. J. Histochem. Cytochem. 29, 663–671.
3. Bendayan, M. and Shore, G.C. (1982) Immunocytochemical localization of mitochondrial proteins in the rat hepatocyte. J. Histochem. Cytochem. 30, 139–147.
4. Goldsmith, P.C. and Ganong, W.F. (1975) Ultrastructural localization of luteinizing hormone-releasing hormone in the median eminence of the rat. Brain Res. 97, 181–193.
5. Coulter, H.D. and Elde, R.P. (1978) Somatostatin radioimmunoassay and immunofluorescence in the rat hypothalamus: Effects of dehydration with alcohol and fixation with aldehydes and OsO_4. Anat. Rec. 190, 369.
6. McNeill, T.H. and Sladek, C.D. (1980) The effect of tissue processing on the retention of vasopressin in neurons of the neurohypophyseal system. J. Histochem. Cytochem. 28, 604–605.
7. Sabatini, D.D., Bensch, K. and Barrnett, R.J. (1963) Cytochemistry and electron microscopy. The preservation of cellular ultrastructure and enzymic activity by aldehyde fixation. J. Cell Biol. 17, 19–58.
8. Altmann, R. (1890) Die Elementaroganismen und ihre Beziehungen zu den Xellen. Leipzig.
9. Sjostrand, F.S. (1943) Electron-microscopic examination of tissues. Nature (London) 154, 725.
10. Sjostrand, F.S. and Andersson-Cedergren, E. (1958) The ultrastructure of the skeletal muscle myofilaments at various states of shortening. J. Ultrastruct. Res. 1, 239–246.

11. Van Harreveld, A. and Crowell, J. (1964) Electron microscopy after rapid freezing on a metal surface and substitution fixation. Anat. Rec. 149, 381–386.
12. Heuser, J.E., Reese, T.S., Dennis, M.J., Jan, Y., Jan, L. and Evans, L. (1979) Synaptic vesicle exocytosis captured by quick freezing and correlated with quantal transmitter release. J. Cell Biol. 81, 275–300.
13. Heuser, J.E. (1981) Preparing biological samples for stereomicroscopy by the quick-freeze, deep-etch, rotary replication technique. Meth. Cell Biol. 22, 97–122.
14. Heuser, J.E. and Kirschner, M.W. (1980) Filament organization revealed in platinum replicas of freeze-dried cytoskeletons. J. Cell Biol. 86, 212–234.
15. Rash, J.E. (1983) The rapid-freeze technique in neurobiology. Trends Neurol. Sci. 6, 208–212.
16. Coulter, H.D. and Terracio, L. (1977) Preparation of biological tissues for electron microscopy by freeze-drying. Anat. Rec. 187, 477–494.
17. Boyne, A.F. (1979) A gentle bounce-free assembly for quick-freezing tissues for electron microscopy: Application to isolated torpedine ray electrocyte stacks. J. Neurosci. Methods 1, 353.
18. Phillips, T.E. and Boyne, A.F. (1984) Liquid nitrogen-based quick freezing: Experiences with bounce-free delivery of cholinergic nerve terminals to a metal surface. J. Electr. Microsc. Technol. 1, 9–29.
19. Hellerstrom, C., Lewis, N.J., Borg, H., Johnson, R. and Freinkel, N. (1979) Method for large-scale isolation of pancreatic islets by tissue culture of fetal rat pancreas. Diabetes 28, 769–776.
20. Dudek, R.W., Childs, G.V. and Boyne, A.F. (1982) Quick-freezing and freeze-drying in preparation for high quality morphology and immunocytochemistry at the ultrastructural level: Application to pancreatic beta cell. J. Histochem. Cytochem. 30, 129–138.
21. Terracio, L. and Schwabe, K.G. (1981) Freezing and drying of biological tissues for electron microscopy. J. Histochem. Cytochem. 29, 1021–1028.
22. De Mey, J., Moeremans, M., Geuens, G., Nuydens, R. and De Brabander, M. (1981) High resolution light and electron microscopic localization of tubulin with the IGS (ImmunoGold Staining) method. Cell Biol. Int. Rep. 5, 889–899.
23. Varndell, I.M., Tapia, F.J., Probert, L., Buchan, A.M.J., Gu, J., De Mey, J., Bloom, S.R. and Polak, J.M. (1982) Immunogold staining procedure for the localisation of regulatory peptides. Peptides 3, 259–272.
24. Tapia, F.J., Varndell, I.M., Probert, L., De Mey, J. and Polak, J.M. (1983) Double immunogold staining method for the simultaneous ultrastructural localization of regulatory peptides. J. Histochem. Cytochem. 31, 977–981.
25. Varndell, I.M., Tapia, F.J., De Mey, J., Rush, R.A., Bloom, S.R. and Polak, J.M. (1982) Electron immunocytochemical localization of enkephalin-like material in catecholamine-containing cells of the carotid body, the adrenal medulla, and in pheochromocytomas of man and other mammals. J. Histochem. Cytochem. 30, 682–690.
26. Ravazzola, M. and Orci, L. (1980) Glucagon and glicentin immunoreactivity are topologically segregated in the alpha granule of the human pancreatic A cell. Nature (London) 284, 66–68.
27. Thim, L. and Moody, A.J. (1980) The primary structure of porcine glicentin (proglucagon). Regulatory Peptides, 2, 139–150.

Immunolabelling for Electron Microscopy (Polak/Varndell, eds.)

CHAPTER 17

Lectin cytochemistry

Marc Horisberger

Research Department, Nestlé Products Technical Assistance Co. Ltd., P.O. Box 88, CH-1814 La Tour de Peilz, Switzerland

CONTENTS

1. INTRODUCTION

Lectins bind specific sugars and are extensively used to investigate properties of cell surface glycoconjugates. One may define lectins as 'proteins of non-immunoglobulin nature capable of specific recognition and reversible binding to carbohydrate moieties of complex carbohydrates without altering covalent structure of any of the recognised glycosyl ligands' [1].

It is now about 20 years since lectins were found to represent a new class of reagents for carbohydrate cytochemistry which exhibit a high level of specificity when the results are carefully interpreted.

2. Lectin specificity

Lectins considered 'identical' in terms of monosaccharide specificity, possess the ability to recognise subtle differences in more complex structures [2]. Indeed the monosaccharide residue in a terminal non-reducing position on a glycan is not the only carbohydrate moiety recognised. Therefore, the best saccharide inhibitor found for a given lectin does not necessarily represent the true glycan site recognised by this lectin as a cell surface glycoconjugate. Furthermore, lectins can react with a broad spectrum of cell surface glycoproteins.

The formation and dissociation of lectin-glycoprotein complexes is dependent upon variables such as temperature, pH, ionic strength, binding affinity, binding capacity and avidity of the lectin, and charge effects [3].

Chemical modification can also alter and restrict the specificity of lectins. Succinylated wheat germ lectin, which is negatively charged at physiological pH, in contrast with the unmodified lectin which is positively charged, does not bind cell surface glycoconjugates containing N-acetylneuraminic acid but does bind cell surface glycoconjugates containing N-acetylglucosamine [4]. A decrease of the apparent number of cell surface binding sites upon succinylation of this lectin has also been observed and attributed to changes in the isoelectric point of the lectin and on the acidic properties of the cell surface [4]. Therefore it is almost needless to stress that cytochemical observations should be carefully interpreted with respect to lectin specificity.

3. Lectins as a tool in cytochemistry

Since lectins are devoid of enzymic activity and are not opaque to the electrons, they must be conjugated either to enzymes that produce electron-dense reaction products or to an electron-opaque molecule or to a metal particle for use in transmission electron microscopy.

A variety of techniques has been proposed. Markers and their preparative procedures fall into different classes: markers are particulate (e.g. colloidal gold, ferritin) or diffuse (e.g. peroxidase). The binding of lectins to markers can be non-covalent in nature (e.g. colloidal gold), covalent (e.g. ferritin, peroxidase) and via specific recognition sites such as avidin-biotin or sugars (e.g. glycosylated ferritin and peroxidase). Some marker systems have a general application (e.g. colloidal gold, ferritin and peroxidase conjugates, avidin-biotin complexes), others have a limited application (e.g. iron-dextran, haemocyanin).

Several reviews have already appeared on various aspects of marking cell surface carbohydrates with lectins [5–7]. This review will focus primarily on lectin-labelled gold markers since they allow a discussion of general principles upon which lectin cytochemistry is based.

4. Colloidal gold

The general principles underlying the preparation of lectin-labelled gold markers are found in an earlier chapter on 'Electron-opaque Markers' (Chapter 2, pp.

17–26). Detailed procedures have been published by Horisberger and Rosset [8] and by Geoghegan and Ackerman [9]. The preparation, labelling, stabilisation, stability, binding characteristics as well as the various applications of lectin-labelled gold markers in transmission and scanning electron microscopy have been discussed extensively in general articles [10, 11].

4.1. Direct procedures

Lectins can be adsorbed as monolayers onto a clean metal surface. For instance, Concanavalin A, a tetramer approximately 8 nm in size, is adsorbed onto nickel with an average layer thickness of 7.4 nm [12]. Colloidal gold, which has a large surface area, also adsorbs a variety of lectins presumably through a non-covalent process [11]. Much evidence indicates that the process is irreversible [11, 13]. All lectins adsorbed onto gold particles have been found to retain their sugar binding capacity although their fine specificity may be modified upon binding [14].

Gold particles labelled with lectins have been used in pre-embedding and post-embedding techniques. In pre-embedding techniques, the number of bound particles decreases, sometimes abruptly when the particle size is increased due to steric hindrance [13]. As a consequence no relationship has been found between the number of lectin binding sites and the number of bound particles since the accessibility of gold markers is strongly regulated by the size of the probe. Indeed binding sites close to the membrane bilayer appear to be less accessible to large probes than binding sites extending from the cell surface. Measurement of the distance between the cell membrane and the gold particles are in agreement with this hypothesis [15]. Therefore, steric hindrance may be turned into an advantage when one wishes to study not only the lateral but also the longitudinal distribution of cell surface lectin binding sites and estimate the size of their crypt. The influence of the size of lectin-labelled gold particles on binding has been demonstrated with red blood cells [13, 16], hepatocytes [15, 17], and platelets [18].

In water, all solids acquire a surface charge due to dissociation of surface groups and adsorption of ions. Therefore, in an aqueous environment electrostatic double layer forces are present in addition to attractive Van der Waal's forces. The mutual interpenetration of double layers results usually in repulsion since the sign of the charge is negative for most materials. As the double layer is compressed by the addition of electrolytes, the binding of lectin-labelled gold particles to cell surface is also dependent upon ionic strength and pH. For instance, the number of Au_{32}* particles labelled with wheat germ lectin bound to red blood cells decreases when the ionic strength is lowered. In water alone, no binding occurs [19]. The agglutination of red blood cells by various lectin-labelled gold particles is also dependent upon ionic strength and size of the probes [20].

The direct procedures can be used in pre-embedding and post-embedding techniques. In the latter, for some unknown reason, the marking density generally increases when the size of the particle is decreased [21, 22]. With particles as small

* The subscript indicates the mean diameter of the particles in nanometers.

as Au_3, the organisation of the marker into groups is found to be more pronounced than with larger lectin-labelled particles [22]. Owing to the fact that monodisperse gold particles can be produced in different sizes, multiple markings are easily obtained by the direct gold method (Fig. 1).

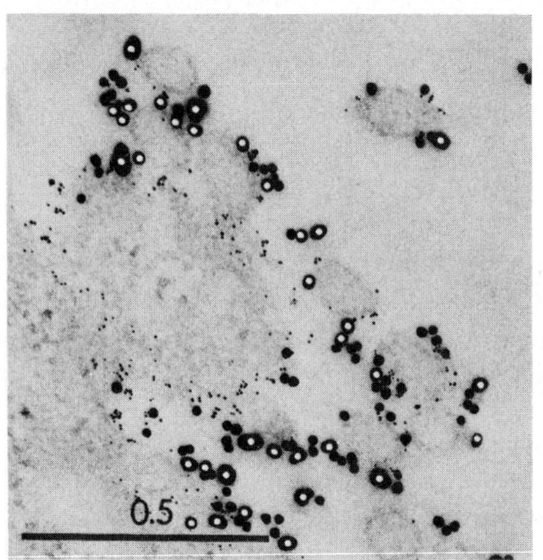

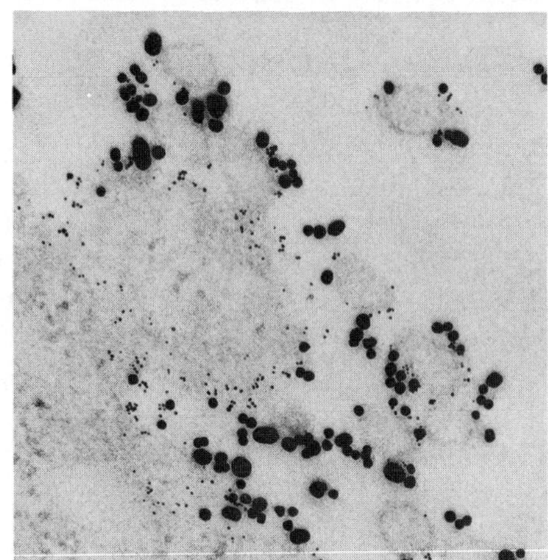

Fig. 1. Stereoscopic transmission electron micrograph of mouse embryo fibroblast marked with three lectins. The cells were successively incubated with Au_5, Au_{17} and Au_{26} particles labelled respectively with Concanavalin A, *Ricinus communis* 120 lectin and wheat germ lectin (white dots). When the stereopair is examined, most of the particles are bound by spatially separated sites. It is best viewed with an optical viewer. However, most people with a little determination can achieve stereopsis by crossing the eyes. The bar represents 0.5 μm (Horisberger and Tacchini-Vonlanthen [13]).

4.2. Indirect procedures

The indirect method (two steps or more) has also been applied in pre-embedding and post-embedding procedures. Bound lectins can be marked with gold particles labelled with polysaccharides [8] or glycoprotein [9] bearing sugars specific for the lectin. Alternatively biotin-labelled lectins are detected with avidin-labelled gold particles [23]. A third method consists of detecting immunoglobulins bound to lectin with protein A-gold particles [13]. When compared to the direct method, the indirect method results generally in a much increased density of marking (see also Chapter 15, pp. 189–233). Although most lectins are tetravalent, it is likely that in the indirect procedure, the method does not always allow visualisation of all lectin molecules since all their binding sites may have reacted with cell surface glycoproteins.

4.3. Binding constants of lectin-labelled gold markers

As the binding constants of lectin-labelled gold markers are several orders of magnitude higher than those of the lectins binding to the cell surface [10, 13], this indicates that they are the result of multivalent interactions. It is possible that secondary interactions between the marker and cell surface such as hydrophobic bonds [2] add to the strength of the binding since bound particles cannot be totally desorbed by the addition of sugar inhibitors [13]. However, in pre- and post-embedding procedures, almost total inhibition of binding is achieved by incorporating specific monosaccharides to the lectin-labelled gold marker. In some instances (e.g. wheat germ lectin, Concanavalin A), total inhibition is only achieved in the presence of specific oligomers or polysaccharides [18].

5. Glycosylated markers

As lectins are multivalent, they can bind both the cellular carbohydrate and a marker carrying a specific sugar for that lectin. Glycosylated derivatives of horseradish peroxidase and ferritin have been prepared for most common lectins [24, 25].

Both markers are valuable. While glycosylated ferritin detects lectins bound only in the superficial area of a cell coat, glycosylated peroxidase can have access to lectin bound in the underlying layer [25] (Fig. 2). Therefore the use of both glycosylated ferritin and glycosylated horseradish peroxidase can give interesting information on the accessibility of cell surface binding sites.

The glycosylated ferritin method alleviates some problems inherent to the preparation of lectin-ferritin conjugates. Indeed the commonly used glutaraldehyde coupling procedure results in a mixture of molecules which must be purified by chromatography to yield a lectin-ferritin complex in a 1:1 ratio, but this is in extremely low quantity.

The glycosylated marker method must be applied with caution. Certain conditions must be met: the marker concentration has to be low enough to avoid a possible displacement of the lectin bound to the cell surface. The presence of endogenous lectin in numerous tissues throughout both animal and plant kingdoms [26] can interfere with the detection of bound exogenous lectins. With glycosylated peroxidase, there may be a reaction due to endogenous peroxidatic activity or pseudoperoxidatic activity. These interfering activities are usually not detected with the post-embedding technique [27]. Finally, the method can detect only cell-bound lectin molecules which still have free sugar binding sites.

6. Avidin-biotin complex

The term affinity cytochemistry was first introduced for a method where biotinylated proteins, such as lectins, are allowed to bind to cells in the first step followed by avidin-ferritin or avidin-peroxidase conjugates [28, 29]. Many variants have been proposed both in histochemistry and in cytochemistry. For instance, carbohydrates can be localised utilising a biotinylated lectin followed by an avidin-biotin-peroxidase complex (ABC) [30]. With peanut lectin, the sensitivity of staining for

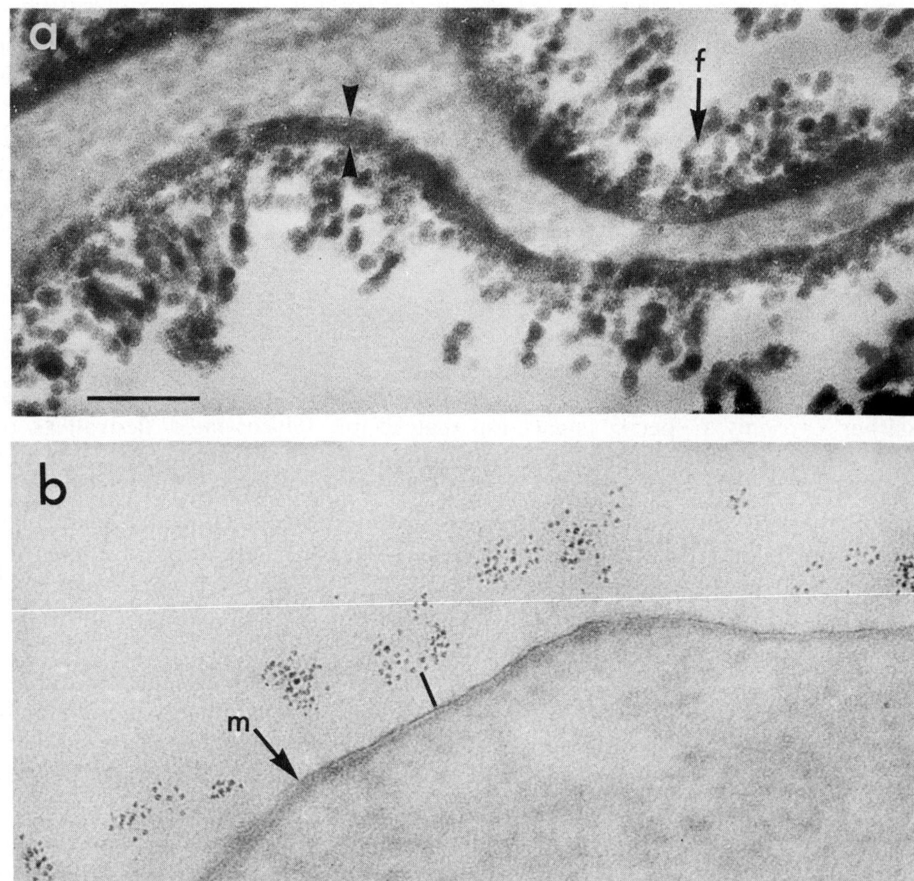

Fig. 2. Concanavalin A binding sites of cell surface of *Amoeba proteus*; (a) the lectin and horseradish peroxidase-diaminobenzidine method stains the amorphous layer (between the two arrows) and the fuzzy layer (f). (b) Concanavalin A and mannosylated ferritin method. The binding sites are detected only in the fuzzy layer and not in the amorphous area (line), near the unit membrane (m). The bars represent 0.2 μm (from Schrevel et al. [25]).

different methods is ABC > PAP > lectin-peroxidase conjugate]30]. When the avidin-biotin method was used in combination with the gold method, both better sensitivity and precise localisation were achieved by the two-step method, via a biotinyl derivative of Concanavalin A and gold particles labelled with avidin, than by the direct gold procedure [23].

The avidin-biotin complex has been used for localising binding sites for a number of lectins and applied to study different pathological alterations [29]. The advantage of the method is that only one conjugate (i.e. ferritin-avidin) must be prepared and characterised for all affinity systems. However, the successful application of the method requires the preparation of biotinylated lectins without affecting their binding characteristics. Most lectins commonly used have been derivatised by biotin, their biological activity being little affected [28].

7. Ferritin and peroxidase conjugates

The use of lectin-peroxidase conjugates and of lectin-ferritin conjugates was first introduced by Avrameas [31] and Nicolson and Singer [32], respectively. The conjugates can be prepared in one [33] or two steps [34] using glutaraldehyde as the coupling agent.

While ferritin-lectin conjugates allow a precise localisation of lectin binding sites, the use of horseradish peroxidase precludes exact determination of binding site distribution and movement due to diffusion of the reaction product.

The ferritin conjugate method can be quantified. However, the data must be carefully interpreted since it has been found that the distribution and/or density of surface-bound lectin is concentration-dependent for lectin-ferritin conjugates [35]. An important application of lectin-ferritin conjugates has been in cell surface marking for immuno-freeze-etching [6]. However, colloidal gold particles are more easily recognised in this application.

To reduce steric hindrance, lectins can be conjugated with smaller enzymes such as microperoxidase or cytochrome but the activity of the conjugates is reduced when compared to that of horseradish peroxidase.

For double marking experiments, peroxidase or ferritin conjugates can be used with lectin-labelled gold particles [36]. However, the combination of peroxidase and ferritin conjugates cannot be recommended since the electron density of the oxidised diaminobenzidine reaction product equals at least that of the ferritin particles [36].

8. Markers with restricted application

Iron-dextran [37] and iron-mannan [38] particles have been used in a two-step procedure to mark Concanavalin A binding sites on cell surface owing to the affinity of the lectin for these polysaccharides. Although procedures are available to obtain covalent iron-dextran conjugates with proteins [39], this method has found limited use. The method suffers from some other limitations owing to the non-covalent nature of the interaction between the marker and the lectin. The lectin may be released to some extent during the marking procedure. Furthermore, only a fraction of the lectin bound by the cell surface may be available to the marker.

9. Application to scanning electron microscopy

In spite of the potential interest in localising lectin binding sites on cell surfaces by scanning electron microscopy, relatively few studies have been reported. Reliable lectin markers for scanning electron microscopy are essentially particulate (e.g. haemocyanin, co-polymer microspheres, colloidal gold [8, 10, 11, 40]). Haemocyanin, with its distinct cylindrical shape (35–50 nm) can be chemically coupled to lectins or used in a two-step cytochemical affinity technique but the latter application is restricted to Concanavalin A. A variety of synthetic macromolecular

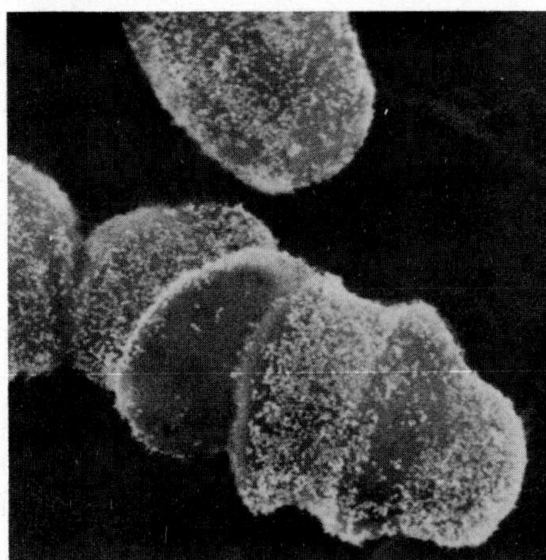
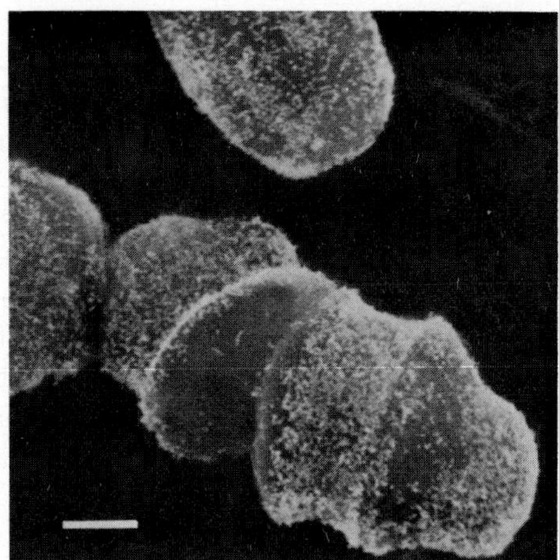

Fig. 3. Stereoscopic scanning electron micrograph of the yeast *Schizosaccharomyces pombe*. The cells were marked for cell wall galactomannan by the direct method with Au_{56} particles labelled with *Ricinus communis* 120 lectin. Crosswalls just established by fission are not marked. However, the growing end reacted with the marker. The bar represents 1 μm (Horisberger and Vauthey, unpublished).

markers has been developed [40]. These microspheres can be chemically conjugated to lectins and used in a direct procedure [40]. However, in all cases, including iron co-polymer microspheres, the marker can be identified only when the cell surface is coated with a 5–15 nm layer of gold or gold-palladium.

In contrast with these markers, lectin-labelled gold particles offer a unique system since (a) the particles can be prepared in a size suitable for scanning electron microscopy (30–50 nm) and (b) their high secondary electron emission enables visualisation on cell surfaces not coated with metal (Fig. 3). The particles are easily detected. The gold method thus provides highly specific and sensitive marking whether applied in one step or more.

10. Acknowledgements

The author thanks Mrs. M. Rouvet and Mrs. M. Beaud for the artwork and Dr. J. Schrevel for providing micrographs of his own work.

11. References

1. Kocourek, J. and Horejsi, V. (1983) A note on the recent discussion on definition of the term 'lectin'. In: T.C. Bog-Hansen and G.M. Spengler (Eds.) Lectins, Vol. 3. Walter de Gruyter, Berlin, pp. 3–6.
2. Debray, H., Decout, D., Strecker, G., Spik, G. and Montreuil, J. (1981) Specificity of twelve lectins towards oligosaccharides and glycopeptides related to N-glycosylproteins. Eur. J. Biochem. 117, 41–55.
3. Monsigny, M., Roche, A.C., Sene, C., Maget-Dana, R. and Delmotte, F. (1980) Sugar-lectin interactions: how does wheat-germ agglutinin bind sialoglycoconjugates? Eur. J. Biochem. 104, 147–153.
4. Monsigny, M., Sene, C., Obrenovitch, A., Roche A.C., Delmotte, F., and Boschetti, E. (1979) Properties of succinylated wheat-germ agglutinin. Eur. J. Biochem. 98, 39–45.
5. Nicolson, G.L. (1978) Ultrastructural localization of lectin receptors. In: J.K. Koehler (Ed.) Advanced Techniques in Biological Electron Microscopy, Vol. 2. Springer-Verlag, New York, pp. 1–38.
6. Roth, J. (1978) The lectins: Molecular probes in cell biology and membrane research. Exp. Pathol. Suppl. 3, 5–186.
7. Lotan, R. (1979) Qualitative and quantitative aspects of labelling cell surface carbohydrates using lectins as probes. In: O. Johari (Ed.) Scanning Electron Microscopy, Vol. 3. SEM Inc., AMF O'Hare (Chicago) IL., pp. 549–564.
8. Horisberger, M. and Rosset, J. (1977) Colloidal gold, a useful marker for transmission and scanning electron microscopy. J. Histochem. Cytochem. 25, 295–305.
9. Geoghegan, W.D. and Ackerman, G.A. (1977) Adsorption of horseradish peroxidase, ovomucoid and antiimmunoglobulin to colloidal gold for the indirect detection of concanavalin A, wheat germ agglutinin and goat anti-human immunoglobulin on cell surfaces at the electron microscopic level: a new method, theory and application. J. Histochem. Cytochem. 25, 1187–1200.
10. Horisberger, M. (1979) Evaluation of colloidal gold as a cytochemical marker for transmission and scanning electron microscopy. Biol. Cell. 36, 253–258.
11. Horisberger, M. (1981) Colloidal gold: a cytochemical marker for light and fluorescent microscopy and for transmission and scanning electron microscopy. In: O. Johari (Ed.) Scanning Electron Microscopy, Vol. 2. SEM Inc., AMF O'Hare (Chicago) IL., pp. 9–31.
12. Horisberger, M. (1980) An application of ellipsometry. Assessment of polysaccharide and glycoprotein interaction with lectin at a liquid/solid interface. Biochim. Biophys. Acta 632, 298–309.
13. Horisberger, M. and Tacchini-Vonlanthen, M. (1983) Stability and steric hindrance of lectin-labelled gold markers in transmission and scanning electron microscopy. In: T.C. Bog-Hansen and G.A. Spengler (Eds.) Lectins, Vol. 3. Walter de Gruyter, Berlin, pp. 189–197.
14. Debray, H., Pierce-Crétel, A., Spik, G. and Montreuil, J. (1983) Affinity of ten insolubilized lectins towards various glycopeptides with the N-glycosamine linkage and related oligosaccharides. In: T.C. Bog-Hansen and G.A. Spengler (Eds.) Lectins, Vol. 3. Walter de Gruyter, Berlin, pp. 335–350.
15. Horisberger, M. (1980) Lectins—ultrastructural localization of receptors using colloidal gold. In: H. Popper, L. Bianchi, F. Gudat and W. Reutter (Eds.) Communications of Liver Cells. MTP Press Limited, Lancaster, England, pp. 25–29.
16. Horisberger, M. and Vonlanthen, M. (1979) Multiple marking of cell surface receptors by gold granules: simultaneous localization of three lectin receptors on human erythrocytes. J. Microsc. 115, 97–102.
17. Horisberger, M., Rosset, J. and Vonlanthen, M. (1978) Location of lectin receptors on rat hepatocytes by transmission and scanning electron microscopy. Experientia 34, 274–276.
18. Nurden, A.T., Horisberger, M., Savariau, E. and Caen, J.P. (1980) Visualization of lectin binding sites on the surface of human platelets using lectins adsorbed to gold granules. Experientia 36, 1215–1217.

19. Horisberger, M. (1980) Adhesion of human and chicken red blood cells to polystyrene: influence of electrolyte and polyethylene glycol concentration. Physiol. Chem. Phys. 12, 195–204.
20. Horisberger, M. (1978) Agglutination of erythrocytes using lectin-labelled spacers. Experientia 34, 721–722.
21. Horisberger, M. and Vonlanthen, M. (1979) Location of mannan and chitin on thin sections of budding yeasts with gold markers. Arch. Microbiol. 115, 1–7.
22. Roth, J., Brown, D. and Orci, L. (1983) Regional distribution of N-acetyl-D-galactosamine residues in the glycocalyx of glomerular podocytes. J. Cell Biol. 96, 1189–1196.
23. Horisberger, M. and Vonlanthen, M. (1979) Fluorescent colloidal gold: a cytochemical marker for fluorescent and electron microscopy. Histochemistry 64, 115–118.
24. Kiéda, C., Delmotte, F. and Monsigny, M. (1977) Preparation and properties of glycosylated cytochemical markers. FEBS Lett. 76, 257–261.
25. Schrevel, J., Kiéda, C., Caigneaux, E., Gros, D., Delmotte, F. and Monsigny, M. (1979) Visualization of cell surface carbohydrates by a general two-step lectin technique: lectins and glycosylated cytochemical markers. Biol. Cell. 36, 259–266.
26. Monsigny, M., Kiéda, C. and Roche, A.C. (1979) Membrane lectins. Biol. Cell. 36, 289–300.
27. Gros, D., Bruce, B., Challice, C.E. and Schrevel, J. (1982) Ultrastructural localization of Concanavalin A and Wheat germ agglutinin binding sites in adult and embryonic mouse myocardium. J. Histochem. Cytochem. 30, 193–200.
28. Bayer, E.A., Wilchek, M. and Skutelsky, E. (1976) Affinity cytochemistry: the localization of lectin and antibody receptors on erythrocytes via the avidin-biotin complex. FEBS Lett. 68, 240–244.
29. Skutelsky, E. and Bayer, E.A. (1979) The ultrastructural localization of cell surface glycoconjugates: affinity cytochemistry via the avidin-biotin complex. Biol. Cell. 36, 237–252.
30. Hsu, S.M. and Raina, L. (1982) Versatility of biotin-labelled lectins and avidin-biotin-peroxidase complex for localization of carbohydrate in tissue sections. J. Histochem. Cytochem. 30, 157–161.
31. Avrameas, S. (1970) Emploi de la Concanavaline-A pour l'isolement, la détection et le résumé des glycoprotéines et glucides extra ou endocellulaire. C.R. Acad. Sci. (Paris) 270, 2205–2208.
32. Nicolson, G.L. and Singer, S.J. (1971) Ferritin-conjugated agglutinins as specific saccharide stains for electron microscopy: application to saccharides bound to cell membranes. Proc. Natl. Acad. Sci. U.S.A. 68, 942–945.
33. Avrameas, S. (1979) Coupling of enzymes to proteins with glutaraldehyde. Use of the conjugates for the detection of antigens and antibodies. Immunochemistry 6, 43–52.
34. Avrameas, S. (1972) Enzyme markers: their linkage with proteins and use in immunohistochemistry. Histochem. J. 4, 321–330.
35. Hixson, D.C., Miller, M.F., Maruyama, K., Walborg, E.F., Wagner, S., Starling, J.J. and Bowen, J.M. (1979) A statistical evaluation of the binding of ferritin-conjugated lectins to the surface of rat cells. Topographical variations as a function of lectin concentration and cell type. J. Histochem. Cytochem. 27, 1618–1629.
36. Roth, J. and Binder, M. (1978) Colloidal gold, ferritin and peroxidase as markers for electron microscopic double labelling lectin techniques. J. Histochem. Cytochem. 26, 163–169.
37. Martin, B.J. and Spicer, S.S. (1974) Concanavalin A-iron dextran technique for staining cell surface mucosubstances. J. Histochem. Cytochem. 22, 206–207.
38. Roth, J. and Franz, H. (1975) Ultrastructural detection of lectin receptors by cytochemical affinity reaction using mannan-iron complexes. Histochemistry 41, 365–368.
39. Dutton, A.H., Tokuyasu, K.T. and Singer, S.J. (1979) Iron-dextran antibody conjugates: general method for simultaneous staining of two components in high-resolution immunoelectron microscopy. Proc. Natl. Acad. Sci., U.S.A. 76, 3392–3396.
40. Molday, R.S. and Maher P. (1980) A review of cell surface markers and labelling techniques for scanning electron microscopy. Histochem. J. 12, 273–315.

Immunolabelling for Electron Microscopy (Polak/Varndell, eds)

CHAPTER 18

Combined use of autoradiography and immunocytochemical methods to show synaptic interactions between chemically defined neurons

Virginia M. Pickel[1] and Alain Beaudet[2]

[1]*Laboratory of Neurobiology, Department of Neurology, Cornell University Medical College, New York, NY 10021, U.S.A. and* [2]*Department of Neurology and Neurosurgery, Montreal Neurological Institute, 3801 University St., Montreal, Quebec, Canada*

Contents

1. Introduction

One of the most promising current uses for autoradiography and peroxidase immunocytochemistry is the identification of two neurotransmitters within single sections of central or peripheral tissues. Using these methods, it is possible to determine whether or not two transmitters co-exist within the same structure or whether different chemically defined neurons are synaptically linked. The radiolabelled compounds may be selectively incorporated into neurons by specific high affinity uptake systems, as demonstrated for [^{3}H]serotonin (5-[^{3}H]HT; for review see Descarries and Beaudet [1]). Alternatively, the endogenous compounds may be recognised by the binding of tritium labelled antisera as recently demonstrated for monoclonal [^{3}H]choline acetyltransferase ([^{3}H]CAT) [2] and substance P [3] antisera. Autoradiographic detection of either marker is compatible with peroxidase immunocytochemistry for the visualisation of other antigens [4]. We will briefly summarise the methods for combining the autoradiographic detection of 5-[^{3}H]HT or [^{3}H]CAT with the immunocytochemical localisation of tyrosine hydroxylase (TH), the enzyme used in the first step of catecholamine synthesis [5]. Examples are described for detection of synapses between serotonergic and catecholaminergic neurons in the medial nuclei of the solitary tracts (m-NTS) and ventral tegmental area (VTA) of rats.

2. Methodology

2.1. *Autoradiographic markers*

2.1.1. *Ventricular uptake of tritiated serotonin*

Serotonin (5-hydroxytryptamine, 5-HT)-containing neurons and their processes are selectively labelled in vivo following prolonged ventricular infusions of tritiated 5-HT (5-[^{3}H]HT) [1]. Rats are pretreated with monoamine oxidase inhibitor (Pargyline, 75 mg/kg, i.p.) 1 h prior to ventricular infusion of the isotope. The infusate (5-[^{3}H]HT-creatine sulphate, specific activity 11 Ci/mmole; Amersham International p.l.c.) is prepared by evaporation from a stock solution and reconstitution at 10^{-4} M in isotonic saline containing 0.1% ascorbic acid and 10^{-3} M non-radioactive noradrenaline. One hundred and fifty microlitres of the diluted isotope is infused into the lateral ventricle of anaesthetised rats over a 2 h period. The animals are sacrificed 15 min following the termination of infusion and processed for immunocytochemical localisation of tyrosine hydroxylase and autoradiography as described in the next section.

2.1.2. *[^{3}H] monoclonal antiserum*

The detection of two antigens by autoradiography and peroxidase immunocytochemistry is limited by the relatively small number of antigens to which tritiated antisera are available. A monoclonal antiserum to rat striatal CAT can be produced by fusion of synthesised mouse lymphocytes with murine plasmacytoma (NS1) cells [2, 6]. The CAT monoclonal antiserum is radiolabelled in culture by incubating hybridomas in a medium containing a mixture of ^{3}H-labelled amino acids (1.5–2.0 μCi [^{3}H]leucine, [^{3}H]lysine, [^{3}H]phenylalanine, [^{3}H]proline and [^{3}H]tyrosine) Amersham International p.l.c., specific activity 70–100 Ci/mmole. Detailed protocols for purification of enzyme, preparation of radiolabelled antiserum and criteria of specificity are given elsewhere [2, 6]. The [^{3}H]CAT can be combined with the immunocytochemical localisation of TH or other neurotransmitter specific antigens using rabbit polyclonal antibodies.

2.2. *Tissue preparation and immunocytochemical labelling*

The brains from animals infused with [^{3}H]serotonin or untreated, were fixed by aortic arch perfusion for 6 min (50 ml/min) with a solution containing 0.2–0.5% glutaraldehyde and 4% paraformaldehyde in 0.1 M phosphate buffer pH 7.2. The lower concentrations of glutaraldehyde are most appropriate when only immunocytochemical markers are used, whereas in situ retention of 5-[^{3}H]HT in axon terminals requires the highest concentration of the aldehyde compatible with immunocytochemistry. In combined studies of TH and 5-[^{3}H]HT, the optimal fixation included a 6 min perfusion and 1-h immersion in 0.5% glutaraldehyde at room temperature. Following fixation, the brains are sectioned on a vibrating microtome (Vibratome®, Oxford Instruments) and processed for immunocytochemistry using the peroxidase-antiperoxidase (PAP) method of Sternberger [7].

When using the [^{3}H]antiserum in combination with peroxidase immunocytochemistry, the tritiated antiserum always precedes the antiserum being localised by the PAP method. Unless this sequence is followed, the second primary antiserum frequently binds non-specifically to the reaction product thus giving a false co-localisation for the two antigens. For example, we found that when [^{3}H]CAT antiserum was applied following immunocytochemical labelling for TH by the PAP method the autoradiographic and peroxidase markers were localised in single catecholaminergic neurons. Moreover, when the first antiserum was directed against enkephalin and the second against [^{3}H]CAT, both markers were found to co-exist in the dorsal horn and other regions of the CNS which normally exhibited immunoreactivity exclusively for enkephalin. However, co-localisation of the two markers did not occur when the [^{3}H]CAT antiserum preceded that of TH or enkephalin. A practical sequence for the combined localisation of CAT and TH includes:

1) 24-h incubation of Vibratome® slices in monoclonal [^{3}H]CAT antiserum (1:10 dilution at 4°C);
2) two 30-min washes;
3) 12-h incubation in rabbit antiserum to TH (1:1,000);
4) two 30-min washes;
5) 45-min incubation in goat anti-rabbit immunoglobulin (1:50, Miles Lab);
6) two 5-min washes;
7) 45-min incubation in a solution containing PAP (1:100);
8) two 30-min washes; and
9) 6 min in a solution containing 3,3′-diaminobenzidine and hydrogen peroxide.

All washes contain 0.1 M Tris-saline pH 7.2, and 1% normal goat serum. All incubations except the first step, are at room temperature. Following the diaminobenzidine reaction, both the slices from animals infused with [^{3}H]serotonin and incubated with [^{3}H]CAT are similarly processed for autoradiography. In the serotonin studies the procedure begins with step 2. The immunocytochemical labelling procedures have been described in detail in two recent reviews [4, 8].

2.3. *Autoradiography*

For light microscope autoradiography, immunolabelled Vibratome® slices were mounted onto gelatin-coated slides. The sections were then dehydrated in graded ethanols, defatted in xylene and rehydrated through an inverse series of ethanols prior to autoradiography. The sections were dried overnight at 37°C and coated by dipping in melted (40°C) Kodak NTB2 emulsion diluted 1:1 with distilled water. The autoradiographs were developed 1–2 weeks later in Kodak D-19 developer (4 min at 17°C), rinsed in distilled water, air-dried and coverslipped. Examination of these autoradiographs by light microscopy enables reactive regions to be selected for ultrastructural analysis.

For electron microscope autoradiography, immunolabelled Vibratome® slices adjacent to those processed for light microscopy were post-fixed for 1 h in 2% neutral-buffered osmium tetroxide (0.12 M phosphate buffer plus 7% sucrose), dehydrated in ethanols and flat-embedded in Epon between two plastic coverslips. After polymerisation, the embedded slices were viewed by light microscopy in order to select regions containing optimal immunocytochemical and autoradiographic labelling. The blocks were then trimmed and ribbons of thin sections were cut from the surface of each block and deposited on parlodion-coated slides. The sections on the slides were stained with uranyl acetate and lead citrate, coated with carbon and dipped in Ilford L-4 emulsion diluted 1:5. After 2–12 months of exposure, the autoradiographs were developed in Kodak Microdol X ($1\frac{1}{2}$ min at 18°C), rinsed for 30 s in distilled water, fixed for 5 min in 30% sodium thiosulphate and washed at 4°C in three consecutive baths of distilled water. The parlodion membranes were then detached by floating on the surface of distilled water and the sections collected on copper grids. A detailed protocol of the autoradiography procedures may be found in the review by Descarries and Beaudet [1].

3. Applications

3.1. *Serotonin-catecholamine interaction in medial nuclei of the solitary tracts*

The medial nuclei of the solitary tracts (m-NTS) contains the catecholaminergic, predominantly noradrenergic, neurons of the A2 group [9]. In addition, the m-NTS contains relatively high endogenous levels of serotonin (5-HT) [10] which is found within varicose axons seen by fluorescence histochemistry [11] or by immunocytochemistry using an antiserum to 5-HT [12]. By combining autoradiography for 5-[^{3}H]HT with immunocytochemistry for tyrosine hydroxylase (TH), direct synaptic relations between the serotonergic terminals and TH-labelled catecholaminergic neurons can be shown [13]. The autoradiographically-labelled terminals contain a mixed population of small round, large dense-cored, and tubular vesicles (Fig. 1). The synaptic contacts formed by these terminals are primarily located on the shafts of dendrites within the m-NTS. These dendrites are both TH-labelled and unlabelled, even at the most superficial portion of the Vibratome® slice. In sections

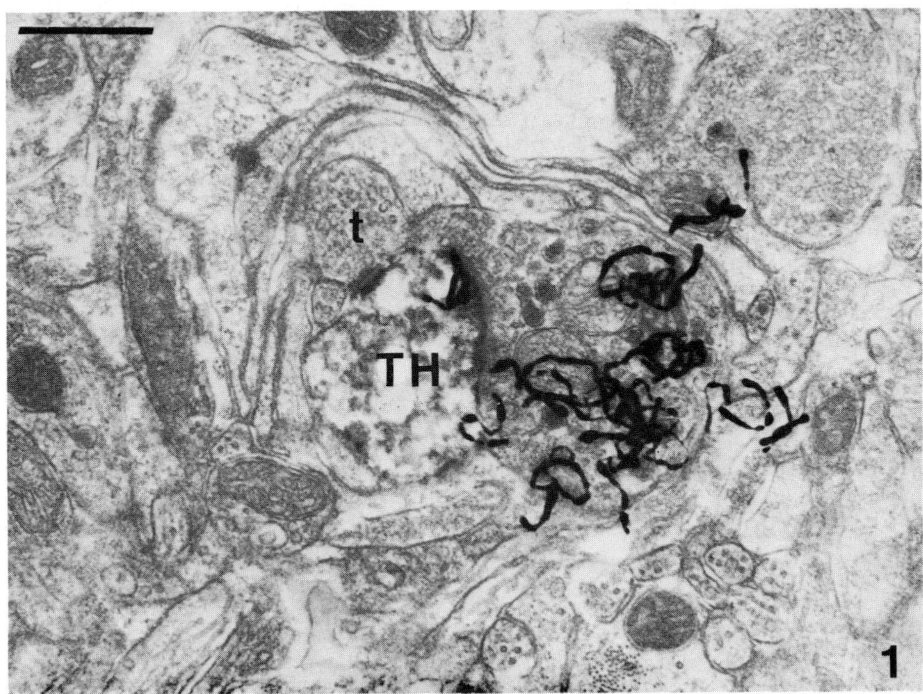

Fig. 1. Combined autoradiographic and immunocytochemical localisation of a serotonergic terminal and tyrosine hydroxylase immunoreactivity in noradrenergic dendrite in m-NTS. Reduced silver grains for 5-[^{3}H]HT are seen over a terminal containing a mixed population of synaptic vesicles. A second unlabelled terminal (t) is seen on same dendrite. Bar = 0.5 μm.

collected more than 1–2 μm from the surface of the Vibratome® slice, the number of TH-labelled profiles diminishes predominantly due to lack of penetration of the antisera. Thus, conclusions regarding whether the recipient dendrites contain TH can be determined only for the most superficial sections. Ironically, near the surface, where immunocytochemical labelling is maximal, the number of terminals labelled with 5-[^{3}H]HT may be reduced by loss of the isotope from cut axons. Thus, the number of synaptic contacts between the immunolabelled processes and those marked autoradiographically probably represents only a small fraction of the total interactions.

3.2. Serotonin-catecholamine interaction in ventral tegmental areas

The ventral tegmental area (VTA), together with the central, and rostral, linear and interfascicular raphe nuclei which lie medial to it, comprise the largest collection of dopamine (DA)-containing nerve cell bodies in the rat brain. These neurons, designated collectively as group A10 by Dahlstrom and Fuxe [9] have been visualised by histofluorescence [14, 15] and immunohistochemistry [16]. The serotonin innervation rat VTA has been documented by microchemical [17],

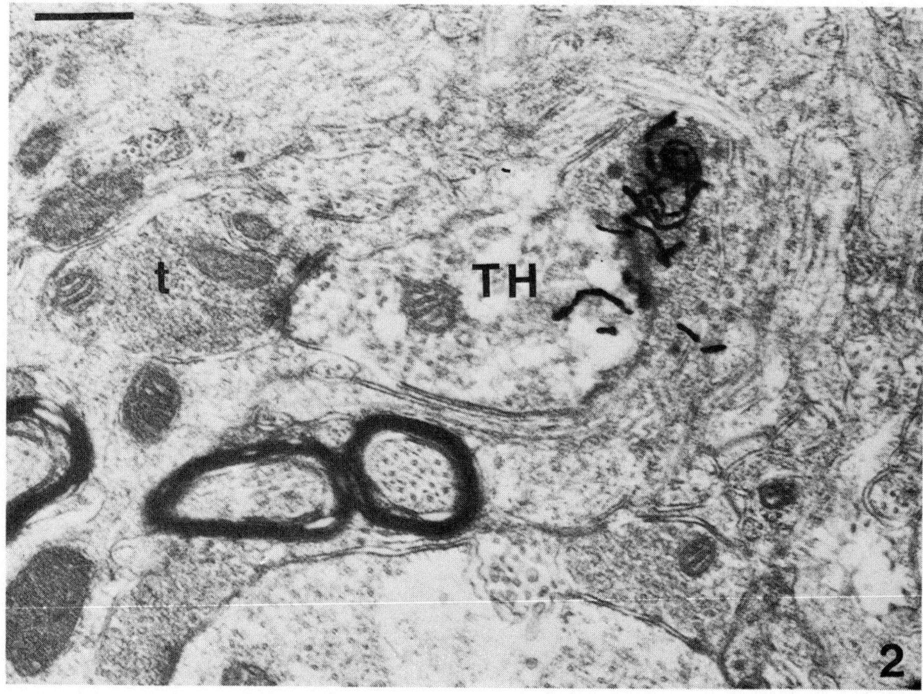

Fig. 2. Combined autoradiographic and immunocytochemical identification of serotonergic and dopaminergic elements in rat ventral tegmental area. A 5-[^{3}H]HT-labelled axon terminal containing numerous small electron-lucent vesicles and several large granular vesicles is seen in synaptic contact with a tyrosine hydroxylase-immunoreactive dendrite. Note that the latter also receives an unlabelled synaptic terminal (t). Bar = 0.5 μm.

immunohistochemical [12] and autoradiographic [18] methods. Moreover, biochemical studies have suggested that ascending serotonergic neurons might be implicated in the regulation of DA metabolism in mesocortical and mesolimbic projections from the VTA [19].

Ultrastructural examination of cellular relationships between 5-[^{3}H]HT-labelled terminals and TH-immunoreactive elements reveals that many of the 5-HT axonal varicosities are directly apposed to TH-immunoreactive dendrites in the VTA and interfascicular nucleus. However, only about 15% of the appositions show differentiated junctional complexes in single thin sections (Fig. 2). These synaptic junctions are all asymmetrical and located on dendritic shafts. Serotonin varicosities apposed to or in synaptic contact with TH-immunoreactive dendrites show the same ultrastructural features as those synapsing on non-immunoreactive elements. They are small (0.6 μm in mean diameter) and characteristically exhibit a mixed population of small (15–25 nm in diameter), pleomorphic, electron-lucent vesicles and a few large granular vesicles (Fig. 2). The results of the combined labelling provide morphological evidence in favour of direct functional interactions between 5-HT and DA-containing neurons in the VTA.

4. Conclusions

Peroxidase or tritiated immunocytochemical markers combined with uptake of 5-[^{3}H]HT and high resolution autoradiography can be used to determine the existence of synaptic interactions between chemically defined neurons in single sections of brain. The dual localisation requires fixation conditions compatible with immunocytochemistry and in situ retention of 5-[^{3}H]HT. Quantitative evaluation of the number of synaptic interactions is limited by partial penetration of antisera. Thus only positive contacts can be determined.

5. References

1. Descarries, L. and Beaudet, A. (1983) The use of autoradiography for investigating transmitter-specific neurons. In: A. Bjorklund and T. Hökfelt (Eds.) Handbook of Chemical Neuroanatomy, Vol. I: Methods in Chemical Neuroanatomy. Elsevier, Amsterdam, pp. 286–364.
2. Ross, M.E., Park, D.H., Teitelman, G., Pickel, V.M., Reis, D.J. and Joh, T.H. (1983) Immunohistochemical localization of choline acetyltransferase using a monoclonal antibody: an autoradiographic method. Neuroscience 10, 907–922.
3. Cuello, A.C., Priestley, J.V. and Milstein, C. (1982) Immunocytochemistry with internally labeled monoclonal antibodies. Proc. Natl. Acad. Sci. USA 79, 665–669.
4. Pickel, V.M. and Teitelman, G. (1983) Light and electron microscopic immunocytochemical localization of single and multiple antigens. In: J. Furness and M. Costa (Eds.) IBRO Handbook Series: Methods in Neuroscience. John Wiley and Sons, New York.
5. Joh, T.H., Gegham, C. and Reis, D.J. (1973) Immunochemical demonstration of increased tyrosine hydroxylase protein in sympathetic ganglia and adrenal medulla elicited by reserpine. Proc. Natl. Acad. Sci. USA 70, 2767–2771.
6. Park, D.H., Ross, M.E., Pickel, V.M., Reis, D.J. and Joh, T.H. (1982) Antibodies to rat choline acetyltransferase for immunochemistry and immunocytochemistry. Neurosci. Lett. 34, 129–135.
7. Sternberger, L.A. (1974) Immunohistochemistry. Prentice-Hall Inc., Englewood Cliffs, New Jersey, 129 pp.
8. Pickel, V.M. (1981) Immunocytochemical methods. In: L. Heimer and M.J. Robards (Eds.) Immunocytochemical Methods in Neuroanatomical Tract-Tracing Methods. Plenum Press, New York, pp. 483–509.
9. Dahlstrom and Fuxe (1964) Evidence for the existence of monoamine-containing neurons in the central nervous system. I. Demonstration of monoamines in the cell bodies of brain stem neurons. Acta. Physiol. Scand. 62, Suppl. 232, 1–55.
10. Palkovits, M., Brownstein, M. and Saavedra, J.M. (1974) Serotonin content of the brain stem nuclei in the rat. Brain Res. 80, 237–249.
11. Fuxe, K. (1965) Evidence for the existence of monoamine neurons in the central nervous system. IV. Distribution of monoamine terminals in the central nervous system. Acta. Physiol. Scand. 64, Suppl. 247, 39–85.
12. Steinbusch, H.W.M. (1981) Distribution of serotonin-immunoreactivity in the central nervous sytem of the rat, cell bodies and terminals. Neuroscience 6, 557–618.
13. Pickel, V.M., Joh, T.H., Chan, J. and Beaudet, A. (1984) Serotonergic terminals: ultrastructure and synaptic interaction with catecholamine containing neurons in the medial nucleus of the solitary tract. J. Comp. Neurol. (in press).
14. Lindvall, O. and Bjorklund, A. (1974) The organization of ascending catecholamine neuron systems in the rat brain as revealed by the glyoxylic acid fluorescence method. Acta. Physiol. Scand., Suppl. 412, 1–48.
15. Fallon, J.H. and Moore, R.Y. (1978) Catecholamine innervation of the basal forebrain. IV. Topography of the dopamine projection to the basal forebrain and neostriatum. J. Comp. Neurol. 180, 545–580.

16. SWANSON, L.A. (1982) The projections of the ventral tegmental area and adjacent regions: a combined fluorescent retrograde tracer and immuno-fluorescence study in the rat. Brain Res. Bull. 9, 321–353.
17. PALKOVITS, M., SAAVEDRA, J.M., JACOBOWITZ, D.M., KIZER, J.S., ZABOREZKY, L. and BROWNSTEIN, M.J. (1977) Serotonergic innervation of the forebrain: Effect of lesions on serotonin and tryptophan hydroxylase levels. Brain Res. 130, 121–134.
18. PARENT, A., DESCARRIES, L. and BEAUDET, A. (1981) Organization of ascending serotonin systems in the adult rat brain. An autoradiographic study after intraventricular administration of (^{3}H)5-hydroxytryptamine. Neuroscience 6, 115–138.
19. HERVÉ, D., SIMON, H., BLANC, G., LE MOAL, M., GLOWINSKI, J. and TASSIN, J.P. (1979) Increased utilization of dopamine in the nucleus accumbens but not in the cerebral cortex after dorsal raphe lesion in the rat. Neurosci. Lett. 15, 127–134.

Immunolabelling for Electron Microscopy (Polak/Varndell, eds.).

CHAPTER 19

Haematological electron immunocytochemistry

Detection of cell surface antigens with monoclonal antibodies

M. De Waele

Department of Haematology, University Hospital of the Free University Brussels (V.U.B.), B-1090 Brussels, Belgium

Contents

1. Introduction

The application of electron immunocytochemistry in haematology has generally been performed to study the presence, the density and the distribution of an antigen or receptor on the surface membrane [1–24] or in the cytoplasm [25–31] of a given cell type. It has also been applied to define the ultrastructural characteristics of the cells identified by a given antibody [32–35]. Most studies have dealt with cell surface antigens and receptors of leukocytes [3, 7–12, 14, 18–24, 32–35], red blood cells [1, 2, 4, 6, 17] or platelets [10, 13, 15, 16]. In leukocyte cell suspensions cell surface immunoglobulin [3, 7, 8, 11, 25, 32, 33], T-lymphocyte antigens [18, 19, 21, 34, 35] and binding sites for lectins [8, 12], insulin [14] or the Fc portion of IgG [9] have been visualised. Mobility and redistribution after binding with ligands [3, 32] and the uptake and intracellular transport [14] have been studied. The labelling procedures used for haematological cells were not significantly different from those described for cells of other sources [36–38]. Ferritin [1–3, 9], horseradish peroxidase [4–7, 11, 20, 22–33, 36] or colloidal gold [8, 10, 12–19, 21, 34, 35, 39, 40] were used as electron-dense markers.

A large proportion of leukocyte cell surface and intracytoplasmic antigens are inactivated by the 'classical' fixation and embedding procedures. Therefore most studies on tissue biopsies have been performed on frozen sections [23, 24] or on sections of fixed non-embedded tissue [22]. The labelled samples were then processed for electron microscopy following the standard techniques.

In most studies antisera containing polyclonal antibodies directed against the antigen have been used. Such antisera are difficult to standardise and to reproduce in different laboratories. One of the major developments in the field of immunology during the last few years is that of hybridoma technology, permitting the production of monoclonal antibodies against cellular antigens [41]. Their high specificity and their availability in almost unlimited amounts makes these antibodies almost ideal reagents for light and electron immunocytochemistry. A wide range of monoclonal antibodies recognising leukocyte cell surface antigens are now commercially available. In this chapter the use of these antibodies for electron immunocytochemistry is reviewed.

The monoclonal antibodies have mainly been applied on cell suspensions in an indirect method with colloidal gold-labelled goat anti-mouse antibodies as second step [15, 18, 19, 21, 34, 35, 39, 40]. This approach will be discussed in detail. In addition, the work on cell suspensions with more step immunoperoxidase techniques [20] and the detection of lymphocyte cell surface antigens in tissue sections [20, 23, 24] will be summarised.

2. Immunogold staining method on cell suspensions

2.1. Description of the procedure

Leukocyte cell surface antigens identified by monoclonal antibodies were visualised in electron microscopy by incubating the cells with colloidal gold-labelled antibodies [15, 18, 19, 21, 34, 35, 39, 40]. The cells were then fixed and processed for transmission electron microscopy following standard techniques.

2.1.1. Preparation of the cell suspensions
Venous blood was collected in 5% EDTA in phosphate buffered saline (PBS) at pH 7.4. Mononuclear cell suspensions were prepared by Ficoll-Hypaque density gradient centrifugation. The cells were washed three times, for 5 min each in PBS at pH 7.4, containing 1% bovine serum albumin (BSA), 1% heat inactivated normal human AB serum, and 0.02 M sodium azide (PBS-BSA 1%-AB 1%-az: washing buffer). Suspensions of 3×10^{10} cells/l were made in PBS-BSA 5% at pH 7.4, containing 4% AB serum and 0.02 M sodium azide (PBS-BSA 5%-AB 4%-az: incubation buffer).

2.1.2. Monoclonal mouse antibodies
Monoclonal mouse antibodies raised against lymphocyte cell surface antigens are now commercially available from different companies. In most studies antibodies of the OKT series (Ortho Pharmaceutical Corp. and Ortho Diagnostic Inc. Raritan, NJ, U.S.A.) were used. More information about the specificity of these antibodies can be found in the literature [42, 43]. OKT3, OKT4 and OKT8, respectively identify the mature T-lymphocytes, T-inducer/helper cells and T-cytotoxic/suppressor cells. OKT6 reacts with the majority of thymocytes but not with peripheral blood lymphocytes. OKM1 recognises an antigen present on monocytes, granulocytes and some null cells. OKIa1 reacts with a HLA-Dr antigen present on B-lymphocytes, activated T-lymphocytes and some monocytes. The antigen recognised by the OKT10 antibody is present on haemopoietic precursor cells, thymocytes, plasma cells, activated T- and B-lymphocytes and some null cells. OKT11 reacts with the E-rosette receptor and is used to identify the peripheral T-cells.

These monoclonal antibodies were applied as dilutions of ascitic fluid or of the commercially available reagents. The lyophilised reagents were reconstituted with PBS-BSA 5%. Each contained small amounts of sodium azide as preservative. The working dilution of these antibodies, as determined by titration on a leukocyte cell suspension, varied between 1 and 5 μg/ml.

2.1.3. Colloidal gold-labelled secondary antibodies
The colloidal gold-labelled secondary antibodies were prepared as described by De Mey [44]. Mainly gold particles with a mean diameter of 30 nm were used. They were prepared by the method of Frens [45]. Just before coupling, the gold sol was brought to pH 9.0 with potassium carbonate. Goat antibodies to mouse immunoglobulin G (GAM) were isolated from goat anti-mouse IgG serum by affinity chromatography on mouse IgG-Sepharose 4B. These antibodies also reacted with molecules of mouse immunoglobulin classes other than IgG, probably through recognition of the common light chains. The purified antibodies were dialysed against 2 mM borax HCl buffer at pH 9.0. Immediately before use, microaggregates were removed by centrifugation at $100{,}000 \times g$ for 1 h at 0°C.

The amount of antibody necessary for optimal protection of the gold sol against flocculation in salt solutions was determined according to Geoghegan and Ackerman [8]. This antibody was mixed with the gold sol by stirring gently. After 2 min BSA (BSA-Sigma) fraction V, 10% in borax buffer pH 9.0 was added to achieve a final concentration of 10 mg/ml. Unstabilised marker and free or loosely bound proteins were removed by three cycles of centrifugation (14,000 × g, 1 h, 0°C) and resuspension in 0.02 M Tris-buffered saline at pH 8.2, containing 10 mg/ml BSA and 0.02 M sodium azide (TBS-BSA-az). The gold probe (GAMG30—Janssen Pharmaceutica, Beerse, Belgium) was then obtained by resuspending the last pellet (mobile red pool) in the appropriate volume of TBS-BSA-az to assure that the spectrophotometric extinction at 520 nm of a 1/20 reagent dilution was 0.350. It was centrifuged at low speed before use to remove microaggregates.

2.1.4. Cell labelling

One hundred microlitres (μl) of the cell suspension (approximately 5×10^6 cells) was incubated for 30 min at room temperature with 100 μl of the monoclonal antibody dilution. The cells were subjected to three 5 min washings with PBS-BSA 1%-AB 1%-az. Thereafter the cells were resuspended in 100 μl of PBS-BSA 5%-AB 4%-az and incubated with 100 μl of the GAMG30 reagent for 1 h at room temperature. During the incubations the cells were agitated every 10 min. After three 5 min washings the cells were fixed in a pellet [46] with 1% glutaraldehyde in 0.1 M cacodylate buffer at pH 7.2 for 30 min at room temperature. Then the fixative was carefully removed from the pellet and replaced with 0.1 M cacodylate buffer. The pellets were generally firm and showed no tendency to disintegrate in this washing solution. They were then postfixed with 1% osmium tetroxide in distilled water for 90 min at room temperature and were stained with 0.5% uranyl acetate in veronal buffer pH 5.3 for 3 h at 37°C. The pellets were then dehydrated in a graded series of ethanol and embedded in Spurr's resin. Ultrathin sections were stained with Reynold's lead citrate for 2 min at room temperature. They were examined with a Zeiss 9S electron microscope.

2.2. Results

2.2.1. Appearance of the preparations

Although the cells were only fixed after the labelling procedure, their ultrastructure in the preparations was well preserved. A good contrast due to the uranyl acetate and lead citrate counterstaining permitted an optimal visualisation of even small cell organelles such as electron-dense granules and parallel tubular arrays. The highly electron-dense colloidal gold particles were easily distinguished and allowed a precise localisation of the antigens on the cell surface membrane to be made.

Granulocytes and monocytes were easily differentiated from the lymphocytes by their morphological characteristics. In the lymphocytes, different morphological subtypes were found. Leukocytes reacting with the monoclonal antibodies showed patches, formed by numerous gold particles, on their surface membrane (Fig. 1). A capping of the label was rarely found. Cells without or with only a few scattered gold particles on the surface membrane were considered to be unreactive with the

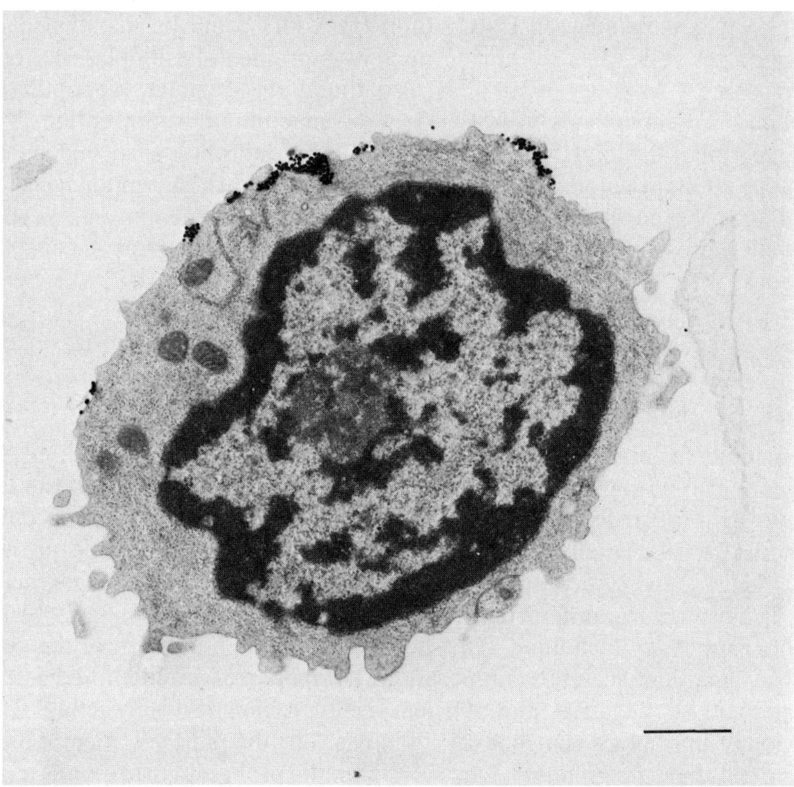

Fig. 1. Peripheral blood lymphocyte labelled with OKT3 and GAMG30. The gold particles are found in patches on the surface membrane. The ultrastructure is well preserved (bar = 1 μm).

monoclonal antibody. Some of these cells may, however, be reactive cells sectioned in a plane not passing through a single patch of gold particles. To evaluate this possibility quantitative studies in comparison with light microscopy should be performed.

A small percentage of the labelled lymphocytes showed pinocytotic vesicles containing a few gold particles. Endocytosis of the labelled membrane in large phagosome-like organelles was occasionally found in monocytes.

2.2.2. Ultrastructural characteristics of normal lymphocyte subpopulations

With this immunogold staining technique, Matutes and Catovsky [34] have analysed the ultrastructural characteristics of the normal lymphocyte subpopulations identified by the OKT3, OKT4, OKT8 and FMC4 monoclonal antibodies. They found that the T-helper (OKT4+) and the T-cytotoxic/suppressor cells (OKT8+) have distinct ultrastructural morphology. The majority of the OKT4+ cells have a high nuclear/cytoplasmic (N/C) ratio and few cytoplasmic organelles, whilst most OKT8+ cells have a low N/C ratio and numerous organelles, notably a well-developed Golgi apparatus, lysosomal structures and parallel tubular arrays.

B-lymphocytes labelled with FMC4 (anti-HLA-Dr) could be distinguished from T-lymphocytes (OKT3+) by having numerous profiles of endoplasmic reticulum and ribosomes. Matutes et al. [35] also found two distinct subpopulations of convoluted T-lymphocytes (OKT3 +) in normal blood, representing 2–4% of normal peripheral blood lymphocytes. The close morphological and membrane phenotype similarities observed between these two types of lymphocytes and the cells of Sezary syndrome and adult T-cell lymphoma/leukaemia suggest that they may well represent the normal counterparts of the malignant T-cells in both conditions.

2.3. Comments

2.3.1. Light microscopical examination of the labelled cells

The outcome of an immunogold labelling experiment for transmission electron microscopy can be rapidly evaluated by processing a small sample of the labelled cells for light microscopy. All details of this approach have been published elsewhere [21, 39]. Briefly, the labelled cells are fixed with low concentrations of glutaraldehyde, and cytocentrifuge preparations are made. The endogenous peroxidase activity in granulocytes and monocytes can be stained with the Graham/Karnovsky technique. This helps with the identification of the cells and permits a rapid and accurate enumeration. Lymphocytes are identified as endogenous peroxidase negative mononuclear cells. Lymphocytes reacting with the monoclonal antibodies have dark granules on the surface membrane. The negative cells have no granules. One such 'granule' probably corresponds to a patch of gold particles seen in electron microscopy.

Lymphocyte subsets enumerated with this method showed a good correlation with those obtained with immunofluorescence microscopy [39]. This proved the reliability of this approach.

2.3.2. Incubation with the monoclonal antibody

The concentration of the monoclonal antibodies in the labelling procedure for electron microscopy was identical to that used for light microscopy. The latter was determined by establishing an antibody dilution curve [21]. Therefore, the same amount of mononuclear cells was incubated with graduated concentrations of the monoclonal antibody and the number of positive cells was determined. In Figure 2 an OKT8 ascites dilution curve is shown. The number of OKT8 positive cells did not vary significantly with ascites dilutions varying between 1/250 and 1/2,000, although the density of the surface labelling on the positive cells decreased. With further dilutions of the monoclonal antibody the number of OKT8 positive cells rapidly fell. The highest dilution of the antibody with a number of positive cells in the horizontal part of the curve, was used as the working dilution in the labelling procedure (e.g. for this OKT8 ascites: 1/2,000). This antibody dilution curve was nearly identical to that found with immunofluorescence microscopy in the same cell suspension (Fig. 2, IF). Similar dilution curves have been found for other monoclonal antibodies [21]. In general, with commercial preparations the working concentrations varied between 1 and 5 μg/ml.

Variation of the incubation time with the OKT3, OKT4 and OKT8 monoclonal antibodies from 15 min to 1 h did not change the number of positive cells in light microscopy [21]. At the electron microscopical level, a slight increase of the density of the gold marker on the surface membrane of the positive cells was observed.

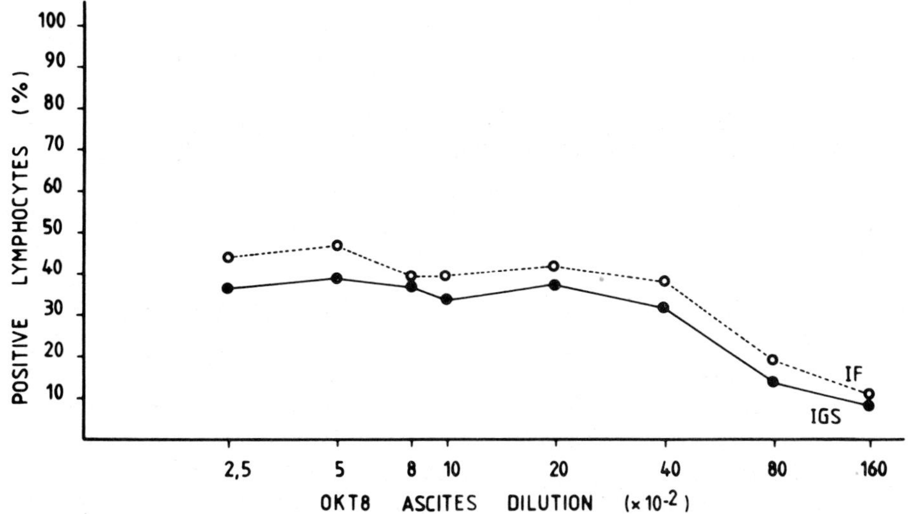

Fig. 2. OKT8 ascites dilution curve as established in light microscopy. Approximately 10^6 mononuclear cells were incubated with graduated concentrations of the monoclonal antibody (horizontal axis). The percentage of OKT8 positive lymphocytes was determined (vertical axis) by immunogold staining (IGS) and immunofluorescence microscopy (IF).

2.3.3. Incubation with the colloidal gold-labelled secondary antibodies

The choice of the mean particle size of the gold probe for electron immunocytochemistry depends upon the magnification desired and upon the steric accessibility of the binding sites [47]. The latter is especially important when direct labelling techniques are used. In indirect techniques the macromolecule from the first layer may protrude from its binding site and provide a greater stereochemical accessibility to the gold probe. According to Horisberger [47], cell surface marking in transmission electron microscopy is best achieved with particles from 5 to 12 nm diameter.

With monoclonal antibodies secondary antibodies coupled to gold particles of 5, 10, 20 or 30 nm were used. For a given concentration of the monoclonal antibody the number of gold particles on the surface membrane of the positive cells increased with decreasing diameter of the gold particles in the probe. This is in accordance with the findings of Horisberger [47] on red blood cells labelled with soyabean lectin directly coupled to gold particles. He hypothesised that the larger size markers can cover many cell surface binding sites or that some binding sites may be inaccessible to the larger particles.

Prolongation of the incubation time of the cells with the gold probe increased the gold density on the surface membrane of the positive cells, especially during the first 30 min. This was accompanied by an increase in the number of positive cells. With incubations longer than 1 h the background staining became more important and significantly affected the data obtained.

2.3.4. Ligand-induced redistribution of cell surface binding sites

When leukocyte surface membrane antigens or receptors are detected on unfixed cells with divalent antibodies and at room temperature, patching occurs [3, 32, 48, 49]. This is due to a passive redistribution of the antigen-antibody-marker complexes in the cell surface membrane. Normally these patches fuse into a single mass at one pole of the cell (= cap formation) [49]. The labelled membrane can also be internalised or shed from the cell [14, 48].

Patching, capping and internalisation can be prevented by fixing the cells before labelling [32, 49] or by the use of monovalent antibody fragments for the detection [3, 32, 48]. In the technique described above prefixation was not performed as it could inactivate the cell surface antigens. Patches of gold particles were found on the positive cells, but capping and internalisation were not prominent. These phenomena require an active cell metabolism [48] and this was inhibited by adding sodium azide to the incubation and washing buffers [3, 48, 49]. Without this reagent capping and internalisation was readily seen. In the lymphocytes a few pinocytotic vesicles containing a small number of gold particles were found. In labelled monocytes and granulocytes however, larger parts of the membrane were internalised and phagosome-like organelles filled with gold particles were present (Fig. 3). Capping and endocytosis can also be reduced by performing the labelling at low temperature [3, 32, 48, 49].

2.3.5. Control procedures

The result of the labelling procedure can be evaluated by performing several control procedures [36]. Two negative controls were regularly performed [21, 34, 39]. The monoclonal antibody was replaced with an antibody of the same species but with unrelated specificity. When normal peripheral blood was examined, the OKT6 antibody was chosen for this purpose. In other studies an ascites dilution from a Balb/c mouse injected intraperitoneally with a non-producing hybridoma clone was used. Another negative control consisted of omitting the monoclonal antibody from the procedure.

In both negative control procedures a small percentage of the lymphocytes showed gold particles on the surface membrane [21, 39]. This background staining was probably due to a non-specific fixation of the colloidal gold probe on to the cell, e.g. on Fc receptors. This non-specific positivity could be reduced by adding normal human AB serum to the incubation media [21, 39].

When surface immunoglobulin is detected, Fc receptor binding of plasma immunoglobulins may significantly affect the data obtained [38]. This phenomenon can be reduced by preincubating the cells for 30 min at 37°C [50], or by fixing the cells before the labelling with low concentrations of formaldehyde [51].

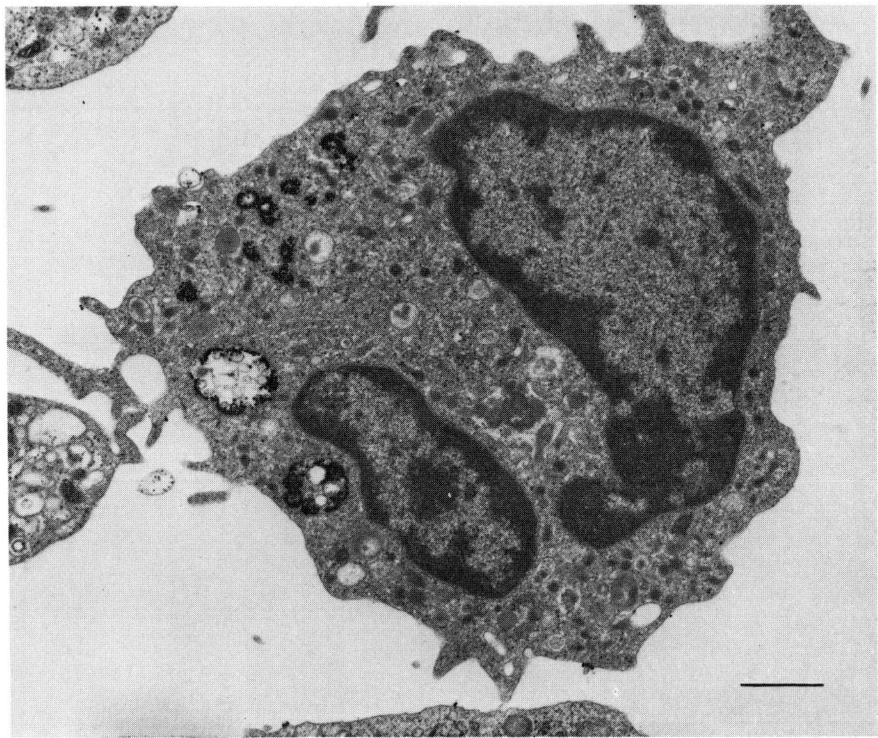

Fig. 3. Peripheral blood monocyte labelled with the OKM1 monoclonal antibody and GAMG30 in a procedure without sodium azide. Most of the labelled membrane has been internalised (bar = 1 μm).

Non-specific positivity may be important when cells of B-lymphoproliferative disorders are examined. Many cells may show a few isolated gold particles or small patches on the surface membrane. In most instances however, a specific labelling of the cells can be distinguished by a higher gold density.

Positive control procedures should also be performed. False-negative results can be due to a loss of reactivity of the monoclonal antibody or of the secondary reagent. The stability of these reagents depends on the way of storage [39]. In addition, fixation of the cells before labelling may denature the cell surface antigens and lead to false-negative results.

2.3.6. Fixation of cells before labelling

Performing the labelling on unfixed cells may not be favourable for an optimal preservation of cellular morphology. However, most leukocyte surface antigens identified by the monoclonal antibodies described above proved to be relatively sensitive to glutaraldehyde fixation. Only very mild conditions of prefixation did not reduce significantly the number of gold particles on the cell surface membrane. In general, these conditions did not improve the ultrastructure of the labelled cells.

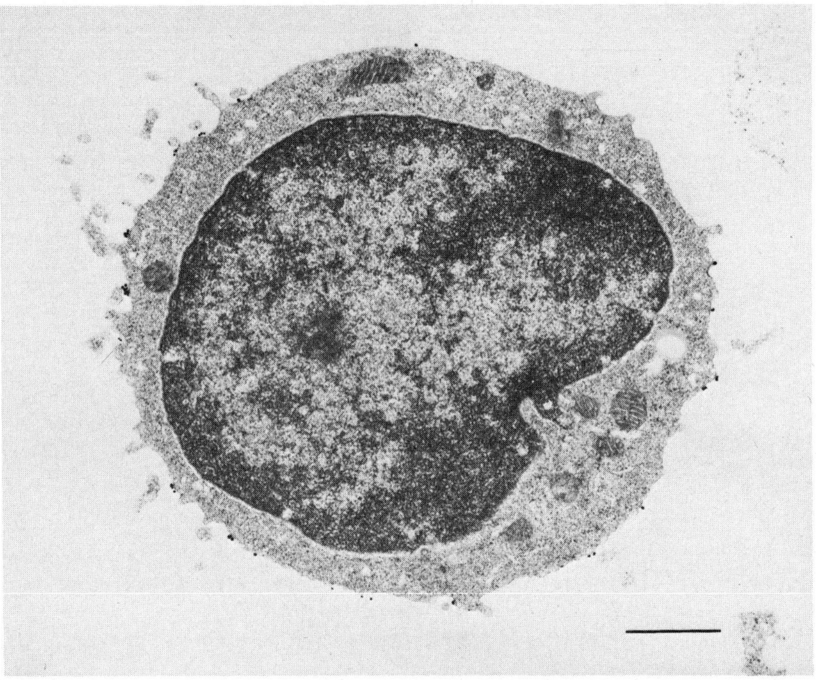

Fig. 4. Peripheral blood lymphocyte labelled with OKT3 and GAMG30. The cells were fixed before the labelling with 5% formaldehyde in 0.1 M cacodylate buffer for 10 min at room temperature. The gold label is distributed homogeneously along the cell surface membrane. The cellular morphology is not as good as without prefixation (bar = 1 μm).

Formaldehyde, freshly prepared from paraformaldehyde, is known to be a less active cross-linking agent [52]. Compared to the same concentration of glutaraldehyde it gave less denaturation of the cell surface antigens but even with high concentrations the cellular morphology was worse than that without prefixation (Fig. 4).

Fixation immobilised the surface antigens so that the gold particles were rather homogeneously distributed along the cell surface membrane. Large patches were not found. In light microscopy, the labelled cells showed a rim of the fine gold granules instead of distinct dark granules on the surface membrane. This made a rapid and accurate enumeration of the positive cells more difficult.

After prefixation of the cells with aldehydes, free aldehyde groups must be blocked since they are reactive with amino groups and could introduce non-specific binding of antibodies [49]. Blocking of reactive aldehyde groups can be done by incubating the cells with ammonium chloride, glycine, lysine or borohydride [49].

2.3.7. Labelling of platelet and red blood cell antigens

In addition to leukocyte surface antigens, platelet and red blood cell surface antigens can also be labelled with this immunogold staining technique. Figure 5

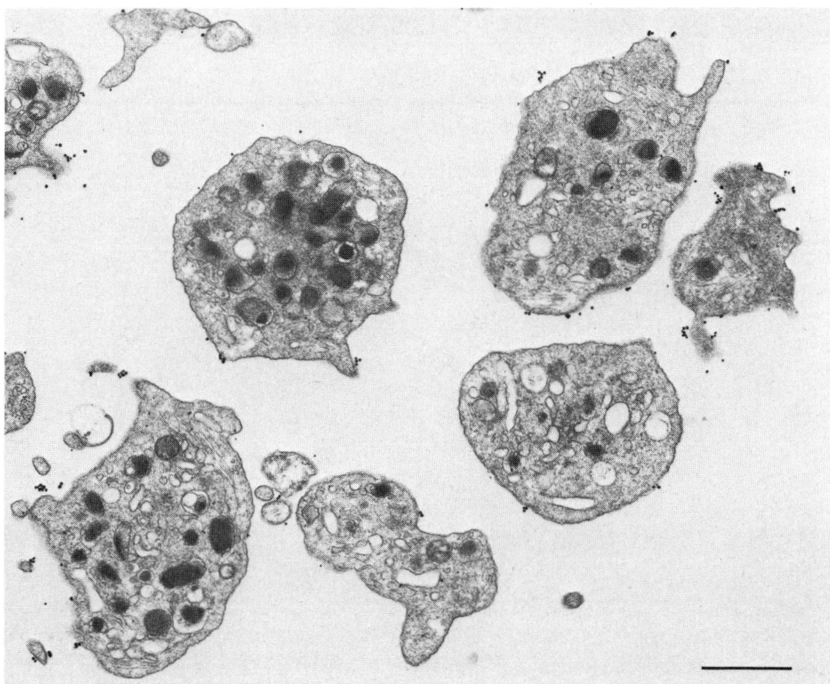

Fig. 5. Blood platelets labelled with a monoclonal antibody directed against β_2-microglobulin and GAMG20. The platelets were fixed before the labelling. The gold label is homogeneously distributed along the surface membrane (bar = 1 μm).

shows normal human platelets labelled with a monoclonal antibody directed against β_2-microglobulin (Becton Dickinson) and goat anti-mouse immunoglobulin antibodies (GAM). On the red blood cells in Figure 6, B blood group antigens were detected with a human antiserum and goat anti-human immunoglobulin antibodies (GAH). Both secondary reagents were coupled to 20 nm gold particles (GAMG20 and GAHG20—Janssen Pharmaceutica, Beerse, Belgium). The cells were fixed before labelling with 1% glutaraldehyde for 30 min at room temperature. The labelling was performed as described for leukocytes. Platelet and red blood cell morphology was well preserved. The prefixation with glutaraldehyde did not completely inactivate the surface membrane antigens. The gold label was homogeneously distributed along the surface membrane in the form of single particles or small patches. There was a striking heterogeneity in density of the label within cells from the same sample.

Without fixation before labelling, most platelets appeared degranulated and a lot of red blood cells had lost their discoid shape. In these conditions, patching of the gold label was observed on the positive cells.

2.3.8. Enzyme cytochemistry on immunogold-labelled cells

The immunological detection of cell surface antigens with immunogold staining can

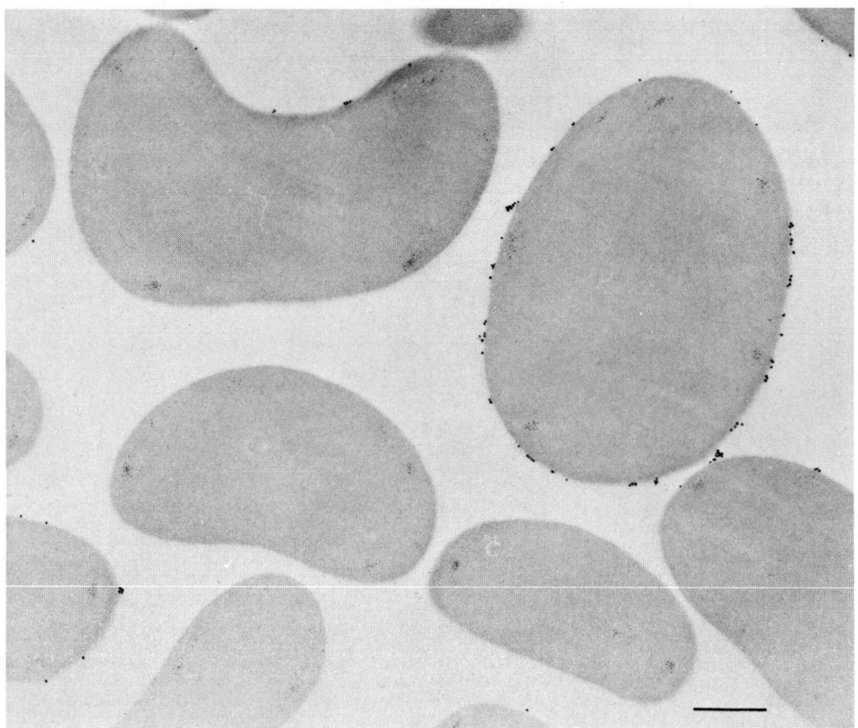

Fig. 6. Red blood cells of the B blood group labelled with a human anti-B serum and GAHG20. The red blood cells were fixed before the labelling with 1% glutaraldehyde for 30 min. Their morphology is well preserved. There is a striking heterogeneity of the labelling density (bar = 1 μm).

be combined with the cytochemical detection of intracellular enzymatic activities. In light microscopical preparations endogenous peroxidase, α-naphthyl acetate esterase, acid phosphatase and β-glucuronidase have been studied with classical cytochemical reactions [19, 21, 40, 53]. With this approach the cytochemical profile of the T- and B-lymphocytes and the T-helper and T-cytotoxic/suppressor subsets as defined by monoclonal antibodies has been determined [21, 40, 53].

Also ultrastructural cytochemistry has been performed on immunogold-labelled cells [15, 21, 40]. Labelling and fixation with glutaraldehyde was carried out at 4°C in order to achieve optimal preservation of the enzymatic activities. The cells were not fixed in a pellet but resuspended in the fixative medium. Myeloperoxidase was stained with the Graham/Karnovsky technique [54]. The cell suspension was incubated with a freshly prepared staining solution, containing 0.5 mg/ml 3,3′-diaminobenzidine (DAB) and 0.0006% H_2O_2 in 0.1 M Tris-HCl buffer pH 7.6 for 20 min at room temperature.

For acid phosphatase, the method of Barka and Anderson [55] was used. The fixed cells were incubated for 1 h at 37°C in Tris-maleate buffer pH 5 containing sodium-β-glycerophosphate as substrate and lead nitrate as coupling agent. After

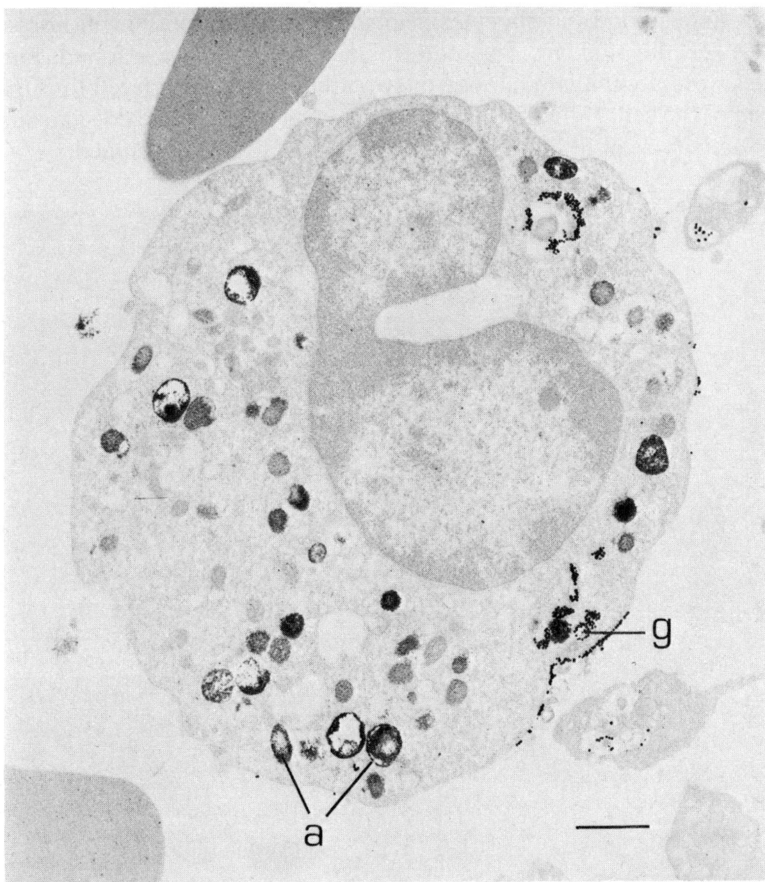

Fig. 7. Peripheral blood neutrophil labelled with OKM1 and GAMG30. The endogenous peroxidase activity is stained with the Graham/Karnovsky technique. The enzymatic activity (a) is present in cytoplasmic granules. Parts of the labelled membrane have already been internalised (bar = 1 μm) (g = gold).

the cytochemical reaction the cells were fixed with osmium tetroxide, dehydrated and embedded in Spurr's resin. In general, ultrathin sections were examined without counterstaining. Only in some of the samples was uranyl acetate and lead nitrate staining performed.

In the preparations, the enzymatic reaction product could easily be distinguished from the gold label. In general, uranyl acetate and lead citrate staining of the samples reduced the contrast in the preparations but enhanced the visibility of the morphological detail.

Myeloperoxidase is present in granulocytes and monocytes and their precursors. It is localised in intracytoplasmic granules (Fig. 7). Staining of this enzyme may help to identify the cells of these lineages in the preparations.

In peripheral blood lymphocytes acid phosphatase is present in lysosome-like organelles [21, 40] (Fig. 8). Two patterns of distribution were found: the stained organelles were accumulated in one or two areas or were scattered throughout the cytoplasm of the cell. Whether these patterns correspond to the dot-like and diffuse granular activity seen in light microscopy remains to be determined.

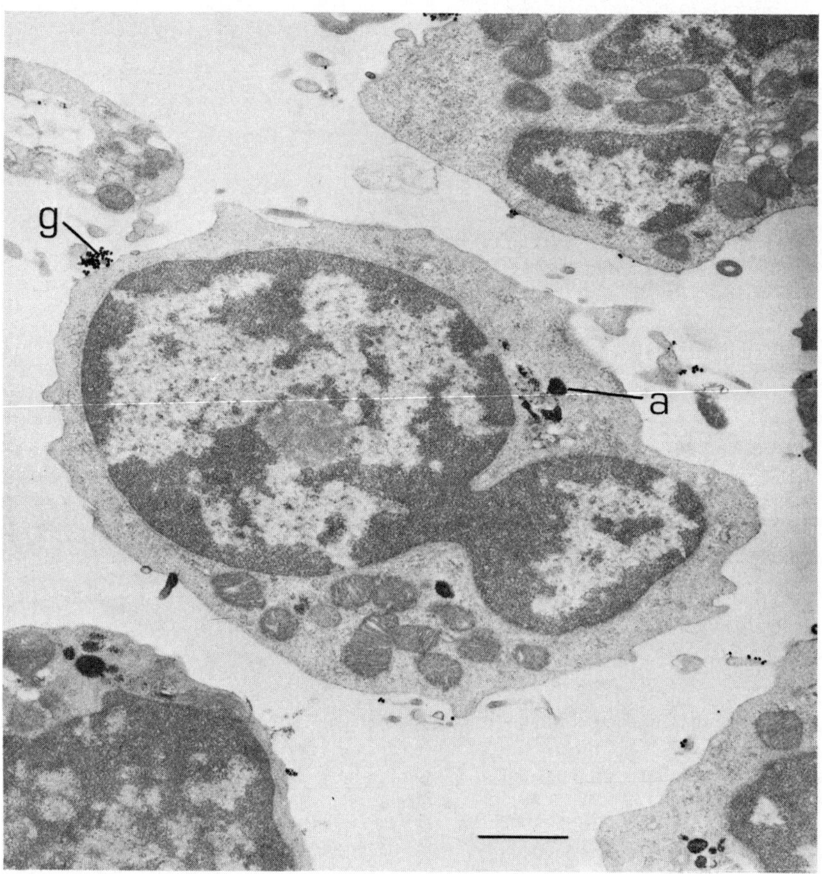

Fig. 8. Peripheral blood lymphocyte labelled with OKT3 and GAMG30. The acid phosphatase activity is stained with the method of Barka and Anderson. The enzymatic activity (a) is present in lysosomal organelles (g = gold) (bar = 1 μm).

Immunogold labelling in combination with enzyme cytochemistry was used by Bain et al. [15] to characterise the abnormal circulating cells in some cases of acute myelofibrosis. These cells were identified as megakaryoblasts by the presence on their surface membrane of a platelet antigen, recognised by the AN51 monoclonal antibody, and by a positive platelet peroxidase reaction in the nuclear envelope and endoplasmic reticulum.

3. Avidin-biotin-peroxidase complex method on cell suspensions

N. Cerf-Bensussan has used an avidin-biotin-peroxidase complex (ABC) method for the electron immunocytochemical detection of leukocyte cell surface antigens with monoclonal antibodies [20]. Mononuclear cells were first treated with Nakane's fixative and then sequentially incubated in dilutions of monoclonal antibody, biotinylated horse anti-mouse IgG and avidin-biotin-peroxidase complex. The cells were fixed with glutaraldehyde and cytosmears were made. The peroxidase activity was revealed with diaminobenzidine and the smears were processed for electron microscopy.

With this method the ultrastructural characteristics of normal peripheral blood lymphocytes were examined. Prefixation with Nakane's fixative did not destroy the antigenic activity and gave a good preservation of cellular morphology. The reaction product on positively stained cells was present as a dense layer around the whole cell surface membrane. However, it had a tendency to diffuse from the surface of stained lymphocytes to the membranes of adjacent unstained cells in areas of close contact. In most samples uranyl acetate and lead citrate staining was omitted to enhance the contrast between the cytoplasm and the reaction product on the cell surface membrane. This may however reduce the visibility of the fine morphological detail. In the preparations the endogenous peroxidase activity in granulocytes and monocytes was also stained. This helped to differentiate these cells from the lymphocytes. With this method no clear morphological differences were found between the T-helper and T-cytotoxic/suppressor cells as defined with the monoclonal antibodies T4 and T8.

Without prefixation and with labelling conditions similar to those described for the immunogold staining procedure, the distribution of the reaction product on the surface membrane of the positive cells was similar to that described for the gold label (Fig. 9). Only a few segments of the surface membrane were covered with the reaction product. This was probably due to a ligand-induced redistribution process.

4. Double labelling

Colloidal gold-labelled reagents form ideal tools to detect two different cell surface antigens on the same cell or on different cells in the same cell suspension [56, 57]. Direct and indirect techniques can be used. For direct techniques, monoclonal antibodies must be coupled to gold particles of different sizes. As these antibody preparations are very homogeneous the coupling can easily be done [18]. However, a large amount of the still very expensive antibody is needed and it may be impossible to avoid any loss of antibody during the coupling procedure.

If the two monoclonal antibodies are mouse immunoglobulins of different classes or subclasses indirect techniques may also be used, but specific secondary reagents coupled to gold particles of different sizes must be available.

It remains to be determined whether the cells are best incubated simultaneously or sequentially with the two colloidal gold reagents. In the latter possibility the size of the gold probes may determine the order of addition [47, 57].

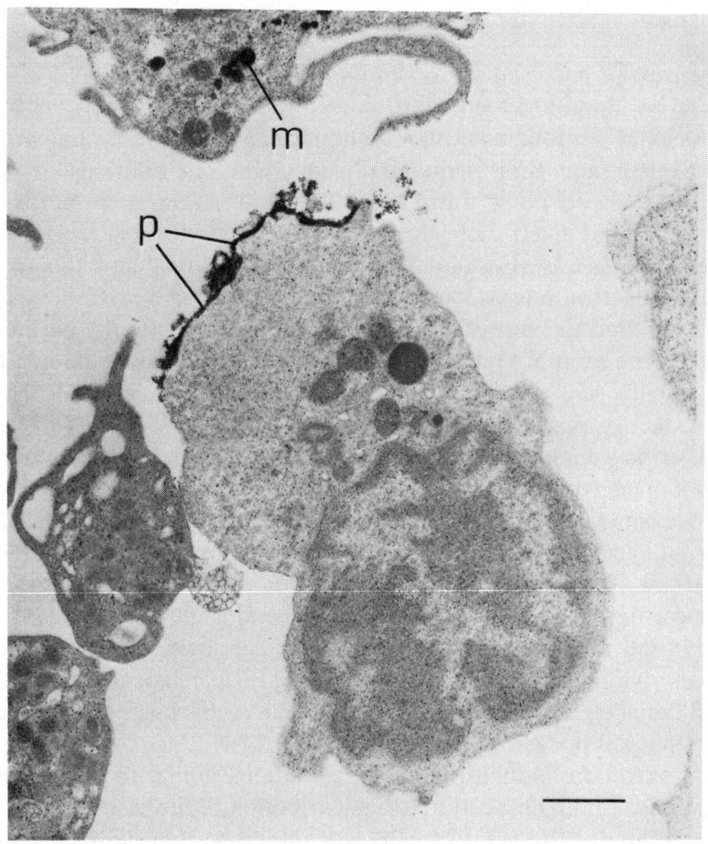

Fig. 9. Peripheral blood lymphocyte labelled with OKT3 in an avidin-biotin-peroxidase complex method. The labelling conditions were similar to those described for immunogold staining. The electron-dense reaction product of DAB (p) is present as a cap at one pole of the cell. The endogeneous peroxidase activity in the monocytes is also stained (m) (bar = 1 μm).

In Fig. 10, results of a double labelling experiment are shown. In peripheral blood lymphocytes, the detection of surface immunoglobulin was combined with that of HLA-Dr antigens. An unfixed mononuclear cell suspension was first incubated with goat anti-human immunoglobulin antibodies coupled to 10 nm gold (GAHG10—Janssen Pharmaceutica, Beerse, Belgium) and then with the OKIa monoclonal mouse antibody followed by goat anti-mouse immunoglobulin antibodies coupled to 30 nm gold. Surface immunoglobulin and HLA-Dr antigens are both present on B-lymphocytes. Therefore, most of the positive lymphoid cells had both gold probes on their surface membrane. The patches of 10 and 30 nm gold

---

Fig. 10 (A, B) Combined detection of surface immunoglobulin (10 nm gold) and HLA-Dr antigens (30 nm gold) in peripheral blood mononuclear cells. Both probes are present in patches on the surface membrane. Ten (a) and 30 (b) nm gold particles can easily be distinguished (B) (bar = 1 μm).

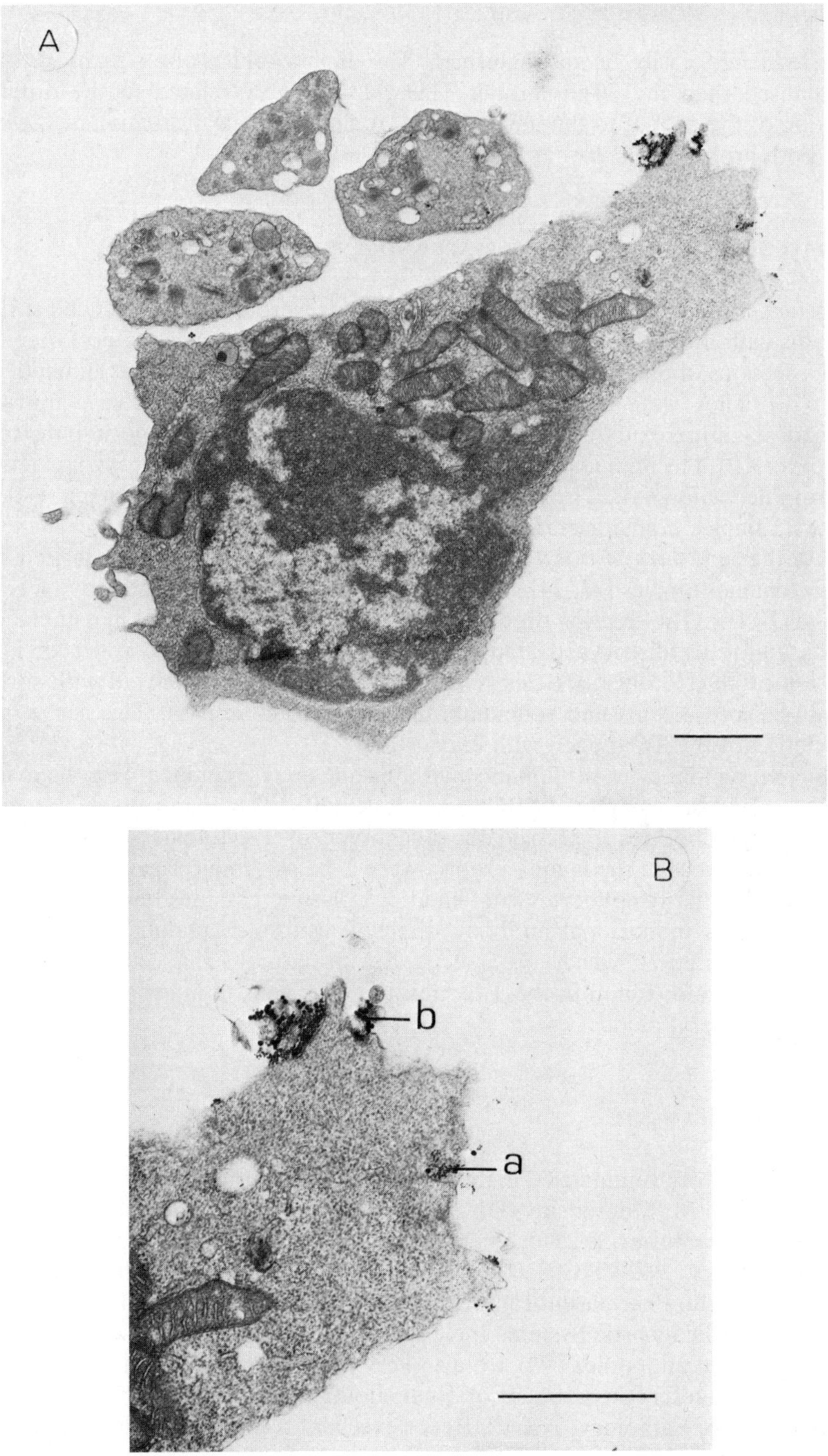

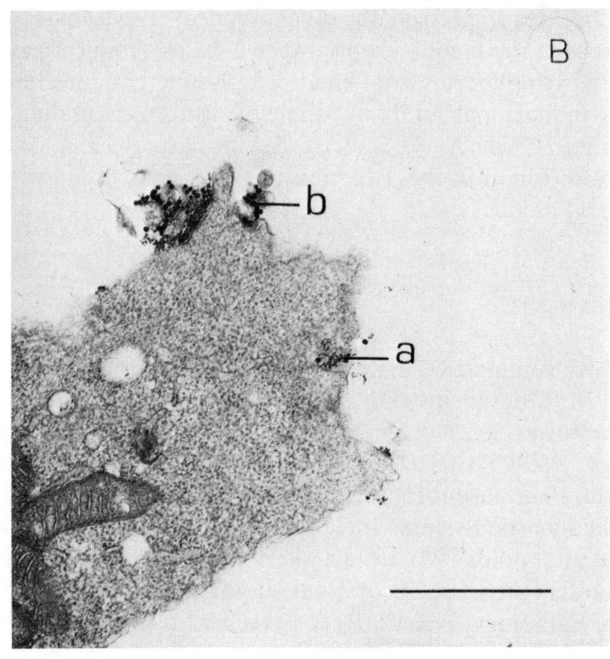
B
b
a

particles could easily be distinguished. The 10 nm gold probe was more often internalised than the 30 nm probe. This was probably related to the order of addition of the probes to the unfixed cells. A simultaneous incubation of the cells with both probes could perhaps limit this phenomenon.

5. Detection of lymphoid cell surface antigens in tissues

In tissues, more step immunoperoxidase methods were used to detect cell surface antigens with monoclonal antibodies [20, 23, 24]. The studies were performed on frozen sections of prefixed material [23, 24] or on sections of material embedded in agar [20]. They were incubated with the monoclonal antibodies in a four-step peroxidase-antiperoxidase (PAP) method [23, 24] or an avidin-biotin-peroxidase complex (ABC) technique [20]. The sections were fixed and the peroxidase activity was revealed with DAB. The sections were then processed for electron microscopy. In most samples uranyl acetate and lead citrate staining was omitted.

With this approach histiocytosis X cells in skin biopsies were characterised with monoclonal antibodies [23, 24]. These cells reacted with the T6 antibody (specificity like OKT6). The reaction product was present only focally while in lymphocytes it was uniformly distributed along the surface membrane. The zones of DAB reaction product on histiocytosis X cells were frequently associated with surface membrane projections and submembrane endocytotic vesicles. This suggests an association of the T6 antigen with endocytosis.

Mucosal lymphocytes of human small intestine were examined with the avidin-biotin-peroxidase complex (ABC) technique [20]. The majority of the intra-epithelial lymphocytes expressed the phenotype of T-cytotoxic/suppressor cells (T8+) while those in the lamina propria were T-helper/inducer cells (T4+). A lot of intra-epithelial lymphocytes contained cytoplasmic granules but none of them reacted with a monoclonal antibody directed against circulating large granular lymphocytes (Leu-7), which are associated with natural killer cell activity. Such granules were also found in the T4+ and the T8+ cells in intestinal mucosa and normal peripheral blood.

6. Acknowledgements

The personal work summarised in this chapter was performed in collaboration with J. De Mey and M. Moeremans (Division of Life Sciences, Janssen Pharmaceutica Research Laboratories, B 2340, Beerse, Belgium) and B. Van Camp (Department of Haematology, AZ-VUB, B 1090 Brussels, Belgium). We are indebted to G. Goldstein (Ortho Pharmaceutical Corporation, Raritan, NJ 08869, U.S.A.) and J. Lifter (Ortho Diagnostic Systems Inc., Raritan, NJ 08869, U.S.A.) for providing the monoclonal antibodies. We are also very grateful to Ph. Reynaerts, L. Smet and L. Broodtaerts (Department of Haematology, AZ-VUB) and H. De Zutter (Department of Pathology, AZ-VUB) for excellent technical assistance and to W.

Gepts for all his support. We would also like to thank M. Borgers and F. Thoné (Division of Life Sciences, Janssen Pharmaceutica Research Laboratories) for performing some of the ultrastructural cytochemistry. The manuscript was typed by M. De Vuyst. This work was supported by grant No. 3.0076.83 from the Fonds voor Geneeskundig Wetenschappelijk Onderzoek, Brussels, Belgium.

7. References

1. Nicolson, G.L. and Singer, S.J. (1971) Ferritin-conjugated plant agglutinins as specific saccharide stains for electron microscopy: Application to saccharides bound to cell membranes. Proc. Natl. Acad. Sci. U.S.A. 68, 942–945.
2. Williams, M.A. and Voak, D. (1972) Studies with ferritin-labelled *Dolichos biflorus* lectin on the numbers and distribution of A Sites on A1 and A2 erythrocytes, and on the nature of its specificity and enhancement by enzymes. Br. J. Haematol. 23, 427–441.
2. De Petris, S. and Raff, M.C. (1973) Normal distribution, patching and capping of lymphocyte surface immunoglobulin studied by electron microscopy. Nature (London) 241, 257–259.
4. Reyes, F., Lejonc, J.L., Gourdin, M.F., Ton That, H. and Breton Gorius, J. (1974) Human Normoblast A Antigen seen by immuno electron microscopy. Nature (London) 247, 461–462.
5. Romano, E.L., Stolinski, C. and Hughes-Jones, N.C. (1975) Distribution and mobility of the A, D and c antigens on human red cell membranes: Studies with a gold-labelled antiglobulin reagent. Br. J. Haematol. 30, 507–516.
6. Reyes, F., Gourdin, M.F., Lejonc, J.L. Cartron, J.P., Breton Gorius, J. and Dreyfus B. (1976) The heterogeneity of erythrocyte antigen distribution in human normal phenotypes: an immuno-electron microscopy study. Br. J. Haematol. 34, 613–621.
7. Borgers, M., Verhaegen, H., De Brabander, M., Thoné, F., Van Reempts, J. and Geuens, G. (1977) Purine nucleoside phosphorylase, a possible histochemical marker for T-cells in man. J. Immunol. Methods 16, 101–110.
8. Geoghegan, W.D. and Ackerman, G.A. (1977) Adsorption of horseradish peroxidase, ovomucoid and antiimmunoglobulin to colloidal gold for the indirect detection of Concanavalin A, wheat germ agglutinin and goat antihuman immunoglobulin G on cell surfaces at the electron microscopic level: A new method, theory and application. J. Histochem. Cytochem. 25, 1187–1200.
9. Itoh, G. and Suzuki, I. (1977) Immunohistochemical detection of Fc receptor. II. Electron microscopic demonstration of Fc receptor by using soluble immune complexes of ferritin-antiferritin immunoglobulin G. J. Histochem. Cytochem. 25, 259–265.
10. Romano, E.L. and Romano, M. (1977) Staphylococcal protein A bound to colloidal gold: A useful reagent to label antigen-antibody sites in electron microscopy. Immunochemistry 14, 711–715.
11. Borgers, M., Verhaegen, H., De Brabander, M., De Cree, J., De Cock, W., Thoné, F. and Geuens, G. (1978) Purine nucleoside phosphorylase in chronic lymphocytic leukemia (CLL). Blood 52, 886–895.
12. Ackerman, G.A. and Freeman, W.H. (1979) Membrane differentiation of developing hemic cells of the bone marrow demonstrated by changes in concanavalin A surface labelling. J. Histochem. Cytochem. 27, 1413–1423.
13. Nurden, A. T., Horisberger, M., Savariau, E. and Caen, J. P. (1980) Visualisation of lectin binding sites on the surface of human platelets using lectins adsorbed to gold granules. Experientia 36, 1215–1217.
14. Ackerman, G.A. and Wolken, K.W. (1981) Histochemical evidence for the differential surface labelling, uptake, and intracellular transport of a colloidal gold-labelled insulin complex by normal human blood cells. J. Histochem. Cytochem. 29, 1137–1149.
15. Bain, B.J., Catovsky, D., O'Brien, M., Prentice, H.G., Lawlor, E., Kumaran, T.O., McCann, S.R., Matutes, E. and Galton, D.A.G. (1981) Megakaryoblastic leukemia presenting as acute myelofibrosis—A study of four cases with the platelet-peroxidase reaction. Blood 58, 206–213.

16. Furlan, M., Horisberger, M., Perret, B.A. and Beck, E.A. (1981). Binding of colloidal gold granules, coated with bovine factor VIII to human platelet membranes. Br. J. Haematol. 48, 319–324.
17. Lafferty, M.D., Ackerman, G.A., Dunn, C.D.R. and Lange, R.D. (1981) Ultrastructural, immunocytochemical localisation of presumptive erythropoietin binding sites on developing erythrocytic cells of normal human bone marrow. J. Histochem. Cytochem. 29, 49–56.
18. De Mey, J., Moeremans, M., De Waele, M., Geuens, G. and De Brabander, M. (1982). The IGS (Immuno gold staining) method used with monoclonal antibodies. In: H. Peeters (Ed.) Protides of the Biological Fluids, Vol. 29. Pergamon Press, Oxford and New York, pp. 943–947.
19. Van Camp, B., Thielemans, C., Dehou, M.F., De Mey, J. and De Waele, M. (1982) Two monoclonal antibodies (OKTa1* and OKT10*) for the study of the final B cell maturation. J. Clin. Immunol. 2, Suppl., 67–74.
20. Cerf-Bensussan, N., Schneeberger, E.E. and Bhan, A.K. (1983) Immunohistologic and immunoelectron microscopic characterization of the mucosal lymphocytes of human small intestine by the use of monoclonal antibodies. J. Immunol. 130, 2615–2622.
21. De Waele, M., De Mey, J., Moeremans, M., De Brabander, M. and Van Camp, B. (1983) Immunogold staining method for the detection of cell surface antigens with monoclonal antibodies. In: G.R. Bullock and P. Petrusz (Eds.) Techniques in Immunocytochemistry, Vol. 2. Academic Press, London and New York, pp. 1–23.
22. Farr, A.G. and Nakane, P.K. (1983) Cells bearing Ia antigens in the murine thymus. An ultrastructural study. Am. J. Pathol. 111, 88–97.
23. Harrist, T.J., Bhan, A.K., Murphy, G.F., Sato, S., Berman, R.S., Gellis, S.E., Friedman, S. and Mihm, M.C. (1983) Histiocytosis-X: In situ characterization of cutaneous infiltrates with monoclonal antibodies. Am. J. Clin. Pathol. 79, 294–300.
24. Murphy, G.F., Harrist, T.J., Bhan, A.K. and Mihm, M.C. (1983) Distribution of cell surface antigens in histiocytosis X cells. Quantitative immunoelectron microscopy using monoclonal antibodies. Lab. Invest. 48, 90–97.
25. Leduc, E.H., Avrameas, S. and Bouteille, M. (1967) Ultrastructural localization of antibody in differentiating plasma cells. J. Exp. Med. 127, 109–118.
26. Kuhlmann, W.D., Avrameas, S. and Ternynck, T. (1974) A comparative study for ultrastructural localization of intracellular immunoglobulins using peroxidase conjugates. J. Immunol. Methods 5, 33–48.
27. Reyes, F., Gourdin, M.F., Farcet, P., Breton-Gorius, J. and Dreyfus, B. (1978) Immunoglobulin production in lymphoma cells: an immunoelectron microscopy study. Recent Results Cancer Res. 64, 176–179.
28. Gourdin, M.F., Farcet, J.P. and Reyes, F. (1982) The ultrastructural localization of immunoglobulins in human B cells of immunoproliferative diseases. Blood 59, 1132–1140.
29. McLaren, K.M. and Pepper, D.S. (1983) The immunoelectron microscopic localisation of human platelet factor 4 in tissue mast cells. Histochem. J. 15, 795–800.
30. Parmley, R.T., Hurst, R.E., Takagi, M., Spicer, S.S. and Austin, R.L. (1983) Glycosaminoglycans in human neutrophils and leukemic myeloblasts: Ultrastructural, cytochemical, immunologic and biochemical characterization. Blood 61, 257–266.
31. Pham, T.D., Kaplan, K.L. and Butler, V.P. (1983) Immunoelectron microscopic localization of platelet factor 4 and fibrinogen in the granules of human platelets. J. Histochem. Cytochem. 31, 905–910.
32. Reyes, F., Lejonc, J.L., Gourdin, M.F., Mannoni, P. and Dreyfus, B. (1975) The surface morphology of human B lymphocytes as revealed by immunoelectron microscopy. J. Exp. Med. 141, 392–410.
33. Biberfeld, P., Mellstedt, H. and Petersson, D. (1977) Ultrastructural and immunocytochemical characterization of circulating mononuclear cells in patients with myelomatosis. Acta Pathol. Microbiol. Scand. Sect. A: 85, 611–624.
34. Matutes, E. and Catovsky, D. (1982) The fine structure of normal lymphocyte subpopulations—a study with monoclonal antibodies and the immunogold technique. Clin. Exp. Immunol. 50, 416–425.

35. Matutes, E., Robinson, D., O'Brien, M., Haynes, B.F., Zola, H. and Catovsky, D. (1984) Candidate counterparts of Sezary cells and adult T-cell lymphoma/leukaemia cells in normal peripheral blood. Leuk. Res. 7, 787–801.
36. Williams, M.A. (1977) Immunocytochemistry at EM level: staining antigens with electron-dense reagents. In: A.M. Glauert (Ed.), Practical Methods in Electron Microscopy, Vol. 6 (1). North-Holland Publishing Company, Amsterdam–New York–Oxford, pp. 41–76.
37. De Mey, J. (1983) A critical review of light and electron microscopic immunocytochemical techniques used in neurobiology. J. Neurosci. Methods 7, 1–18.
38. Sternberger, L.A. (1979) Immunocytochemistry. Wiley, London and New York, pp. 170–174.
39. De Waele, M., De Mey, J., Moeremans, M., De Brabander, M. and Van Camp, B. (1983) Immunogold staining method for the light microscopic detection of leukocyte cell surface antigens with monoclonal antibodies. Its application to the enumeration of lymphocyte subpopulations. J. Histochem. Cytochem. 31, 376–381.
40. De Waele, M., De Mey, J., Moeremans, M., Smet, L., Broodtaerts, L. and Van Camp, B. (1983) Cytochemical profile of immunoregulatory T-lymphocyte subsets defined by monoclonal antibodies. J. Histochem. Cytochem. 31, 471–478.
41. Mason, D.Y., Cordell, J.L. and Pulford, K.A.F. (1983) Production of monoclonal antibodies for immunocytochemical use. In: G.R. Bullock and P. Petrusz (Eds.) Techniques in Immunocytochemistry, Vol. 2. Academic Press, London and New York, pp. 175–216.
42. Kung, P.C. and Goldstein, G. (1980) Functional and developmental compartments of human T-lymphocytes. Vox Sang. 39, 121–127.
43. Reinherz, E.L. and Schlossman, S.F. (1981) The characterization and function of human immunoregulatory T-lymphocyte subsets. Immunology Today (April), 69–75.
44. De Mey, J. (1983) Colloidal gold probes in immunocytochemistry. In: J.M. Polak and S. Van Noorden (Eds.) Immunocytochemistry: Practical Applications in Pathology and Biology. John Wright and Sons, Bristol–London–Boston, pp. 82–112.
45. Frens, G. (1973) Controlled nucleation for the regulation of the particle size in monodisperse gold suspensions. Nature Phys. Sci. 241, 20–22.
46. Glauert, A.M. (1978) Fixation methods. In: A.M. Glauert (Ed.) Practical Methods in Electron Microscopy, Vol. 3 (1). North-Holland Publishing Company, Amsterdam–New York–Oxford, pp. 73–110.
47. Horisberger, M. (1981) Colloidal gold: A cytochemical marker for light and fluorescent microscopy and for transmission and scanning electron microscopy. In: O. Johari (Ed.) Scanning Electron Microscopy, Vol. II. SEM Inc., AMF O'Hare (Chicago) IL. pp. 9–31.
48. Loor, F. (1977) Structure and dynamics of the lymphocyte surface, in relation to differentiation, recognition and activation. In: Progress in Allergy, Vol. 23. Karger, Basel, pp. 1–153.
49. Roth, J. (1983) The colloidal gold marker system for light and electron microscopic cytochemistry. In: G.R. Bullock and P. Petrusz (Eds.) Techniques in Immunocytochemistry, Vol. 2. Academic Press, London and New York, pp. 217–284.
50. Kumagai, K., Abo, T., Sekizawa, T. and Sasaki, M. (1975) Studies of surface immunoglobulins on human B-lymphocytes. I. Dissociation of cell-bound immunoglobulins with acid pH or at 37°C. J. Immunol. 115, 982–987.
51. Schuit, H.R.E. and Hijmans, W. (1980) Identification of mononuclear cells in human blood. II. Evaluation of morphological and immunological aspects of native and formaldehyde fixed cell populations. Clin. Exp. Immunol. 41, 567–574.
52. Glauert, A.M. (1978) Fixatives. In: A.M. Glauert (Ed.) Practical Methods in Electron Microscopy, Vol. 3 (1). North-Holland Publishing Company, Amsterdam—New York—Oxford, pp. 5–72.
53. Crockard, A. and Catovsky, D. (1983) Cytochemistry of normal human lymphocyte subsets defined by monoclonal antibodies and immunocolloidal gold. Scand. J. Haematol. 80, 433–443.
54. Graham, R.C. and Karnovsky, M.J. (1966) The early stages of absorption of injected horseradish peroxidase in the proximal tubules of mouse kidney: ultrastructural cytochemistry by a new technique. J. Histochem. Cytochem. 14, 291–302.
55. Barka, T. and Anderson, P.J. (1962) Histochemical methods for acid phosphatase using hexazonium pararosanilin as a coupler. J. Histochem. Cytochem. 10, 741–743.

56. Roth, J. and Binder, M. (1978) Colloidal gold, ferritin and peroxidase as markers for electron microscopic double labelling lectin techniques. J. Histochem. Cytochem. 26, 163–169.
57. Horisberger, M. and Vonlanthen, M. (1979) Multiple marking of cell surface receptors by gold granules: simultaneous localization of three lectin receptors on human erythrocytes. J. Microsc. 115, 97–102.

Immunolabelling for Electron Microscopy (Polak/Varndell, eds)

CHAPTER 20

Recent advances in microbiological immunocytochemistry

Julian E. Beesley

Wellcome Research Laboratories, Langley Court, Beckenham, Kent BR3 3BS, U.K.

CONTENTS

1. INTRODUCTION

The localisation of tissue antigens by electron microscope immunocytochemistry is a compromise between retaining sufficient antigenic activity within the specimen to bind with antibody and maintaining adequate morphological detail within the tissue for reliable identification of the site of the immunological reactions.

High contrast and good resolution are very important factors in microbiological studies because the objects of interest, which may be viruses or bacterial pili, are so small. Pre-embedding labelling techniques, in which the tissue is incubated with antibody and electron-dense probe after initial aldehyde fixation, permit further fixation and contrasting with osmium tetroxide and uranyl acetate. This adequately preserves the ultrastructure of the specimen and good contrast can be obtained but the technique is limited to the localisation of surface antigens. Localisation of internal antigens by this technique necessitates disrupting the limiting cell membranes, to enable antibody and probe to penetrate into the cell, most commonly with detergents such as saponin [1, 2] or Triton X [3, 4]. This treatment disrupts the organisation of the cells and might also induce antigen migration, or false-negative results might be obtained because of incomplete penetration of antibodies into the cells. Consequently, although this technique is suitable for the high resolution localisation of surface antigens it is not entirely suitable for the localisation of internal antigens.

Labelling procedures performed on ultrathin sections of resin embedded tissue, the post-embedding techniques, enable the primary antibody to react equally with all areas of the tissue exposed at the face of the section. Unfortunately, routine fixation and embedding decreases the sensitivity of immunocytochemical methods [5] and thus processing techniques for post-embedding immunocytochemistry usually necessitate the omission of osmium tetroxide and uranyl acetate contrasting. Ultrastructural preservation and contrast are therefore not optimised in resin sections of tissue prepared for immunocytochemistry. Recently, osmium tetroxide has been included in the fixation schedule [6–8] but even then, absence of the triple fixation routine limits the contrast and resolution required for microbiological examination. Furthermore, the absence of uranyl acetate block staining is unacceptable for studies involving viruses or bacterial pili since the on-grid staining of resin sections with uranium and lead salts contrasts only those structures exposed on the cut surface of the section. In general, viruses are very small (many are smaller than 100 nm diameter) in relation to section thickness (usually less than 100 nm) and hence most of the organism will be embedded in the section and thus not available to the stain.

Microbiological immunocytochemistry is also beset by other problems. The antigenic mass of viruses and structures such as bacterial pili is so low that each particle will bind only very few antibody molecules and labelling with the electron-dense probe will be low [9]. Immunolabelling on resin sections, like staining with uranium and lead salts, occurs only on antigens exposed at the cut surface of a section [10, 11] and thus the low labelling of viruses and bacterial appendages will be further reduced by the inaccessibility of the antigen. These constraints do not apply to bacterial cells since the organism is large in relation to section thickness and many antigens will be exposed at the surface of the section.

Finally, some bacteria, such as *Pasteurella haemolytica* are surrounded by a capsule. The capsule of *P. haemolytica*, which is important in the serological specificity and pathogenicity of the organism [12], is fragile and easily removed from the organism by centrifugation. However, centrifugation is a necessary step during the preparation of micro-organisms for embedding and the amount of capsular antigen remaining after processing may be only a very small proportion of the original [13].

These introductory remarks show that many potential immunocytochemical studies in microbiology are hindered by the small size of the organism studied. A desirable preparative technique to enable the application of post-embedding immunocytochemical procedures to microbiological studies would also include, if possible, high density staining of the specimen within the section without masking the antigens.

Recently, cryotechniques developed for immunocytochemistry have been applied to microbiological problems. Immuno-negative stain techniques and an immuno-replica technique have been designed specifically for microbiological applications and are proving to be successful. The principles of electron immunocytochemistry are adequately explained elsewhere in this book and only immunocytochemical techniques applicable to microbiological problems will be discussed in this chapter.

2. Immunolabels for microbiological immunocytochemistry

Over the last few years there have been substantial advances in the development of electron-dense markers for electron microscope immunocytochemistry. The use of these markers, mainly ferritin [14, 15], peroxidase [16, 17] and colloidal gold [10, 12, 13, 18–21] in microbiology, has reflected the more general trend in the development of different labels for immunocytochemistry.

An ideal immunocytochemical probe for microbiological studies would be particulate, small and extremely dense so that it could be used for labelling small structures of high density. Ferritin is a particulate probe and this has been used for immunolabelling bacterial pili [15] and vesicular stomatitis viruses budding from the cell membrane of infected Chinese hamster ovary cells [22] but it has such a low density that it is unsuitable for many microbiological studies, especially on ultrathin cryosections [5, 19, 23].

Colloidal gold [24, 25] closely approaches the ideal immunocytochemical probe for microbiological applications. The gold probe is extremely dense and has such a characteristic appearance, quite unlike other structures normally found in biological tissues, that it can be used on very dense samples. It is particulate and therefore, unlike the peroxidase marker, does not obliterate fine appendages such as bacterial pili or virus fringes [26] (Fig. 1). The colloidal gold probes are easy to use and prepare [24]. Different size probes may be used for double labelling experiments (see Chapter 12, pp. 143–154 and Chapter 13, pp. 155–177). The probes are readily quantified and they can be used for all immunocytochemical applications [24].

3. Cryotechniques

Ultrathin cryosections are extensively used by some groups for immunocytochemistry, and there is now a substantial amount of technical data on this subject [23]. Cryoultramicrotomy has now reached the stage where it satisfies the demand for high resolution immunoelectron microscopy in studies of cell and tissue ultrastructure [27, 28].

Despite the early descriptive studies of viruses and bacteria in ultrathin frozen sections [29–32] there are only a few reports of ultrathin cryosections used for microbiological immunocytochemistry. This is somewhat surprising as ultrathin cryosections possess many advantageous features when compared with resin

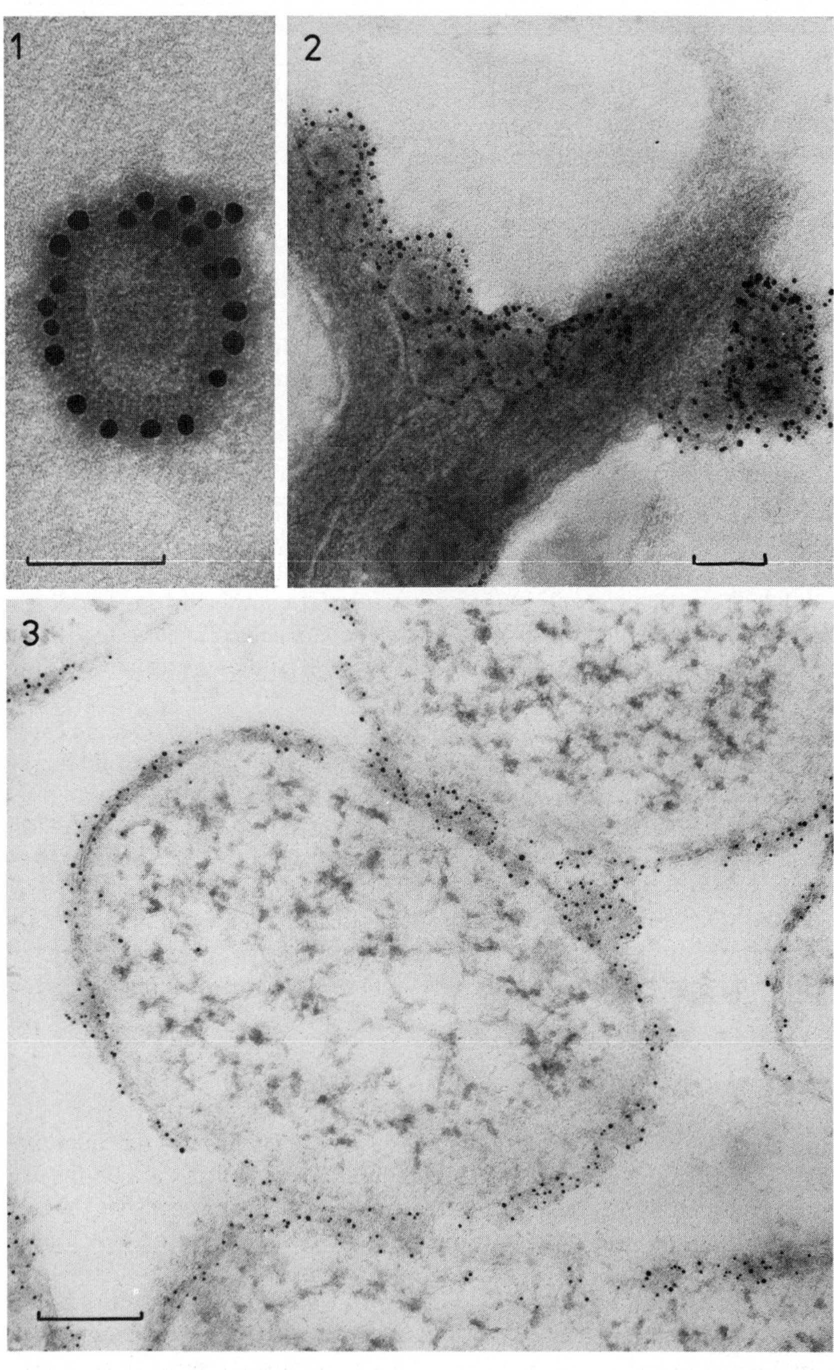
1
2
3

sections for microbiological immunocytochemistry although it must be reiterated that the latter have provided much useful information [12, 17, 18]. Cryosection immunocytochemistry has revealed viral antigens to be located in the Golgi apparatus [22, 33–35] and on budding viruses [34–36]. Moreover, Geuze and Slot [37] have demonstrated tumour IgG in virus-containing smooth endoplasmic reticulum of mineral oil induced plasmacytomas of the mouse, and Beesley and Campbell have studied the location of several influenza virus antigens in infected Vero cell cultures [21].

There are fewer reports of the localisation of bacterial antigens using ultrathin cryosections. Surface antigens of *Bacillus subtilis* cells have been described by Fournier-Laflèche et al. [38] and intracellular antigens of the bacterium *Escherichia coli* have been reported [39]. There have also been several studies on the localisation of capsular antigen in *Pasteurella haemolytica* [10, 13, 19].

Of those authors who have reported the use of ultrathin cryosections for microbiological immunocytochemistry Geuze and Slot [37], Fournier-Laflèche [38] and Bernadac and Lazdunski [39] pre-embed the samples before freezing; Bergmann et al. [22], Green et al. [33], Griffiths et al. [34, 35] and Bourguignon and Butman [36] embed the sections after cryoultramicrotomy; whilst Beesley and Adlam [10, 13], Beesley et al. [19] and Beesley and Campbell [21] use no ancillary embedding procedures. The latter technique permits the attainment of high contrast by a negative stain technique for full visualisation of very small structures such as the units in the virus fringe [21] and allows visualisation of the bacterial capsule with a shadowing technique [13].

In this laboratory specimens are routinely fixed with 2.5% (v/v) glutaraldehyde. This relatively high concentration of glutaraldehyde is used because many samples encountered in microbiological studies are very fragile. Microbiological specimens are often monolayer cultures of cells, weakened by infection and some cells are already partially disrupted. These cells are relatively unprotected against osmotic changes as compared with cells in tissues [40] and they need the maximum stabilising effect of the fixative. The specimens are cryoprotected with 2.3 M sucrose for 15–30 min [41], mounted on LKB (Bromma, Sweden) specimen pins and frozen in liquid nitrogen slush. The pins are then mounted in the cryoultramicrotome. Sections are cut at a specimen temperature of −110°C and a knife temperature of −90°C on a modified LKB Ultratome III [42] (Bromma, Sweden).

Fig. 1. Influenza virus, dried down onto an electron microscope grid and labelled with anti-monomer haemagglutinin antibody and the 20 nm colloidal gold probe before staining with ammonium molybdate. The virus fringe is visible beneath the colloidal gold probe. Bar = 0.1 μm.

Fig. 2. Ultrathin cryosection of feline infectious peritonitis virus budding from cultured feline embryonic lung cell, labelled with antibody, colloidal gold probe and finally stained with ammonium molybdate. The antigenic sites on the virus are not masked and immunolabelling of external antigens can occur throughout the thickness of the section. Bar = 0.1 μm.

Fig. 3. Ultrathin methacrylate section of *P. haemolytica* labelled with antibody against capsular polysaccharide and the colloidal gold probe. The labelling is confined to a thin ring of probe on the surface of the section. Bar = 0.2 μm.

Sections are collected with drops of sucrose solution [23] and mounted on coated grids, washed in water then immunolabelled with antiserum and colloidal gold according to standard procedures [19, 20]. The ultrathin cryosections are then negatively stained with 1.5% ammonium molybdate containing 0.05% bovine serum albumin, which is added to prevent the negative stain accumulating at the edge of the section.

Influenza virus antigens have been located in infected Vero cells using this technique [21]. In this study we were able to show that ultrathin cryosections of cultured monolayer Vero cells infected with influenza virus produce good viral morphology and adequate fine structural preservation of host cells. Specific immunolabelling of influenza virus antigens both on the virus and within the host cells was achieved in conjunction with the protein A-gold probe (see Chapter 9, pp. 113–121).

Insoluble embedding media are not used in the production of ultrathin cryosections according to our methodology with the result that the negative stain can contrast throughout the section thickness. The virus fringe in ultrathin cryosections of influenza virus is sufficiently distinct to be a reliable diagnostic feature, since all the viral fringe subunits will be stained. Identification of virus particles is thus much easier in ultrathin cryosections than in resin sections in which only those few subunits exposed at the cut surface of the section will be stained [21].

Unembedded ultrathin cryosections, unlike resin sections, also allow immunolabelling of external antigens throughout the thickness of the section [10, 21]. This increased labelling is valuable in the study of external antigens on small organisms (Fig. 2) but careful interpretation is needed when comparing the amount of labelling of external antigens with those inside the tissue section since it is not known whether or not the immunological reagents penetrate sections of the actual tissue.

Bacterial studies can involve both resin and cryotechniques, one technique complementing knowledge gained by the other. The bacterium *P. haemolytica* is a causative agent of respiratory diseases in sheep and shipping fever in cattle and studies are now centred on the bacterial capsular substance [12]. Colloidal gold immunocytochemistry applied to sections of methacrylate-embedded bacteria shows the presence of capsular antigen as a thin ring of gold probe at the periphery of the bacterium [10] (Fig. 3). The gold probes do not penetrate the resin. Variations in thickness of the sections, therefore, do not affect the amount of labelling. The labelling on these sections can be quantified by counting the number

---

Fig. 4. Ultrathin cryosection of *P. haemolytica* labelled with antibody against capsular polysaccharide and the colloidal gold probe. The label is spread across the bacterial membranes and in the background of the section. Bar = 0.5 μm.

Fig. 5. Ultrathin cryosection of *P. haemolytica* immunolabelled with antibody against capsular polysaccharide and colloidal gold probe. The section was then shadowed with platinum and carbon to reveal the presence of an irregular matrix (C) possibly capsule that has been labelled. The positions of the bacterial membranes (M) are visible beneath the shadow. Bar = 0.2 μm.

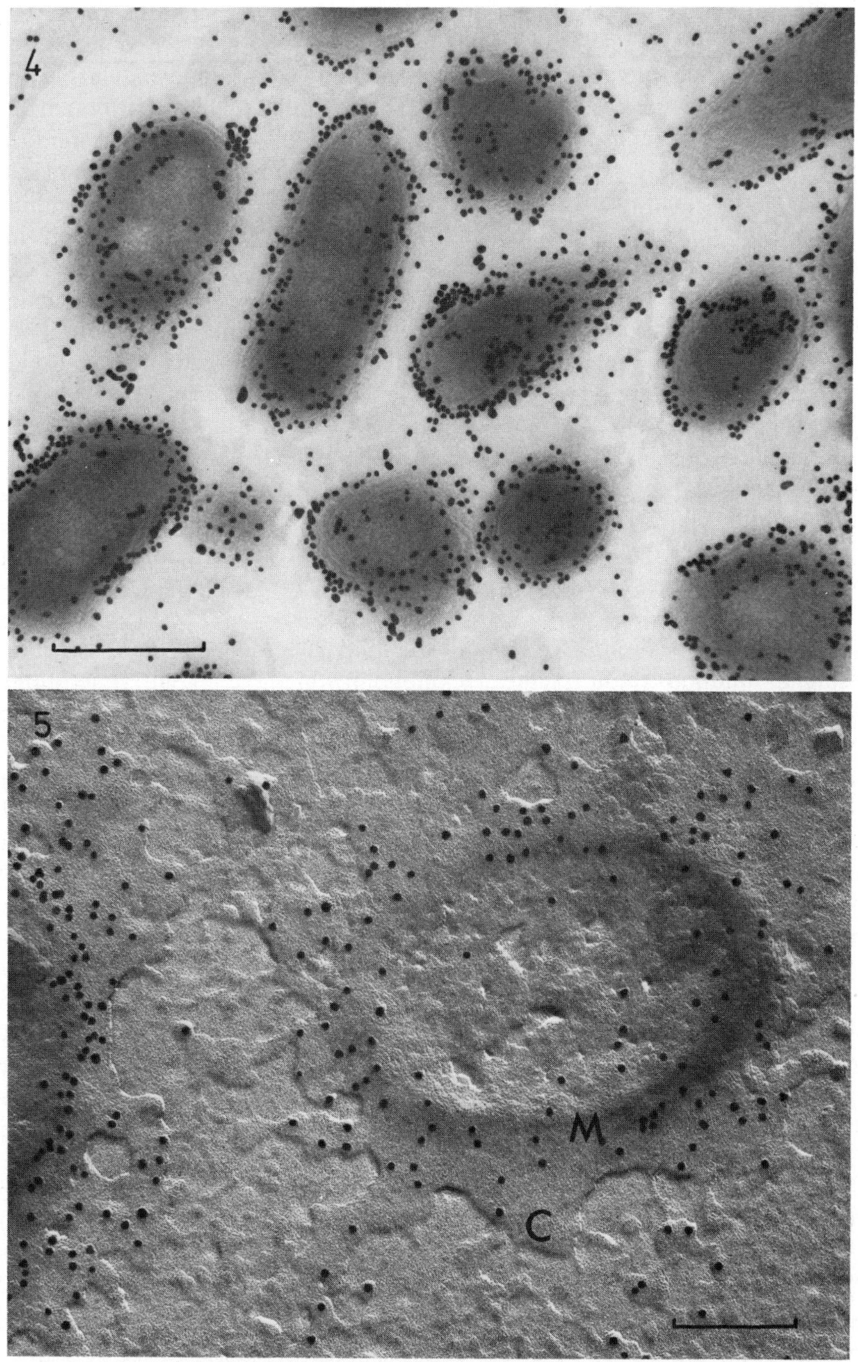
4
5
M
C

of probes obtained after different treatments. The cross-reactions between an antiserum raised against one bacterial serotype and other bacterial serotypes have been quantified in this manner [20].

Ultrathin cryosections of *P. haemolytica* labelled with antibody against capsular polysaccharide and the colloidal gold probe show probe widely distributed across the bacterial membranes and in the background of the section around each bacterium [10, 13] (Fig. 4). Measurements from stereo pairs of photographs show the wide band of gold probe as labelling associated with the bacterial membranes throughout the thickness of the section; the thicker the section, the wider the band of label [10]. The cause of labelling around the bacterium in the background of the section was indicated by shadowing ultrathin cryosections of bacteria with platinum and carbon. The bacteria were surrounded by an irregular matrix, possibly capsule. Shadowing after immunolabelling with antibody against capsular polysaccharide and the colloidal gold probe revealed that the labelling was specific to this irregular matrix [13] (Fig. 5).

This degree of capsular preservation can also be detected in samples in which the capsule is stabilised in situ by infiltrating bacterial pellets with 15% (w/v) bovine serum albumin, which is then lightly cross-linked with 2.5% glutaraldehyde. The specimens are labelled with antibody and colloidal gold probe before embedding in resin. Sections of these preparations confirm the presence of a wide band of capsular antigen (Fig. 6).

A more precise localisation of immunolabelling bacterial membranes is obtained by using an immuno-freeze fracture technique in which the cryoultramicrotome is used as a fracturing apparatus. *P. haemolytica* are pelleted and infiltrated with 15% (w/v) bovine serum albumin which is cross-linked with 2.5% (v/v) glutaraldehyde. After cryoprotection with 2.3 M sucrose, the specimens are affixed to the specimen holder of the cryoultramicrotome and frozen in liquid nitrogen slush. Ultrathin cryosections are repeatedly taken from this block and examined until a suitable region of the block is found. These sections can be labelled with antibody and colloidal gold probe to determine the optimum reagent dilutions. The formation of a section from the block is a fracturing process and this produces a fracture face on the specimen block. The specimen block is removed from the cryoultramicrotome and thawed in 2.5% glutaraldehyde. After a brief rinse in buffer it is incubated with antibody and then a colloidal gold probe. The block is treated with 2% osmium tetroxide, dehydrated in ethanol and embedded in Araldite. Thin sections are cut vertically through the fractured face of the block and are stained with uranyl acetate and lead citrate before examination. The presence, or absence, of antigen

---

Fig. 6. Pre-embedding labelling of the stabilised capsule of *P. haemolytica* with antibody against capsular polysaccharide and colloidal gold probe. The presence of an irregular wide band of capsule confirms the results of shadowing ultrathin cryosections with platinum and carbon. Bar = 0.2 μm.

Fig. 7. Freeze fracture face of *P. haemolytica* labelled with antibody against capsular polysaccharide and the colloidal gold probe. Antigen is associated with the bacterial membranes, not the cytoplasm. Bar = 0.5 μm.

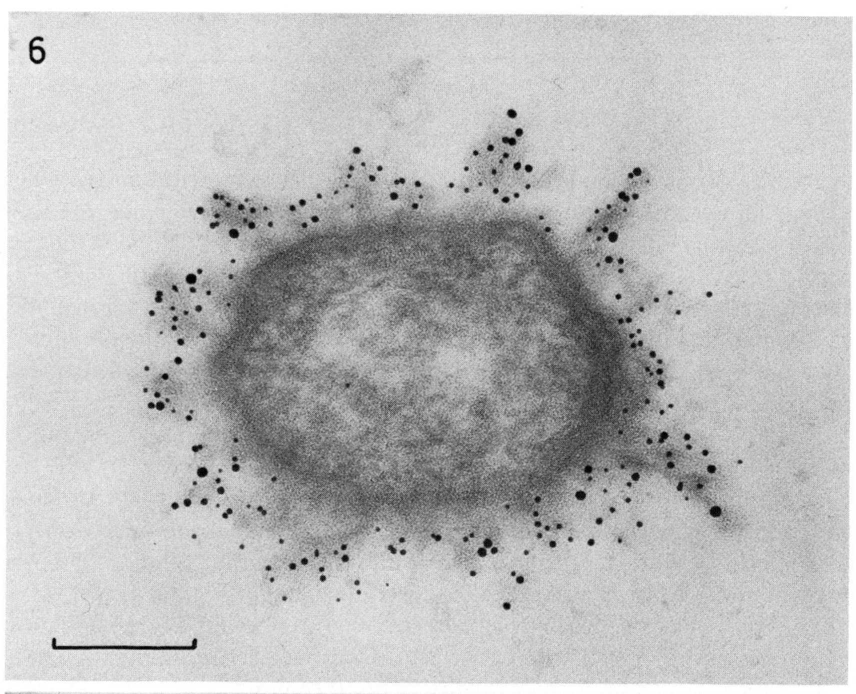
6

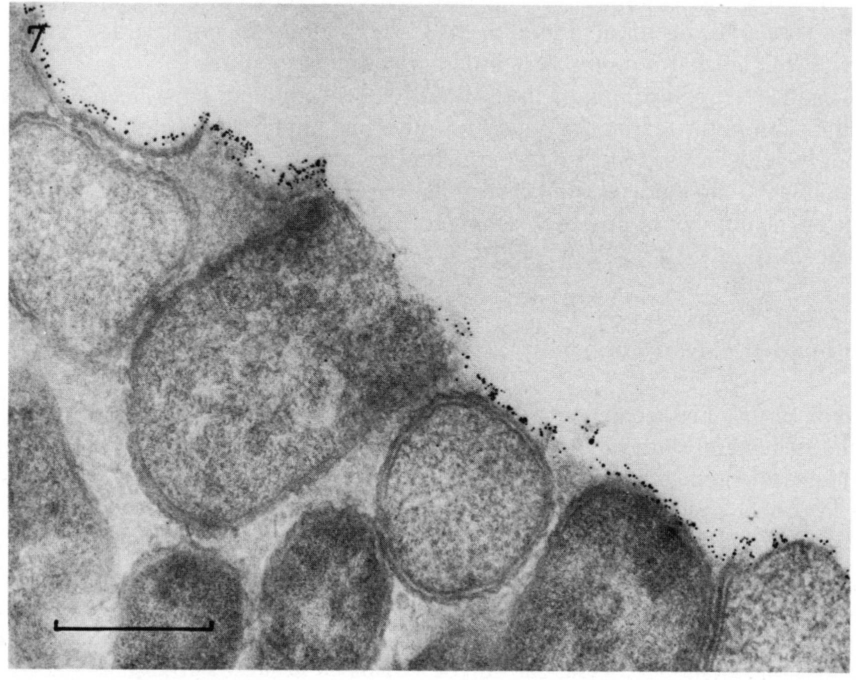
7

associated with labelling on each fractured face of the membranes can then be determined from these sections (Fig. 7).

4. Immuno-negative stain technique

Electron microscope techniques for the detection of surface viral antigens necessitate incubating virus suspensions with antibody and then screening the sample for large aggregates of viruses [43–45]. Alternatively, viruses can be serologically trapped to microscope grids [46–48] (see Chapter 22, pp. 323–340). Bacterial pili have been similarly incubated with antibody alone, or antibody and a ferritin probe before being dried down onto an electron microscope grid and negatively stained to visualise decorating antibody or probe [15, 49]. These methods are time consuming and need relatively large volumes of suspension of organisms. These techniques also depend on consistently high quality negative stain and abundant decoration by antibody.

The colloidal gold probe has proved useful in detecting surface antigens of structures that can be obtained in suspension. These antigens are dried down onto electron microscope grids, labelled in situ with an appropriate dilution of antibody followed by the gold probe and finally negatively stained (Figs. 1 and 8).

The gold probe is so dense and can detect such low amounts of antibody that the immunocytochemical reaction is easily visualised over a wide range of antibody concentration and negative stain quality. This technique is now employed as a routine screen for bacterial pilus antigens [50] and viral antigens [51]. It is a simple, short technique that requires very little antigen in suspension.

Two further applications of the colloidal gold technique have been employed with this immuno-negative stain method. Five nanometer and 20 nm colloidal gold probes have been used in conjunction with two antibody preparations to label the same pilus [50] and morphologically similar virus types have been differentiated by counting the number of probes attaching to each virus after treatment with different specific antisera [51].

5. Immuno-replica technique

An immuno-replica technique has been described to detect measles virus antigen on the cell membrane of measles infected cultured HeLa cells [52–57]. In this technique infected cell cultures are incubated with anti-measles serum, the protein A-gold probe, then dehydrated, critical point dried and shadowed with platinum and carbon. The gold probe is so dense that it can be distinguished easily. Labelled areas can be differentiated from unlabelled areas and even probes positioned in the evaporation shadow of large structures can be accurately identified. This is a pre-requisite for an exact quantification and mapping of antigen. In addition, the ultrastructure of labelled areas can still be visualised because of the small size of the marker [52].

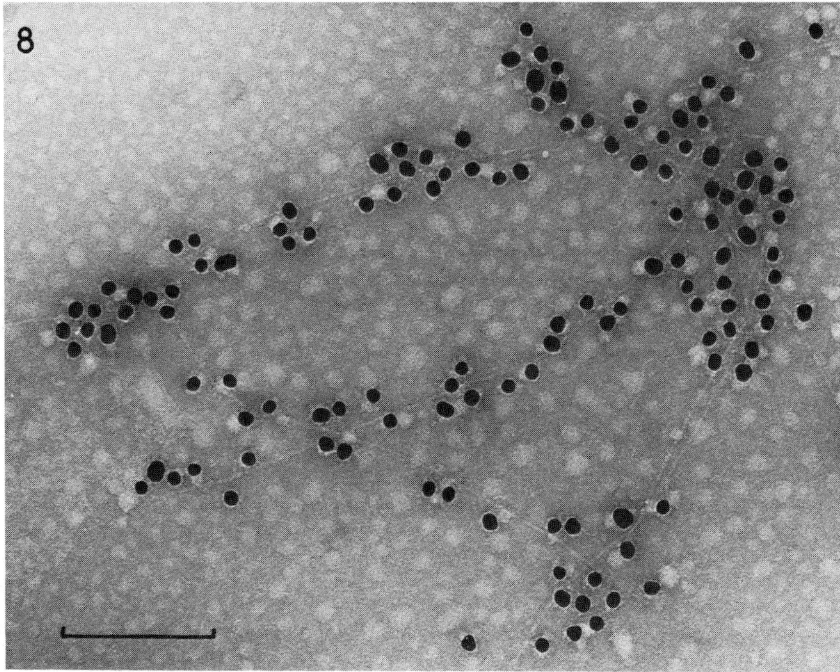

Fig. 8. Isolated *Bacteroides nodosus* pili, dried down onto an electron microscope grid and labelled with specific antibody and colloidal gold probe, and finally stained with ammonium molybdate. The labelling is specific to pili and is very distinctive. Bar = 0.2 μm.

6. Conclusions

Microbiological immunocytochemistry presents special problems because the objects under study are often extremely small and require good contrast and resolution for detailed study. These problems can now be overcome by the correct choice of preparative techniques. The size of the organism is the critical factor for determining which of the techniques described above will be applied.

If the object is large in comparison with the section thickness, such as a bacterium, then resin sections will usually suffice since many of the antigens within the bacterium will be exposed at the cut surface.

If the object of interest is small in relation to the section thickness, such as a virus, then cryotechniques are applicable. Cryotechniques fulfill many of the criteria of the desirable preparation technique described in the introduction to this chapter. They permit high density staining to obtain the resolution needed to visualise small structures, they permit immunolabelling of surface antigens throughout the thickness of the section and also, as in resin sections, expose intracellular antigens. These advantages are combined with the delineation of structures by negative staining comparable to that found in positively stained conventional sections [23].

If only surface antigens are of interest and the sample can be dried down onto a grid, then the immuno-negative stain technique is a simple, rapid technique to use. The immuno-negative stain technique is especially applicable in those studies where it would be difficult to obtain sufficient material to embed, such as in the study of isolated bacterial pili or isolated viruses.

If the appearance of groups of antigens on the host cell surface is sought then the immuno-replica technique is the method of choice since it allows scanning of wide areas of host surface for the identification of the sites of the immunological reactions.

The development of all these techniques was greatly facilitated by the development of the colloidal gold probes. Whilst the techniques described here could be repeated with other electron-dense probes, gold produces superior results and has enabled these techniques to be established as routine procedures, producing a wealth of useful information.

7. Acknowledgements

I gratefully acknowledge the continued technical assistance from the staff of the Electron Microscope Suite, The Wellcome Foundation, Beckenham, Kent, U.K., especially from Miss A. Orpin and M.P. Betts. Thanks are also expressed to all those who have collaborated in these studies. I finally thank Mrs. E.A. Fellowes for typing this manuscript.

8. References

1. Bohn, W. (1978) A fixation method for improved antibody penetration in electron microscopical immuno-peroxidase studies. J. Histochem. Cytochem. 26, 293–297.
2. Bohn, W. (1980) Electron microscopic immunoperoxidase studies on the accumulation of virus antigens in cells infected with shope fibroma virus. J. Gen. Virol. 46, 439–447.
3. De Mey, J., Moeremans, M., Geuens, G., Nuydens, R. and De Brabander, M. (1981) High resolution light and electron microscopic localization of tubulin with the IGS (Immuno Gold Staining) method. Cell Biol. Int. Rep. 5, 889–899.
4. De Mey, J. (1983) A critical review of light and electron microscopic immunocytochemical techniques used in neurobiology. J. Neurosci. Methods 7, 1–18.
5. Griffiths, G.W. and Jockusch, B.M. (1980) Antibody labelling of thin sections of skeletal muscle with specific antibodies: A comparison of bovine serum albumin (BSA) embedding and ultracryomicrotomy. J. Histochem. Cytochem. 28, 969–978.
6. Bendayan, M. and Zollinger, M. (1982) Protein A-gold immunocytochemical technique: Ultrastructural localisation of antigens on postosmicated tissues. Proc. 10th Int. Cong. Electron Microsc. 3, 267–268.
7. Schwendemann, G., Wolinsky, J.S., Hatzidimitriou, G., Merz, D.C. and Waxham, M.N. (1982) Postembedding immunocytochemical localisation of paramyxovirus antigens by light and electron microscopy. J. Histochem. Cytochem. 30, 1313–1319.
8. Walker, P.D. and Beesley, J.E. (1983) Trends in the localisation of bacterial antigens by immunoelectron microscopy. Immunol. Commun. 12, 75.
9. Oram, J.D. and Crooks, A.J. (1979) A comparison of labelled antibody methods for the detection of virus antigens in cell monolayers. J. Immunol. Methods 25, 297–310.
10. Beesley, J.E. and Adlam, C. (1982) The protein A-gold technique: a comparison between the

staining mechanisms on methacrylate and cryosections. Proc. 10th Int. Cong. Electron Microsc. 3, 265–266.

11. Bendayan, M. (1983) Ultrastructural localisation of actin in muscle, epithelial and secretory cells by applying the protein A-gold technique. Histochem. J. 15, 39–58.
12. Adlam, C., Knights, J.M., Mugridge, A., Lindon, J.C., Baker, P.R.W., Beesley, J.E., Spacey, B., Craig, G.R. and Nagy, L.K. (1984) Purification, characterisation and immunological properties of the serotype specific capsular polysaccharide of *Pasteurella haemolytica* (serotype A1) organism. J. Gen. Microbiol. (In press).
13. Beesley, J.E. and Adlam, C. (1983) Capsular antigen in ultrathin cryosections of *Pasteurella haemolytica*. Immunol. Commun. 12, 69.
14. Takamiya, H., Batsford, S., Gelderblom, H. and Vogt, A. (1979) Immuno-electron microscopic localisation of lipopolysaccharide antigens on ultrathin sections of *Salmonella typhimurium*. J. Bacteriol. 140, 261–266.
15. Walker, P.D., Short, J., Thomson, R.O. and Roberts, D.S. (1973) The fine structure of *Fusiformis nodosus* with special reference to the location of antigens associated with immunogenicity. J. Gen. Microbiol. 77, 351–361.
16. Short, J.A. and Walker, P.D. (1975) The location of bacterial antigens on sections of *Bacillus cereus* by use of the soluble peroxidase-anti-peroxidase complex and unlabelled antibody. J. Gen. Microbiol. 89, 93–101.
17. Raybould, T.J.G., Beesley, J.E. and Chantler, S. (1981) Ultrastructural localization of characterized antigens of *Brucella abortus* and distribution among different biotypes. Infect. Immun. 32, 318–322.
18. Garzon, S., Bendayan, M. and Kurstak, E. (1982) Ultrastructural localization of viral antigens using the protein-A gold technique. J. Virol. Methods 5, 67–73.
19. Beesley, J.E., Orpin, A. and Adlam, C.A. (1982) A comparison of immunoferritin, immunoenzyme and gold-labelled protein A methods for the localization of capsular antigen on frozen thin sections of the bacterium *Pasteurella haemolytica*. Histochem. J. 14, 803–810.
20. Beesley, J.E., Orpin, A. and Adlam, C. (1983) An evaluation of the conditions necessary for optimal protein A-gold labelling of capsular antigen in ultrathin methacrylate sections of the bacterium *Pasteurella haemolytica*. Histochem. J. 16, 151–163.
21. Beesley, J.E. and Campbell, D.A. (1983) The protein A-gold (PAG) technique in association with ultrathin frozen sections for the ultrastructural localisation of influenza virus antigens. Proc. R. Mic. Soc. 18, 74.
22. Bergmann, J.E., Tokuyasu, K.T. and Singer, S.J. (1981) Passage of an integral membrane protein, the vesicular stomatitis virus glycoprotein, through the Golgi apparatus en route to the plasma membrane. Proc. Natl. Acad. Sci. USA 78, 1746–1750.
23. Tokuyasu, K.T. (1980) Immunocytochemistry on ultrathin frozen sections. Histochem. J. 12, 381–403.
24. Roth, J. (1982) The protein-A gold (pAg) technique—a qualitative and quantitative approach for antigen localisation on thin sections. In: G.R. Bullock and P. Petrusz (Eds.) Techniques in Immunocytochemistry, Vol. 1. Academic Press, London and New York, pp. 107–133.
25. De Mey, J. (1983) Colloidal gold probes in immunocytochemistry. In: J.M. Polak and S. Van Noorden (Eds.) Immunocytochemistry. Practical Applications in Pathology and Biology, John Wright and Sons, Bristol – London – Boston, pp. 82–112.
26. Gelderblom, H., Pauli, G., Schäuble, H. and Ludwig, H. (1980) Detection of a cross-reacting antigen between herpes simplex virus type 1 and bovine herpes mammillitis virus by immunoelectron microscopy. Proc. 7th European Cong. Electron Microsc. 2, 482–483.
27. Singer, S.J., Tokuyasu, K.T., Dutton, A.H. and Wen-Tien Chen. (1982) High resolution immunoelectron microscopy of cell and tissue ultrastructure. In: J.D. Griffith (Ed.) Electron Microscopy in Biology, Vol. 2, J. Wiley, London and New York, pp. 55–105.
28. Tokuyasu, K.T. (1983) Present state of immunocryoultramicrotomy. J. Histochem. Cytochem. 31, 164–167.
29. Bernhard, W. and Leduc, E.H. (1967) Ultrathin frozen sections. I. Methods and ultrastructural preservation. J. Cell Biol. 34, 757–771.

30. Bernhard, W. and Viron, A. (1971) Improved techniques for the preparation of ultrathin frozen sections. J. Cell Biol. 49, 731–746.
31. Novak, M., Benda, R. and Jelinkova, A. (1976) Electron microscopy of virus particles in ultrathin frozen sections. Acta Virol. 20, 159–161.
32. Novak, M., Jelinkova, A. and Benda, R. (1977) Preparation of ultrathin frozen sections in electron microscopy. Specimen supported by polyethylene glycol. Microsc. Acta 89, 61–64.
33. Green, J., Griffiths, G., Louvard, D., Quinn, P. and Warren, G. (1981) Passage of viral membrane proteins through the Golgi complex. J. Mol. Biol. 152, 663–698.
34. Griffiths, G., Brands, R., Burke, B., Louvard, D. and Warren, G. (1982) Viral membrane proteins acquire galactose in trans Golgi cisternae during intracellular transport. J. Cell Biol. 95, 781–792.
35. Griffiths, G., Quinn, P. and Warren, G. (1983) Dissection of the Golgi complex. I. Monensin inhibits the transport of viral membrane proteins from medial to trans Golgi cisternae in baby hamster kidney cells infected with semliki forest virus. J. Cell. Biol. 96, 835–850.
36. Bourguignon, L.Y.W. and Butman, B.T. (1982) Intracellular localisation of certain membrane glycoproteins in mouse T-lymphoma cells using immunoferritin staining of ultrathin frozen sections. J. Cell. Physiol. 110, 203–212.
37. Geuze, H.J. and Slot, J.W. (1980) The subcellular localisation of immunoglobulin in mouse plasma cells as studied with immunoferritin cytochemistry on ultrathin frozen sections. Am. J. Anat. 158, 161–169.
38. Fournier-Laflèche, D., Chang, A., Benichou, J-C. and Ryter, A. (1975) Immuno-labelling in frozen ultrathin sections of bacteria. J. Microsc. Biol. Cell. 23, 17–28.
39. Bernadac, A. and Lazdunski, C. (1981) Immunoferritin labelling of ultrathin frozen sections of gram-negative bacterial cells. Biol. Cell. 41, 211–216.
40. Glauert, A.M. (1974) Fixation methods. In: A.M. Glauert (Ed.) Practical Methods in Electron Microscopy, Vol. 3. Elsevier North-Holland, Amsterdam – Oxford – New York, p. 90.
41. Geuze, H.J., Slot, J.W., Van de Ley, P.A., Scheffer, R.C.T. and Griffith, J.M. (1981) Use of colloidal gold particles in double-labelling immunoelectron microscopy of ultrathin frozen tissue sections. J. Cell Biol. 89, 653–665.
42. Beesley, J.E. (1983) A new microscope stand for use with the LKB Ultratome III and cryokit. J. Microsc. 129, 111–112.
43. Almeida, J.D. and Waterson, A.P. (1969) The morphology of virus-antibody interaction. Adv. Virus Res. 15, 307–338.
44. Mandel, B. (1971) Methods for the study of virus complexes. In: K. Maramorosch and H. Koprowski (Eds.) Methods in Virology, Vol. 5. Academic Press, New York, pp. 375–397.
45. Doane, F.W. (1974) Identification of viruses by immuno-electron microscopy. In: E. Kurstak and R. Morriset (Eds.) Viral Immunodiagnosis. Academic Press, New York. pp. 237–255.
46. Derrick, K.S. (1973) Detection and identification of plant viruses by serologically specific electron microscopy. Phytopathology 63, 441.
47. Shukla, D.D. and Gough, K.H. (1979) The use of protein A from *Staphylococcus aureus* in immune electron microscopy for detecting plant virus particles. J. Gen. Virol. 45, 533–536.
48. Nicolaieff, A., Obert, G. and Van Regenmortel, M.H.V. (1980) Detection of rotavirus by serological trapping on antibody-coated electron microscope grids. J. Clin. Microbiol. 12, 101–104.
49. Short, J.A., Thorley, C.M. and Walker, P.D. (1976) An electron microscope study of *Bacteroides nodosus*: Ultrastructure of organisms from primary isolates and different colony types. J. Appl. Bacteriol. 40, 311–315.
50. Beesley, J.E., Day, S.E.J., Betts, M.P. and Thorley, C.M. (1984) Immunocytochemical labelling of *Bacteroides nodosus* pili using an immunogold technique. J. Gen. Microbiol. (In press.)
51. Beesley, J.E. and Betts, M.P. (1984) Virus diagnosis: A novel use for the protein A-gold probe. Med. Lab. Sci. (In press.)
52. Hohenberg, H., Mannweiler, K., Rutter, G. and Bohn, W. (1981) Topographic and antigenic demonstration of characteristic ultrastructures at the surfaces of virus-infected cells by comparative high resolution E.M. studies of surface replicas with large areas. Med. Microbiol. Immunol. 169, 155.

53. Bohn, W., Rutter, G. and Mannweiler, K. (1981) Distribution of viral-antigens and viral structures on HeLa-cells lytically and persistently infected with measles virus. Med. Microbiol. Immunol. 169, 149.

54. Mannweiler, K., Hohenberg, H., Bohn, W., Rutter, G., Andresen, I. and Ueckermann, C. (1981) Protein-A gold particles as markers for cell-surface labelling in replica immunocytochemistry (RIC). Acta Anat. 111, 95.

55. Mannweiler, K., Hohenberg, H., Bohn, W. and Rutter, G. (1982) Protein-A gold particles as markers in replica immunocytochemistry: high resolution electron microscope investigations of plasma membrane surfaces. J. Microsc. 126, 145–149.

56. Mannweiler, K., Hohenberg, H., Bohn, W. and Rutter, G. (1982) E.M. cytochemical investigations of virus-induced plasma membrane alterations in large-sized replicas of CP-dried and freeze-fractured cultured cells. Eur. J. Cell Biol. 27, 124.

57. Mannweiler, K., Bohn, W., Rutter, G. and Hohenberg, H. (1982) Replica-immunocytochemical data at the plasma-membrane of HeLa cells after infection with measles virus. Zb. Bakt. A. 251, 445.

Immunolabelling for Electron Microscopy (Polak/Varndell, eds)

CHAPTER 21

Phylogenetic conservation of hormone-related peptides: Immunocytochemical and physico-chemical studies in non-vertebrates and unicellular organisms

Derek Le Roith

University of Cincinnati, College of Medicine, Division of Endocrinology and Metabolism, Mail Location 547, Cincinnati, OH 45267, U.S.A.

CONTENTS

1. INTRODUCTION

Intercellular communication in mammals and other vertebrates has traditionally been restricted to the endocrine system and nervous system. Moreover, the messenger molecules were considered to be unique to each system; hormones being associated with endocrine glands and neurotransmitters with nervous tissue.

Furthermore, since the glandular structures first appeared at the level of vertebrates, hormones were considered to be unique to vertebrates. In contrast, neurotransmitters were found in lower metazoa at the level where nerves first appear phylogenetically.

Over the last decade or so, numerous reports have described the presence of materials very similar to classic vertebrate hormone-related peptides in multicellular non-vertebrate tissues [1–13] as well as in unicellular organisms [14–25]. The presence of these peptides has been demonstrated by both immunocytochemical techniques as well as by specific immuno- and bioassays on purified extracts from these tissues. These findings strongly suggest that classic vertebrate peptide hormones and neuropeptides have early phylogenetic origins [26], and in addition that functionally important regions of these molecules have been highly conserved during evolution.

This chapter outlines the evidence for the phylogenetic origins of these hormone-related peptides, drawing on examples where immunocytochemical techniques have been utilised. In addition, studies on other highly conserved and ubiquitous cellular proteins are described for comparison.

2. Hormonal peptides in multicellular non-vertebrates

With the discovery that hormone-related peptides were produced by non-endocrine tissues, for example, cancer-, nerve-, and gut-related cells, investigators began looking for similar materials in nervous tissue and gut-related cells of multicellular non-vertebrates. The bulk of the evidence supporting the presence of these vertebrate hormone-related peptides in non-vertebrate tissues comes from immunocytochemical studies. The studies utilised heterologous antisera raised against mammalian hormones and included immunocytochemical staining of the tissues as well as immunoassay of purified extracts from these tissues (Table 1). At least 18 vertebrate hormone-related peptides have been identified in non-vertebrate tissues, and each major phyletic group is represented in these studies [7]. Regarding the existence of insulin-related material, for example, numerous investigators have independently demonstrated its presence in at least eighteen separate species of Insecta [7]. The insulin-related material has been localised to the brain of these insects using immunocytochemical staining and its physico-chemical characteristics established from studies of purified tissue extracts using immunoassays and bioassays.

2.1. Insulin-related material in Insecta

Following the initial reports that suggested the presence of insulin-related material in insect brains [2], cellular localisation was demonstrated in the median neurosecretory cells (MNC) in the blowfly ([27], Fig. 1), hoverfly, silkworm and tobacco hornworm moth [28]. Extirpation of the median neurosecretory cells from the blowfly caused elevation of glucose and trehalose levels (carbohydrate

TABLE 1

Vertebrate-type hormone-related peptides found in multicellular non-verbebrate tissues [7]

Hormone	*Species* *Ch.*	*E.*	*I.*	*C.*	*M.*	*A.*	*F.*	*H.*
Insulin	+	+	+	+	+	+		
Glucagon	+	+	+	+	+			
Somatostatin	+		+	+	+		+	
Cholecystokinin	+		+		+	+		+
Pancreatic polypeptide	+		+		+	+		
Substance P	+		+	+	+			+
Vasotocin			+		+			+
ACTH	+		+	+	+			
Endorphin	+		+		+	+		
Calcitonin	+				+			+
TRH					+			
Bombesin					+			+
Neurotensin								+
Secretin	+				+			
Vasoactive Intestinal Peptide	+					+		
LHRH	+							

Chordates (Ch), Echinoderms (E), Insects (I), Crustaceans (C), Molluscs (M), Annelids (A), Flatworms (F), *Hydra* (H).

intermediates); hyperglucosaemia and hypertrehalosaemia were reversed by injection of MNC homogenates and brain extracts (Fig. 2). Similar elevations of trehalose levels have been noted after selective removal of the corpus cardiacum of the blowfly (*Phormia regina*) [29], and corpus cardiacum—corpus allatum of the housefly, *Musca domestica* [30]; organs previously shown to contain insulin-related material [2, 8]. These findings strongly suggest that the insulin-related material in the brain plays an important role in regulating carbohydrate metabolism in insects.

Physico-chemical studies on purified insulin-related material from brain extracts suggest that the material is very similar to classic mammalian insulins. Thus the material chromatographed on Sephadex G-50 gel columns with an approximate molecular weight of 6,000 [4, 5], and elution patterns on ion-exchange chromatography columns as well as polyacrylamide gel electrophoresis were similar to those of mammalian insulins [5]. In addition, the purified material demonstrated full biological activity in insulin bioassays [4, 5]. Furthermore, the amino acid composition of the insulin-related material from the tobacco hornworm moth (*Manduca sexta*) demonstrates only about 14 amino acid residue differences when compared to porcine insulin (Table 2).

2.2. *Summary*

Numerous classical vertebrate hormone-related peptides are present in non-vertebrate tissues, especially nervous tissue and gut-related cells. Most studies have

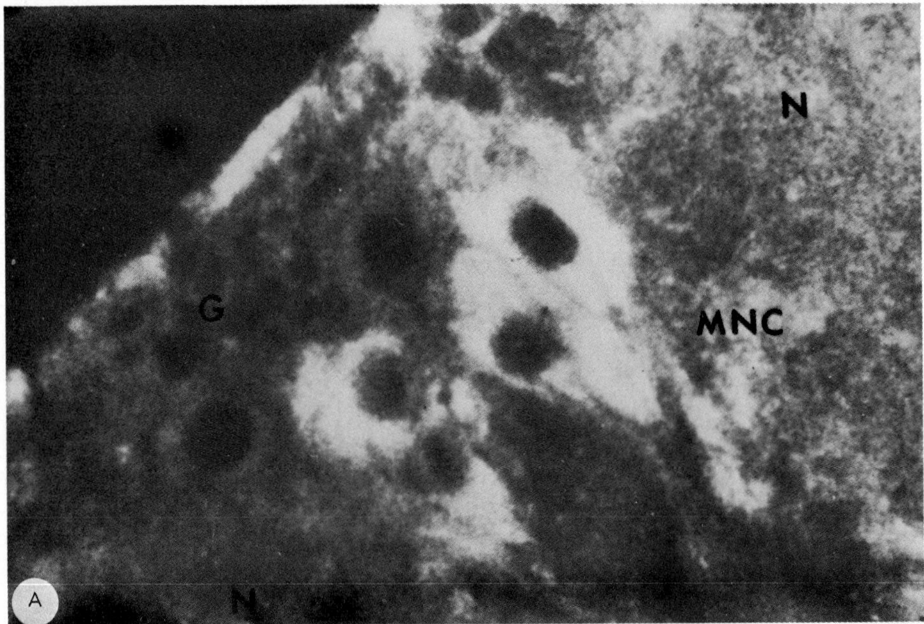

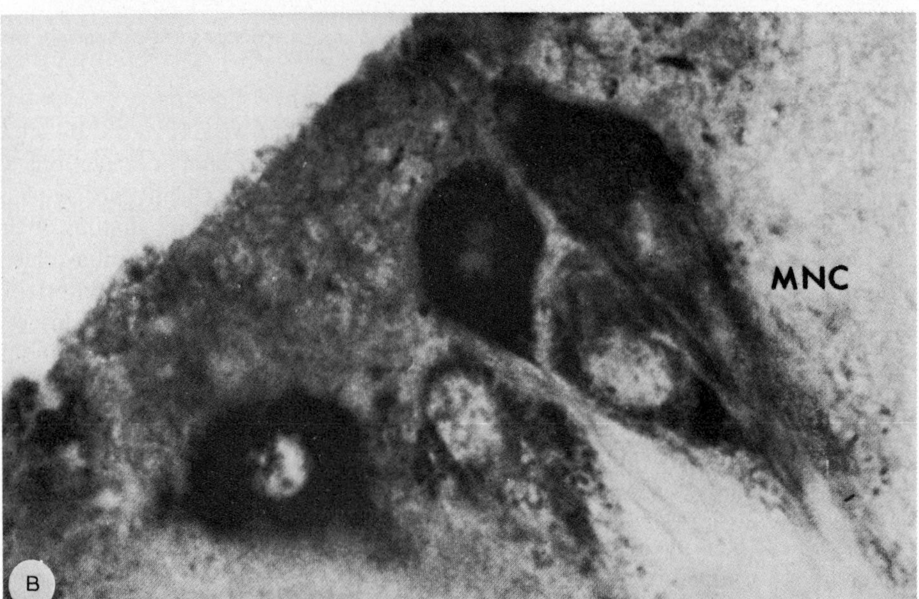

Fig. 1. Insulin-related material in the blowfly. Median neurosecretory cells together with surrounding brain tissue were excised from adult blowflies (*Calliphora vomitoria*—Diptera) during ether anaesthesia. The tissues were fixed in Bouin's fluid, embedded in paraffin and sectioned at 6 μm. Indirect immunofluorescence was used to stain insulin-related material. The sections were initially incubated with a bovine insulin antiserum produced in guinea pigs. The sections were then treated with fluorescein isothiocyanate (FITC)-conjugated porcine anti-guinea pig serum. Photographs were taken using epi-fluorescence photomicroscopy (A). The sections were then washed, stained for neurosecretory material with paraldehyde fuchsin and re-photographed (B). MNC = median neurosecretory cells, N = neuropil. (Reproduced from reference [27] with permission of the authors.)

been immunological; immunocytochemistry using heterologous antibodies and sensitive radioimmunoassays of purified tissue extracts. With regard to insulin in insects, in addition to extensive physico-chemical and structural analyses, a definite physiological function has been attributed.

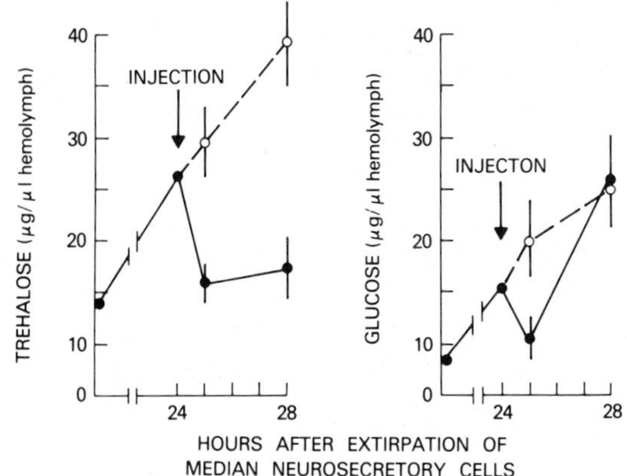

Fig. 2. Effect of insulin on carbohydrate metabolism in the blowfly. Following the extirpation of median neurosecretory cells, glucose and trehalose concentrations in haemolymph rose. Partially purified extracts of the MNC cells (arrow) normalised the concentration of these carbohydrate intermediates (●); the effect on glucose levels was more transient when compared to the trehalose levels. In saline-treated controls (○) the glucose and trehalose levels continued to rise.

TABLE 2

Amino acid composition of porcine insulin and insulin-related material from *Manduca sexta* [8]

Amino acid	*Number of residues* *Manduca*	*Porcine*
Aspartate	3.1	3
Threonine and glycine	8.0	6
Serine	6.8	3
Glutamic Acid	6.2	7
Proline	–	1
Alanine	2.0	2
Valine	2.9	3
Methionine	nd	0
Isoleucine	2.1	2
Leucine	1.8	6
Tyrosine	2.8	4
Phenylalanine	1.0	3
Histidine	2.1	2
Lysine	0.9	1
Arginine	1.1	1
Tryptophan	nd	0
Cysteine	nd	6

Threonine and glycine were not resolved.
nd = methionine, tryptophan and cysteine not stable to acid hydrolysis.

3. Hormonal peptides in unicellular organisms

Material very similar to thyroid-stimulating hormone (TSH) was found in *Clostridium perfringens* by Pastan and co-workers [20]. The TSH-like material had an apparent molecular weight of 30,000, was pronase sensitive suggesting it was protein, and exhibited TSH-like biological activity. Subsequently material resembling human chorionic gonadotropin (hCG) was found in a number of bacteria isolated from patients suffering from cancer (see section 3.2).

3.1. Our findings

In our search for the phyletic origins of peptide hormones, we extended our studies to unicellular organisms. *Tetrahymena pyriformis*, a ciliated protozoan, was grown in defined, synthetic medium in the absence of serum or macromolecules (to exclude possible exogenous contamination). The cells were harvested at the end of

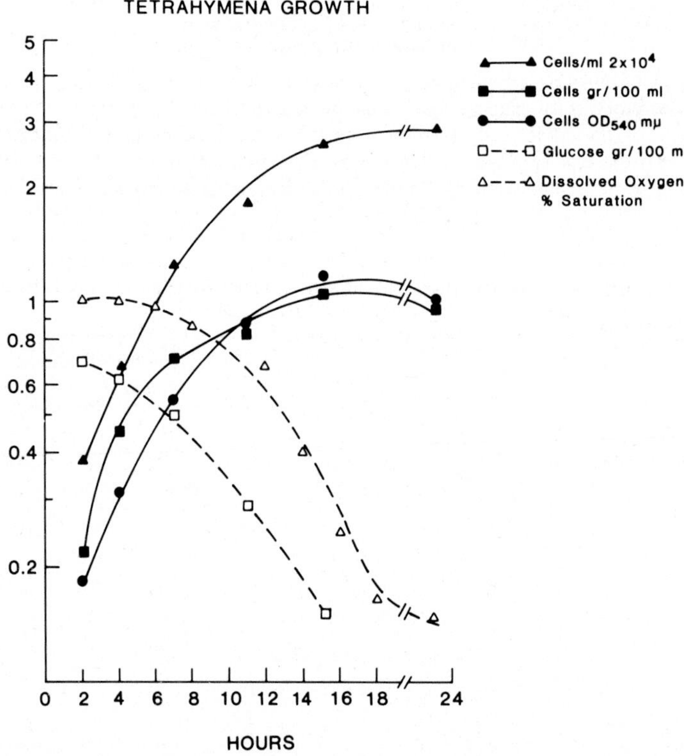

Fig. 3. Large batch fermentation of *Tetrahymena pyriformis*. *Tetrahymena pyriformis* was grown in a 100 litre fermenter at 27°C, with an aeration rate of 0.03 v/v/min and agitation of 50 rpm. Following inoculation with 10 litres of a 48 h culture, the cell growth was followed by measuring optical density (OD) at 540 nm, cell counts and wet weight. After a lag period of 12 h, logarithmic growth phase (OD 0.2) was reached and lasted for about 12 h. The cells were then harvested and concentrated 20-fold by tangential flow filtration using a Millipore® cassette system which utilises a 15 square feet Durapore® membrane of 0.45 μm pore size.

the logarithmic growth phase (24–30 h) and separation of the cells from the medium was achieved using a Millipore® cassette system (Fig. 3). Following homogenisation of the cells, hormone-related peptides were extracted and purified using techniques previously described for extracting these peptides from vertebrate tissues [14–19].

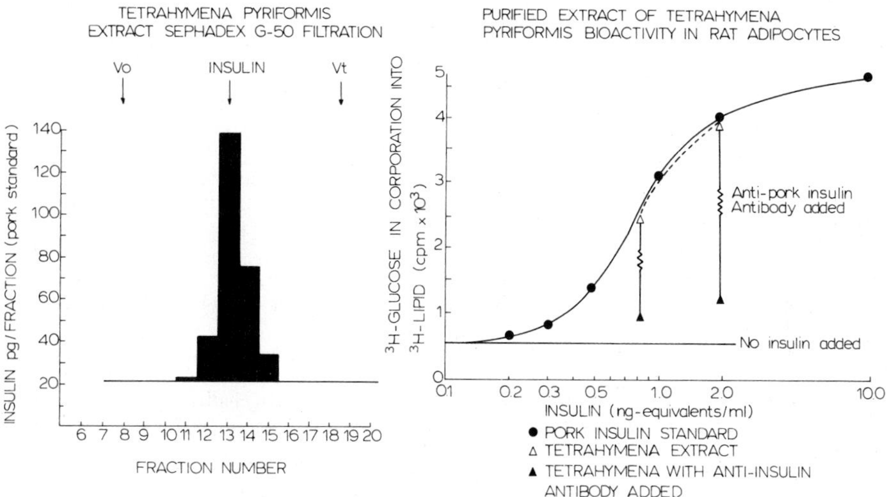

Fig. 4. Insulin-related material in *Tetrahymena pyriformis*. Following separation of *Tetrahymena* cells from the growth medium, the cells were homogenised. Insulin-related material was then extracted in acid-ethanol and purified by several chromatography techniques. A peak of insulin immunoreactivity was recovered on Sephadex G-50 gel filtration, in the region where mammalian insulins migrate (left panel). When the purified *Tetrahymena* insulin-related material was tested in a bioassay for insulin, the bioactivity (glucose incorporation into lipid by isolated adipocytes from young male rats) was predicted by the amount of immunoreactive insulin (right panel). Most of the bioactivity was neutralised in the presence of porcine insulin antibody.

Materials closely resembling insulin were found in *Tetrahymena* (Fig. 4), *Neurospora crassa* and *Aspergillus fumigatus* (unicellular fungi) as well as in five strains of *Escherichia coli* (Table 3). The insulin-related materials behaved like vertebrate insulins immunologically and biologically, as well as in several chromatographic systems including Sephadex gel filtration, ion-exchange and high performance liquid chromatography (HPLC) (unpublished observations). In addition, extracts of *Tetrahymena* contain materials closely resembling ACTH (1–39) and beta-endorphin as well as larger molecular weight material that had both ACTH and beta-endorphin-related moieties on the same molecule suggestive of proopiomelanocortin, the biosynthetic precursor of ACTH and beta-endorphin of mammals [16]. Somatostatin-like, relaxin-like and thyrocalcitonin-related materials have also been demonstrated in extracts of unicellular organisms (Table 3).

3.2. *Chorionic gonadotropin (CG)-like factor production by bacteria*

Human chorionic gonadotropin (hCG) is normally produced by the trophoblastic cells of the placenta during pregnancy as well as by trophoblastic and non-

TABLE 3
Hormone-related peptides in unicellular organisms

	Organisms	Reference
Eukaryotes		
Insulin	*Tetrahymena, Neurospora, Aspergillus*	14
Somatostatin	*Tetrahymena*	17
ACTH, beta-endorphin	*Tetrahymena*	16
Relaxin	*Tetrahymena*	18
Thyrocalcitonin (salmon-type)	*Tetrahymena*	19
(human-type)	*Candida*	25
Prokaryotes		
Insulin	*E. coli*	15
Somatostatin	*E. coli*	–
Neurotensin	many strains	24
hCG	many strains	21–23, 34–36
TSH	*Clostridium perfringens*	20
Thyrocalcitonin (human-type)	*E. coli*	25

In addition, other peptides have been described in these organisms including arginine vasotocin in *Tetrahymena* (Collier et al., personal communication) and cholecystokinin (J. Taylor et al., personal communication).

trophoblastic neoplasms [31, 32]. More recently, studies have demonstrated the production of hCG-like material by normal non-neoplastic human tissues including testes, foetal lung and liver [33]. Wolfsen and Odell have suggested that the production of hCG-like material is ubiquitous in vertebrate tissues [32]. Interestingly, a number of investigators have independently demonstrated the presence of chorionic gonadotropin (CG)-like material in aerobic and anaerobic bacteria isolated from patients suffering from cancer [21–23, 34–36]. The following subsections will outline the results from the more pertinent of these reports.

3.2.1. Immunocytochemistry

Using indirect fluorescein-labelled, as well as indirect peroxidase-antiperoxidase labelled antibody techniques, Acevedo and co-workers [34–36] demonstrated the presence of CG-like materials in subcultures of 16 strains of bacteria (Table 4).

TABLE 4
Chorionic gonadotropin-like material in bacteria isolated from cancer patients [21–23, 34–41]

Positive	Absent
Progenitor cryptocides	*Corynebacterium parvum*
Eubacterium lentum	*Agrobacterium tumefaciens*
Staphylococcus epidermidis	*Pseudomonas aeruginosa*
Escherichia coli	*Streptococcus faecalis*
Pseudomonas maltophilia	
Streptococcus faecalis	
Staphylococcus simulans	
Staphylococcus aureus	

Localisation of the CG-like material by light microscopy demonstrated its association with the bacteria cell wall (Fig. 5). Using these techniques, a large number of non-pathogenic organisms as well as some cancer-associated bacterial failed to demonstrate the presence of CG-like material suggesting that the phenomenon may

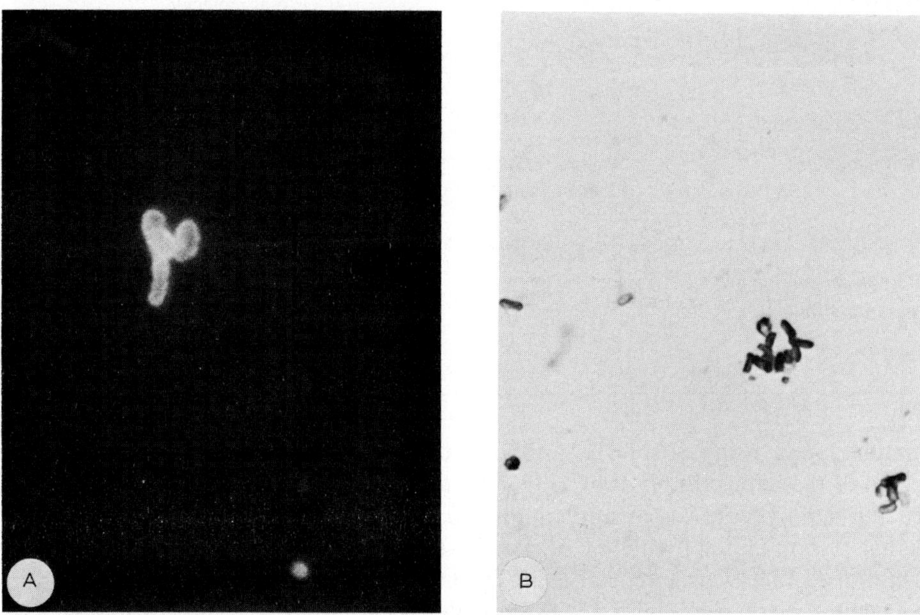

Fig. 5. Light microscopical evidence for CG-like material in a bacterium *Eubacterium lentum*. *E. lentum* (ATCC 25559) an obligate anaerobe, was isolated from a rectal tumour. The organism was grown on 5% sheep blood agar with Columbia base (BBL) for 48 h at 35°C in an anaerobic jar. Several colonies were then smeared on a glass microscope slide, air-dried and stained for chorionic gonadotropin (CG)-like material using rabbit CG antiserum as first antibody and fluorescein-labelled or peroxidase-antiperoxidase-labelled goat anti-rabbit sera as second layer. *E. lentum* demonstrated positive reactions using indirect immunofluorescence (A) as well as with the peroxidase-antiperoxidase technique (B). The highest intensity of labelling was demonstrated by the cell wall. In contrast, *Corynebacterium parvum* showed no positive reaction (negative control, data not shown). Figure reproduced from reference [35] with permission.

not be ubiquitous. With the aid of immuno-electron microscopy these investigators confirmed the intimate association of the CG-like material with the cell wall and further speculated that the CG-like material may in fact be an integral part of the cell wall membrane [36].

3.2.2. Physico-chemical studies

The presence of CG-like material in these cancer-associated bacteria has been confirmed by demonstrating CG-like immunoreactivity in a number of in vitro assays (Table 5). Acetone-dried powder of the cells was prepared from cultures of the bacteria and CG-like material extracted from the powder as previously described [23]. The extracted material demonstrated cross-reactivity in an hCG

TABLE 5

Physico-chemical characteristics of chorionic gonadotropin-like material from *Progenitor cryptocides* [23]

Immunoactivity:	Antisera used:	hCG total molecule hCG β-subunit hCG β-subunit carboxy terminal hCG α-subunit
Radioreceptor assay:		bovine corpus luteum membranes
Chromatography:	Sephadex G-100:	molecular weight similar to hCG
	Concanavalin A Sepharose:	suggests presence of glucose and mannose moieites
	DEAE-Sephadex A-50:	similar pattern to hCG
SDS polyacrylamide gel electrophoresis:		two bands similar to α- and β-subunits of hCG standard
Bioassay:		stimulation of rat uterine and ovarian weight similar to hCG standard

immunoassay, using a specific antibody directed towards the carboxy-terminal region of the beta-subunit of hCG [23]. In addition, the material cross-reacted in a specific radioreceptor assay and stimulated rat uterine growth in a manner similar to that demonstrated by hCG [23]. Purification of the CG-like material from these bacterial extracts was achieved using Sephadex G-100 gel chromatography, Concanavalin A-Sepharose and DEAE-Sephadex A-50 chromatography (Table 5). The CG-like material eluted with an apparent molecular weight (M_r) similar to hCG; adsorbed to Concanavalin A-Sepharose and was eluted using methyl α-D-glucopyranoside, strongly suggesting the presence of glucose or mannose moieties. Furthermore, when electrophoresed on sodium dodecyl sulphate (SDS)-polyacrylamide gel the bacterial-derived CG-like material separated into two major bands with mobilities corresponding to α- and β-subunits of hCG [23].

3.3. Summary

Using sensitive and specific immunoassays and bioassays, several classical vertebrate hormone-related peptides have been demonstrated in extracts of unicellular organisms. Extensive studies in our laboratory and in laboratories of other investigators have excluded the possibility of exogenous contamination of these extracts, strongly suggesting that these hormone-related peptides are native to unicellular organisms.

Immunocytochemical studies have identified and localised CG-like material in several bacteria isolated from patients with cancer. Since not all bacteria studied demonstrated CG-like material using this technique, this would suggest that the phenomenon is not universal amongst bacteria; though it may be a question of quantity since adequate immunocytochemical staining requires the presence of large amounts of antigen.

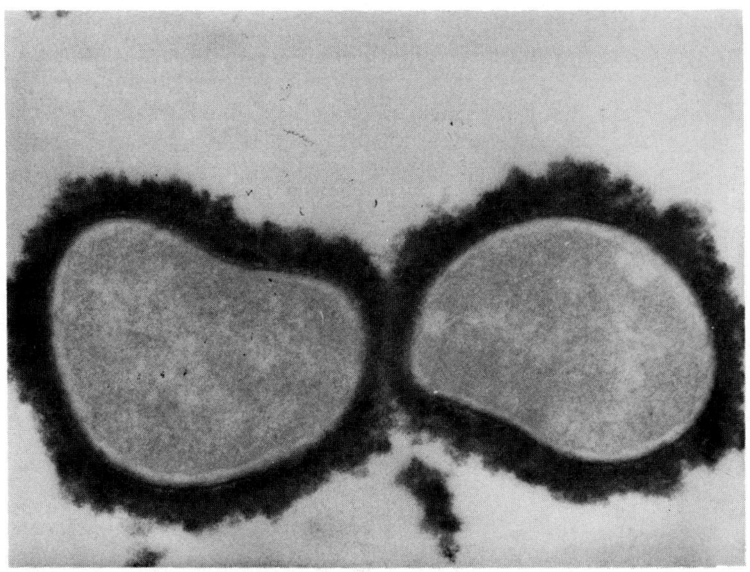

Fig. 6. Immuno-electron microscopic localisation of CG-like material in *E. lentum*. Cultures of *E. lentum* were prepared for immuno-electron microscopy. Cells were fixed in a mixture of glutaraldehyde and periodate-lysine-paraformaldehyde. The fixed cells were then washed and labelled with rabbit CG antiserum (first antibody) followed by peroxidase-antiperoxidase-labelled goat anti-rabbit serum (second layer). Following repeated washing, the peroxidase-antiperoxidase complex was stained with 3,3′-diaminobenzidine tetrahydrochloride (DAB) with hydrogen peroxide substrate in the dark. Post-fixation was performed using 2% osmium tetroxide, followed by ethanol dehydration and embedding in epoxy resin. Ultrathin sections were cut using an ultramicrotome and examined with an electron microscope at 60 KV. Both inner and outer membranes of the bacterial wall display abundant immunostaining for CG-like antigen. Figure reproduced from reference [34] with permission. Magnification = ×48,000.

4. Phylogeny of other cellular proteins

A number of proteins are ubiquitous to most eukaryote cells. These include essential structural proteins, e.g. actin, myosin, troponin as well as other proteins which are vital for other cellular mechanisms, e.g. calmodulin.

4.1. Presence of calmodulin in unicellular organisms

The calcium-binding protein, calmodulin, is ubiquitous in all eukaryote cells which have been tested [37, 38]. First described by Cheung in 1970 [39] as an activator of cyclic nucleotide phosphodiesterase, calmodulin has 148 amino acid residues, an approximate molecular weight of 17,000 and an isoelectric point of 3.9–4.3. In addition, it activates a number of other important intracellular processes including adenylate cyclase, myosin kinase, phosphorylase kinase and microtubule disassembly.

The presence of calmodulin was first reported in non-vertebrate tissues by Waisman et al. [40], who described its existence in at least seven non-vertebrate

TABLE 6

Amino acid compositions of calmodulin-like proteins

	Vertebrates		*Non-vertebrates*		
	Bovine brain	*Rat testis*	*Earthworm*	*Renilla*	*Slime mold*
Lysine	8	7	7	9	6
Histidine	1	1	1–2	1–2	1
Me_3-lysine[a]	1	1	–	1	0
Arginine	7	6	5	6	3
Aspartate	24	22	24	26	18
Threonine	12	11	12	13	9
Serine	5	4	6	6	6
Glutamate	29	28	29	31	43
Proline	2	2	6	2	4
Glycine	12	12	13	13	10
Alanine	12	11	11	11	9
Cysteine	0	0	1	0	0
Valine	8	8	7	7	5
Methionine	10	9	9	9	5
Isoleucine	8	9	8	9	6
Leucine	10	10	10	11	8
Tyrosine	82	2	2	2	5
Phenylalanine	8	8	8	9	8
Tryptophan	0	0	0	0	0

[a]trimethyl-lysine.
Adapted from references [46, 48].

species, including molluscs, annelids and arthropods [40]. Subsequently calmodulin has been identified in sea urchins [41], unicellular eukaryotes [42, 43] higher plants [44] as well as in the prokaryote *Escherichia coli* [45]. In evolutionary terms calmodulin also appears to be a very highly conserved protein. No significant differences in amino acid composition are demonstrable between calmodulin from bovine brain, rat testes, earthworms and a marine coelenterate (Table 6). One exception is calmodulin from slime molds (*Dictyostelium discoideum*), whose amino acid composition is especially distinctive by the absence of trimethyl-lysine (Table 6). Despite these differences calmodulin from unicellular organisms demonstrates full biological activity in activating brain cyclic nucleotide phosphodiesterase and in competing in sensitive and specific immunoassays [46].

Fig. 7. Localisation of calmodulin in *Paramecium tetraurelia. Paramecium* were harvested during early stationary phase, fixed in formalin, post-fixed in acetone and preincubated in 0.01% bovine serum albumin before incubation with the first antibody, goat anti-calmodulin (produced in goats following injection of purified rat testis calmodulin). After repeated washings, the cells were then incubated with rhodamine conjugated rabbit anti-goat IgG and examined using an epi-illumination fluorescence microscope. Anti-calmodulin staining was seen in vacuolar inclusions (A), cilia (B) and in a linear punctuate array along the kineties (basal bodies) in deciliated cells (C). Figure reproduced from reference [47] with permission.

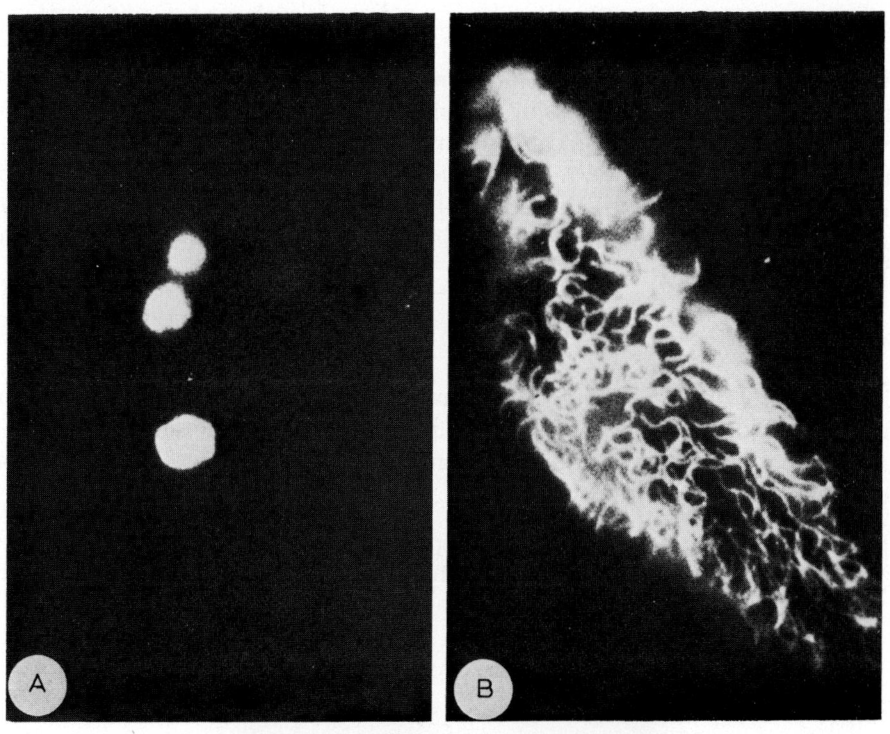

A
B

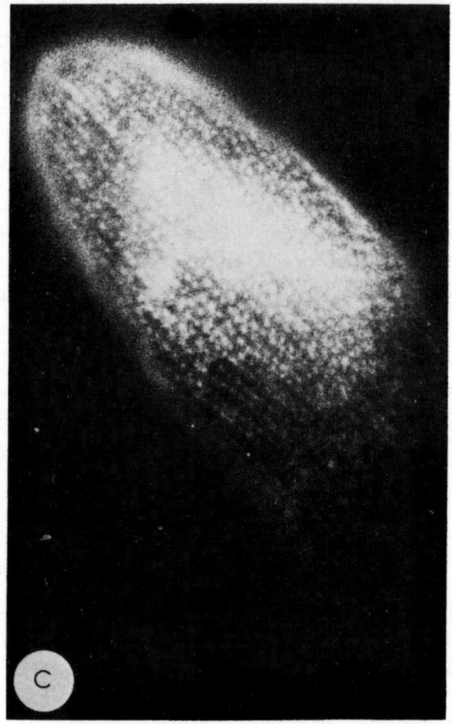

C

4.1.1. *Immunofluorescence studies*

The subcellular distribution of calmodulin in *Paramecium tetrauneria*, identified by indirect immunofluorescence using a calmodulin antibody (Fig. 7), is appropriate for the known functions of calmodulin in protozoa. Thus vacuolar inclusions probably represent ingested food vacuoles and calmodulin in basal bodies (kineties) probably represents their important role in endocytosis. Furthermore, since calcium is vital for ciliary movement the presence of calmodulin in the cilia is not surprising.

4.2. *Summary*

Using recently developed and highly sensitive immunoassays, proteins important for cellular function have been identified widely in vertebrate as well as in non-vertebrate tissues. Calmodulin, in particular, has been extensively studied and demonstrates two important features:

a) the ubiquity of this essential cellular protein, and
b) strong phylogenetic conservation of the structural and functional aspects of the molecule.

5. Conclusions

Over the past two decades, a number of investigators have demonstrated the presence of classic vertebrate peptide hormones and neuropeptides in multicellular non-vertebrate tissues. As outlined in this chapter, the bulk of the evidence has been provided by immunological studies, primarily immunocytochemistry and immunoassays. More recent studies have demonstrated the presence of many of these peptides in unicellular organisms. Since the concentrations of these peptides in the cells is low, their detection was achieved by sensitive immunoassays after many fold concentration and partial purification of the cellular extracts. Thus far, immunocytochemistry has only been successful in identifying and localising peptides in unicellular organisms where these peptides are produced in large amounts, e.g. CG-like material in certain bacteria and calmodulin in *Paramecium*.

With the improvements that are occurring in biotechnology and immunocytochemistry, it could be predicted that immunocytochemical studies will be capable of identifying and localising these peptides in unicellular organisms and eventually the genes or the messenger RNA molecules coding for these peptides.

6. References

1. Falkmer, S., Emdin, S., Havu, N., Lundgren, G., Marques, M., Ostberg, Y., Steiner, D.F. and Thomas N.W. (1973) Insulin in invertebrates and cyclostomes. Am. Zool. 13, 625–638.
2. Tager, H.S., Markese, J., Kramer, K.J., Spiers, R.D and Childs, C.N. (1976) Glucagon-like and insulin-like hormones of the insect neurosecretory system. Biochem. J. 156, 515–520.

3. Plisetskaya, E., Kazakov, V.K., Solititskaya, L. and Leibson, L.G. (1978) Insulin producing cells in the gut of freshwater bivalve molluscs *Anodonta cygnea* and *Unio pictorum* and the role of insulin in the regulation of their carbohydrate metabolism. Gen. Comp. Endocrinol. 35, 133–145.
4. LeRoith, D., Lesniak, M.A. and Roth, J. (1981) Insulin in insects and annelids. Diabetes 30, 70–76.
5. Duve, H., Thorpe, A and Lazarus, N.R. (1979) Isolation of material displaying insulin-like immunological and biological activity from the brain of the blowfly, *Calliphora vomitoria.* Biochem. J. 184, 221–227.
6. Fritsch, H.A.R., Van Noorden, S. and Pearse, A.G.E. (1976) Cytochemical and immunofluorescence investigations of insulin-like producing cells in the intestine of *Mytilus edulis* (Bivalvia). Cell Tissue Res. 165, 365–369.
7. Kramer, K.J., Childs, C.N., Spiers, R.D. and Jacobs, R.M. (1982) Purification of insulin-like peptides from insect hemolymph and royal jelly. Insect Biochem. 12, 91–98.
8. Kramer, J.J. (1984) Vertebrate hormones in insects. In: Comprehensive Insect Physiology, Biochemistry and Pharmacology, Vol. 7. Endocrinology (in press).
9. El Salhy, M., Abou-El-Ela, R., Falkmer, S., Grimelius, L. and Wilander, E. (1980) Immunohistochemical evidence of gastro-enteropancreatic neurohormonal peptides of vertebrate type in the nervous system of the larva of a dipteron insect, the hoverfly, *Eristalis aeneus*. Regul. Pept., 1, 187–204.
10. Grimmelikhuijzen, C.J.P., Carraway, R.E., Rokaeus, A. and Sundler, F. (1981) Neurotensin-like immunoactivity in the nervous system of hydra. Histochemistry 72, 199–209.
11. Iwanaga, T., Fujita, T., Nishiitsutsuji, U.J. and Endo, Y. (1981) Immunocytochemical demonstration of PP, somatostatin, and enteroglucagon immunoreactivities in the cockroach midgut. Biomed. Res. 2, 202–207.
12. Sanders, B. (1983) Insulin-like peptides in the lobster *Homarus americanus*. I. Insulin immunoreactivity. Gen. Comp. Endocrinol. 50, 366–373.
13. Larson, B.A. and Vigna, S.R. (1983) Species and tissue distribution of cholecystokinin/gastrin-like substances in some invertebrates. Gen. Comp. Endocrinol. 50, 469–475.
14. LeRoith, D., Shiloach, J., Roth, J. and Lesniak, M.A. (1981) Insulin or a closely related molecule is native to *Escherichia coli*. J. Biol. Chem. 256, 6533–6536.
15. LeRoith, D., Shiloach, J., Roth, J. and Lesniak, M.A. (1980) Evolutionary origins of vertebrate hormones: substances similar to mammalian insulins are native to unicellular organisms. Proc. Natl. Acad. Sci. USA, 77, 6184–6188.
16. LeRoith, D., Liotta, A.S., Roth, J., Shiloach, J., Lewis, M.E., Pert C.B. and Krieger, D.T. (1982) Corticotropin and β-endorphin-like materials are native to unicellular organisms. Proc. Natl. Acad. Sci. USA 79, 2086–2090.
17. Berelowitz, M., LeRoith, D., Von Schenk, H., Newgard, C., Szabo, M., Frohman, L.A., Shiloach, J. and Roth, J. (1982) Somatostatin-like immunoactivity and biological activity is present in *T. pyriformis* a ciliated Protozoan. Endocrinology 110, 1939–1944.
18. Schwabe, C., LeRoith, D., Thompson, R.P., Shiloach, J. and Roth, J. (1983) Relaxin extracted from protozoa (*Tetrahymena pyriformis*) J. Biol. Chem. 258, 2778–2781.
19. Deftos, L., LeRoith, D., Shiloach, J. and Roth, J. (1984) Salmon calcitonin-like immunoactivity in extracts of *Tetrahymena pyriformis*. Hormone Metab. Res. (in press).
20. Macchia, V., Bates, R.W. and Pastan, I. (1967) Purification and properties of thyroid stimulating factor isolated from *Clostridium perfringens*. J. Biol. Chem. 242, 3726–3730.
21. Acevedo, H.F., Slifkin, M., Pouchet, G.R. and Pardo, M. (1978) Immunocytochemical localization of a choriogonadotropin-like protein in bacteria isolated from cancer patients. Cancer 41, 1217–1219.
22. Backus, B.T. and Affronti, L.F. (1981) Tumor-associated bacteria capable of producing a human choriogonadotropin-like substance. Infect. Immunol. 32, 1211–1215.
23. Maruo, T., Cohen, H., Segal, S.J. and Koide, S.S. (1979) Production of choriogonadotropin-like factor by a microorganism. Proc. Natl. Acad. Sci. USA, 76, 6622–6626.
24. Bhatnagar, Y.M. and Carraway, R. (1981) Bacterial peptides with C-terminal similarities to bovine neurotensin. Peptides 2, 51–59.

25. PEREZ-CANO, R., MURPHY, P.K., GIRGIS, S.I., ARNETT, T.R., BLENKHARN, I. and MACINTYRE, I. (1982) Unicellular organisms contain a molecule resembling human calcitonin. Endocrinology 110, 673A.
26. LEROITH, D., SHILOACH, J. and ROTH, J. (1982) Is there an earlier phylogenetic precursor that is common to both the nervous and endocrine systems? Peptides 3, 211–215.
27. DUVE, H. and THORPE, A. (1979) Immunofluorescent localization of insulin-like material in the median neurosecretory cells of the blowfly *Calliphora vomitoria* (Diptera). Cell Tissue Res. 200, 187–191.
28. EL-SALHY, M., FALKMER, S., KRAMER, K.J. and SPIERS, R.D. (1984) Immunohistochemical investigations of neurohormonal peptides in the brain, corpora cardiaca and corpora allata of an adult Lepidopteran insect, *Manduca sexta*. Cell Tissue Res. (in press).
29. CHEN, A.C. and FRIEDMAN, S. (1977) Hormonal regulation of trehalose metabolism in the blowfly, *Phormia regina*: interaction between hypertrehalosemic and hypotrehalosemic hormones. J. Insect. Physiol. 23, 1223–1232.
30. LIU, T.P. (1973) The effect of allatectomy on blood trehalose in the female housefly, *Musca domestica*. Comp. Biochem. Physiol. 46A, 109–113.
31. BRAUNSTEIN, G.D., KANDAR, V., RASOR, J., SWAMINATHAN, N. and WADE, M.E. (1979) Widespread distribution of chorionic gonadotropin-like substance in normal human tissues. J. Clin. Endocrinol. 49, 917–925.
32. O'DELL, W.D. and WOLFSON, A.R. (1980) Hormones from tumors. Are they ubiquitous? Am. J. Med. 68, 317–318.
33. MCGREGOR, W.G., KUHN, R.W. and JAFFE, R.B. (1983) Biologically active chorionic gonadotropin synthesis by the human fetus. Science 220, 306–308.
34. SLIFKIN, M., PARDO, M., POUCHET-MELVIN, G.R. and ACEVEDO, H.F. (1979) Immuno-electron microscopic localization of a choriogonadotropin-like antigen in cancer-associated bacteria. Oncology 36, 208–210.
35. ACEVEDO, H.F., SLIFKIN, M., POUCHET-MELVIN, G.R. and CAMPBELL-ACEVEDO, E.A. (1979) Choriogonadotropin-like antigen in an anaerobic bacterium, *Eubacterium lentum*, isolated from a rectal tumor. Infect. Immunol. 24, 920–924.
36. ACEVEDO, H.F., KOIDE, S.S., SLIFKIN, M., MARUO, T. and CAMPBELL-ACEVEDO, E.A. (1981) Choriogonadotropin-like antigen in a strain of *Streptococcus simulans*: detection, identification and characterization. Infect. Immunol. 31, 487–494.
37. KLEE, C.B., CROUCH, T.H. and RICHMAN, P.G. (1980) Calmodulin. Ann. Rev. Biochem. 49, 489–515.
38. MEANS, A.R. and CHAFOULEAS, J.G. (1982) Calmodulin in endocrine cells. Ann. Rev. Physiol. 44, 667–682.
39. CHEUNG, W.Y. (1970) Cyclic 3′:5′ nucleotide phosphodiesterase: Demonstration of an activator. Biochem. Biophys. Res. Commun. 38, 533–538.
40. WAISMAN, D., STEVENS, F.C. and WANG, J.H. (1975) The distribution of the Ca^{++}-dependent protein activator of cyclic nucleotide phosphodiesterase in invertebrates. Biochem. Biophys. Res. Commun. 65, 975–982.
41. HEAD, J.F., MADER, S. and KAMINER, B. (1979) Calcium binding modulator protein from the unfertilized egg of the sea urchin, *Arbacia punctulata*. J. Cell. Biol. 80, 211–218.
42. KUZNICKI, J., KUZNICKI, L. and DRABIKOWSKI, W. (1979) Ca^{2+}-Binding modulator protein in protozoa and myxomycete. Cell Biol. Int. Rep. 3, 17–23.
43. CHAFOULEAS, J.G., DEDMAN, J.R., MUNJAAL, R.P. and MEANS, A.R. (1979) Development and application of a sensitive immunoassay. J. Biol. Chem. 254, 10262–10267.
44. ANDERSON, J.M. and CORMIER, M.J. (1978) Calcium-dependent regulator of NAD kinase in higher plants. Biochem. Biophys. Res. Commun. 84, 595–602.
45. IWASA, Y., YONEMITSU, K., MATSUI, K., FUKUNAGA, K. and MIYAMOTO, E. (1981) Calmodulin-like activity in the soluble fraction of *Escherichia coli*. Biochem. Biophys. Res. Commun. 98, 656–660.
46. BAZARI, W.L. and CLARKE, M. (1981) Characterization of a novel calmodulin from *Dictyostelium discoideum*. J. Biol. Chem. 256, 3598–3603.

47. Mahile, N.J., Dedman, J.R., Munjaal, R.P. and Means, A.R. (1979) Development and application of a sensitive immunoassay. J. Biol. Chem. 254, 10262–10267.
48. Jones, H.P., Mathews, J.C. and Cormier, M.J. (1979) Isolation and characterization of Ca^{2+}-dependent modulator protein from the marine invertebrate, *Renilla reniformis*. Am. Chem. Soc. 18, 55–60.

Immunolabelling for Electron Microscopy (Polak/Varndell, eds)

CHAPTER 22

Botanical immunocytochemistry

Daphne M. Wright

Plant Virus Unit, Ministry of Agriculture Fisheries and Food, Agricultural Development and Advisory Service, Block C, Brooklands Avenue, Cambridge, U.K.

CONTENTS

---

1. Introduction

Although the presence of viruses in crop plants is obviously not a recent phenomenon the first records, of necessity, are those which were described unwittingly since the existence of viral disease was not realised. The most famous 'descriptions' perhaps, are paintings by the Dutch masters of the early seventeenth century who included in their studies of still life, tulips whose petals were not uniform in colour but striped with contrasting colours (Fig. 1). The bulbs of such tulips were highly prized, but their owners little realised that they produced quite beautifully classic symptoms of tulip breaking virus. Other paintings showing variegated patterns and colours on ripe plums, an ideal subject for the brush of the artist, were in fact infected with plum pox virus.

The Dutch growers, in their efforts to increase the availability of valuable bulbs, found that the striping symptoms could be transferred to a uniformly coloured bulb by grafting, but it was not until 200 years later in 1886 that the results of the first experiment on the transmission of a plant virus were published. Adolf Mayer [1] found that by injecting sap taken from infected plants into the veins of healthy tobacco plants he induced a mosaic disease. However, he wrongly concluded from these and other experiments that the disease was caused by a bacterium. In 1892 Ivanowski [2] confirmed Mayer's experiments but also found that the causal agent of the disease could pass through a filter impenetrable to bacteria. In 1898, Beijerinck [3] also confirmed Mayer's experiments and found like Ivanowski that the causal agent of the disease could pass through a porcelain filter. From this, and other experiments, he concluded that the pathogen was not a bacterium and he introduced the description '*contagium vivum fluidum*'.

By 1929, Holmes [4, 5] had discovered that tobacco mosaic produced necrotic spots on *Nicotiana glutinosa* whose frequency corresponded with the concentration of virus particles in the original inoculum and in 1931, Smith [6] pioneered the use of 'indicator plants' as a means of distinguishing and separating plant viruses. This, together with the work by Purdy in 1929 [7] on serological methods formed the basis for the diagnosis of plant viruses. It was in the 1930s, however, with the invention of the electron microscope that a whole new world was opened up for the researcher of plant viruses as new techniques were discovered enabling virus particles to be seen and related to the symptoms in plants.

However, despite this early indication of the potential of the electron microscope as a diagnostic tool, technique development was slow and during the '40s and '50s use of the electron microscope was confined to the research laboratory.

2. Development

The National Agricultural Advisory Service (NAAS) latterly known as the Agricultural Development and Advisory Service (ADAS) employs amongst others, plant pathologist advisers, who work with farmers and chemical companies to improve crop production and yield. As part of a disease control strategy pathologists undertake diagnosis of plant pathogens but traditionally have limited specialism in virus diagnosis. In the early days this was largely left to the research stations who had the facilities for such work plus the relevant knowledge and expertise.

By the mid 1960s it was realised that specialist virology expertise within ADAS was required. A small diagnostic unit was set up where virological knowledge could

Fig. 1. Flowers in a Vase by Jan van Huysum including a tulip with tulip breaking virus symptoms (courtesy Wallace collection).

be concentrated. However, routine diagnosis was still hampered by lack of technique and pathologists were confined to using indicator plants, species known to give particular reactions to certain viruses. Indicator plant or sap transmission testing involves grinding a piece of affected material with a little abrasive and buffer in a mortar with a pestle. The ground sap is rubbed on to marked leaves of specific indicator plants, the virus if present enters the plant through wounds made by the abrasive. Excess sap and abrasive is washed off and the plant put into an aphid-free glasshouse for a suitable length of time to produce symptoms or 'reactions'.

Some viruses produce diagnostic symptoms on the marked or 'local' leaves of the indicator plants. Other viruses cause lesions on leaves in addition to those which have been inoculated and this is referred to as a 'systemic' reaction. These reactions can be diagnostic but in some circumstances a further test may be necessary. Such supporting tests usually involve basic serodiagnosis and gel diffusion was traditionally used. 'Infected' and healthy plant sap diffuses from wells in an agar layer to meet antiserum similarly diffusing from a central well. Positive reactions are indicated by a precipitation line where the infected sap and the specific antiserum meet. As can readily be appreciated these methods were time consuming since reactions could occasionally take weeks to appear. Despite its duration sap transmission testing remains a necessary component in the diagnostic repertoire of the plant virologist.

Serodiagnosis has advanced considerably as techniques from medical immunology have been modified for the plant virologist's use. However, perhaps the most significant technical achievement in plant diagnostic virology during the 1960s and '70s was the development of quick methods for the preparation of plant sap extracts for examination in the electron microscope. This coupled with the demonstration that the concentration of many of the most important crop viruses was such that they could be visualised in quick preparation paved the way for the routine use of the electron microscope in virus diagnosis. Perhaps the best known of such quick methods for preparation of leaf material for examination in the transmission electron microscope is that known as the 'quick leaf dip'. The affected leaf is cut, the freshly exposed interior passed through a drop of negative stain on a filmed grid, which is then drained and can be immediately examined.

This quick and easy method resulted in a surge of interest amongst diagnosticians and the establishment of closer contact with the research stations and universities who were prepared to examine specimens or provide viewing time on their electron microscopes. Interest in electron microscopical diagnosis was further stimulated by the discovery that viruses in cultivated mushrooms could also be quite easily and quickly diagnosed. Preparation of mushroom samples involves squeezing a piece of sporophore in muslin, adding negative stain, placing a drop of this mixture on a carbon-coated grid, draining and examining in the electron microscope—in short a very simple process [8]. In the early years this was supported by a mycelial growth rate test on agar which took 3 weeks. Eventually, however, the electron microscope diagnosis of these viruses was found to be sufficiently reliable, and the correlation between mushroom virus and yield loss so close, that it obviated the need for the growth rate test which was thus abandoned [9]. With greater use and reliance, emphasised by the parallel development of serodiagnostic tests, came a fuller appreciation of the limitations of quick methods in electron microscope diagnosis.

These limitations can be summarised as follows:

a) 'Quick' electron microscopy is comparatively insensitive and it became apparent that even relatively high virus concentrations may be missed. Thus the technique cannot be used reliably for viruses which occur in low concentrations and not at all for those which are restricted in distribution within the plant, for example, the luteoviruses which are restricted to phloem. Recent infections in which virus concentration has not built up may also remain undetected by quick tests.
b) Whilst virus particles can often be visualised in the electron microscope it is impossible using quick preparation methods to distinguish quite different viruses which share a common particle size and shape. Thus, where two such viruses are common to a particular host, specific diagnosis is impossible and mixtures of two viruses may pass undetected.
c) Negative stains vary in their effects on virus particles. Under certain circumstances virus particles may lengthen or become aggregated together, or may disintegrate completely. Particular viruses may be more sensitive to damage in quick preparations than others.

Thus the main disadvantages of quick preparation methods for electron microscope diagnosis of plant viruses are; insensitivity, lack of specificity, and sometimes actual physical damage. These problems have been largely overcome using a combination of electron microscopy and serology.

The use of specific antisera to attract virus particles to a filmed grid was first reported in 1941 [10] and is therefore far from new. However, routine use of electron microscopes was mostly restricted to research establishments and the potential of the technique for diagnosis was not realised until the early 1970s.

3. Immunosorbent electron microscopy (Fig. 2)

Immuno Electron Microscopy (IEM) may be divided into three basic methods. The first of these is variously known as the 'antiserum coated-grid method', 'serum activated grids', 'trapping' or the 'Derrick method' after its innovator [11], but may more properly be referred to as Immuno Sorbent Electron Microscopy (ISEM).

As the name suggests this involves coating or 'activating' a filmed grid with a specific antiserum, loading with an extract of virus infected material and finally staining. The antiserum attracts specific virus particles concentrating them on the treated grid and thereby markedly increasing the chances of a diagnosis.

3.1. Use of ISEM

This method has proved to be invaluable for the detection of members of the luteovirus group (*luteus*–yellow) which are phloem restricted and occur in low

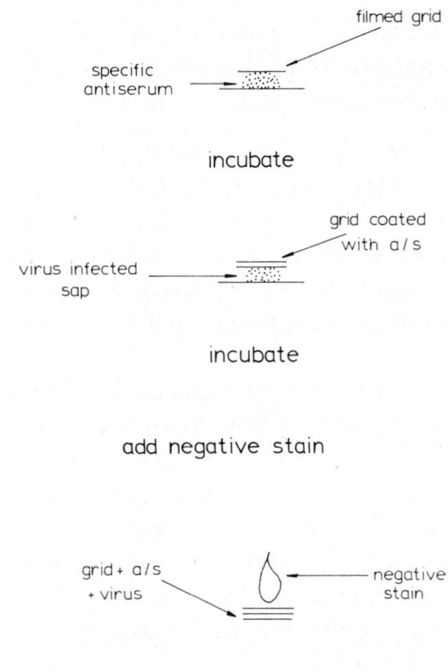

Fig. 2. Diagram of immunosorbent electron microscopy procedure.

concentration. Until the development of ISEM the diagnosis of important members of this group such as potato leaf roll [12] (Fig. 3) and barley yellow dwarf could only be done using an aphid transmission test. This involved the feeding of vector aphids on infected material for a predetermined acquisition period (24–48 h) followed by an inoculation feed on healthy indicator plants (again 24–48 h). Reactions on inoculated indicator plants could take from 2 to 6 weeks to develop and in the case of potato leaf roll, if potato virus Y was also present the indicator was killed. With the use of ISEM, time taken for an accurate diagnosis has been reduced from a possible 6 weeks to just $3\frac{1}{2}$ h using the minimum incubation period (Fig. 4). The great sensitivity achieved by ISEM has made it possible to extract the virus not only from infected plants but also from the virus vector, i.e. the aphid which infects through the leaves and the nematode which infects through the roots [13, 14].

ISEM has been used to illustrate the presence of mixed antibody populations in an antiserum. Whilst using Enzyme Linked Immuno Sorbent Assay (ELISA), a sensitive serological test (see Chapter 10, pp. 123–128) to investigate the incidence and distribution of ryegrass mosaic virus (RMV), variable reactions were recorded even with RMV free material. It was suspected that the antiserum being used in ELISA also contained antibodies to ryegrass seed-borne virus (RGSV). This virus may often be present in ryegrass without showing symptoms and may have been difficult to detect in plants used for purification for RMV antiserum production. When using the 'leaf squash' method, squashing a piece of RMV free ryegrass

Fig. 3. Potato cv. Record showing symptoms of potato leaf roll virus (Crown copyright).

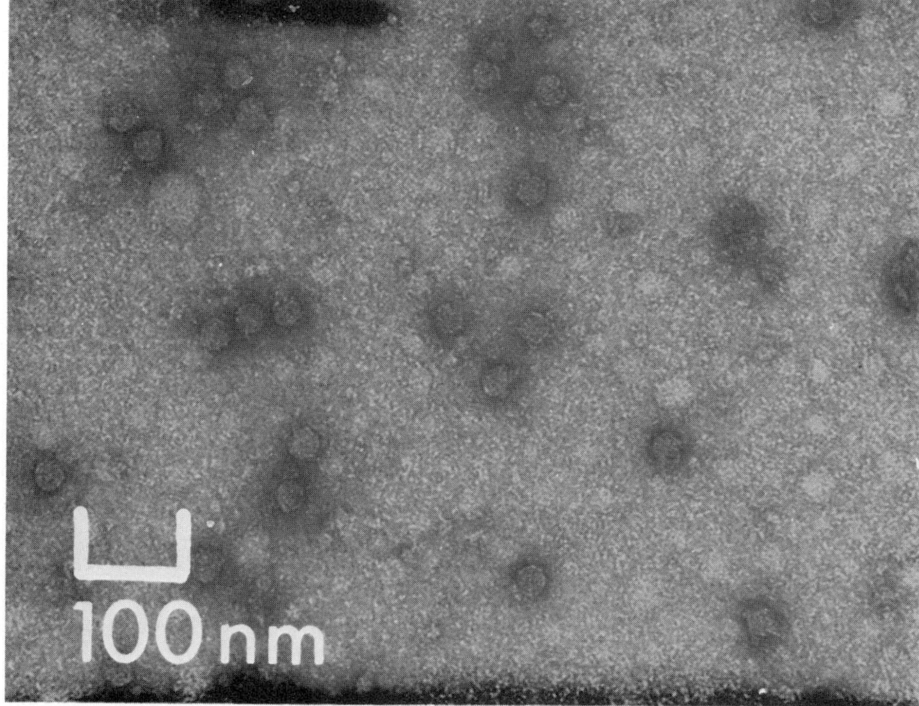

Fig. 4. Electron micrograph of potato leaf roll virus particles using ISEM (Crown copyright).

material in negative stain and placing one drop on a filmed grid, a few spherical virus particles 25 nm in diameter were seen suggesting infection with RGSV (Fig. 5). The ISEM method using the suspect antiserum significantly increased the frequency of RGSV particles observed in the electron microscope (Fig. 6) from which it was concluded that the antiserum contained antibodies to both viruses [15].

3.2. *Variations in the ISEM method*

A standard method for preparation of material for ISEM is appended (section 7.1). However, as with most immunoelectron microscope techniques, opinion as to ideal component stages in the method varies with requirement and from one laboratory to another. There are differing opinions as to the best incubation times and methods. Although the initial incubation of antisera of $\frac{1}{2}$–3 h at 37°C is most frequently used, tests have proved that for some antisera 5 min at room temperature provides enough time for sufficient adsorption of antibody to the grid [16]. For the second stage of the process, the adsorption of the virus to the antibody, a shorter time of 15 min may in some cases be sufficient for rapid diagnosis, but an increase in incubation time of 24–36 h or overnight is necessary to bind the maximum number of virus particles and thus convey greater sensitivity.

Room temperature may be used at all stages in the process. However, for the virus adsorption on to treated grids, 4°C has been widely used since it appears that lower temperatures give greater uniformity of trapping as well as reducing the likelihood of evaporation of the sap. Precoating the grid with Protein A derived from *Staphylococcus aureus*, has been said to increase sensitivity [17] by aligning the antibody molecule's Fab binding regions outwards. Further work has suggested that the increase may not be sufficient to justify the additional step in the test. The potential however, cannot be dismissed since it has been shown that increase in trapping may be more evident with one virus than another [18]. ISEM increases sensitivity by increasing the number of virus particles seen in the electron microscope, and is ideal for viruses which could not be seen with routine methods.

4. Antiserum virus mixture/clumping (Fig. 7)

The second technique may in one process protect fragile viruses, concentrate them and under some circumstances label them specifically. Of particular value when only small quantities of antiserum may be available or when there is no available indicator plant this technique is called 'antiserum virus mixture' (AVM) or 'clumping'.

The purpose is to link virus particles with antibody bridges so that the resultant clumps can be seen in the electron microscope. The antiserum, which is believed to be specific to the suspect virus, should not be over-diluted and the virus particles not too abundant since the objective is for excess antibodies to coat the virus particles and thus specifically label them.

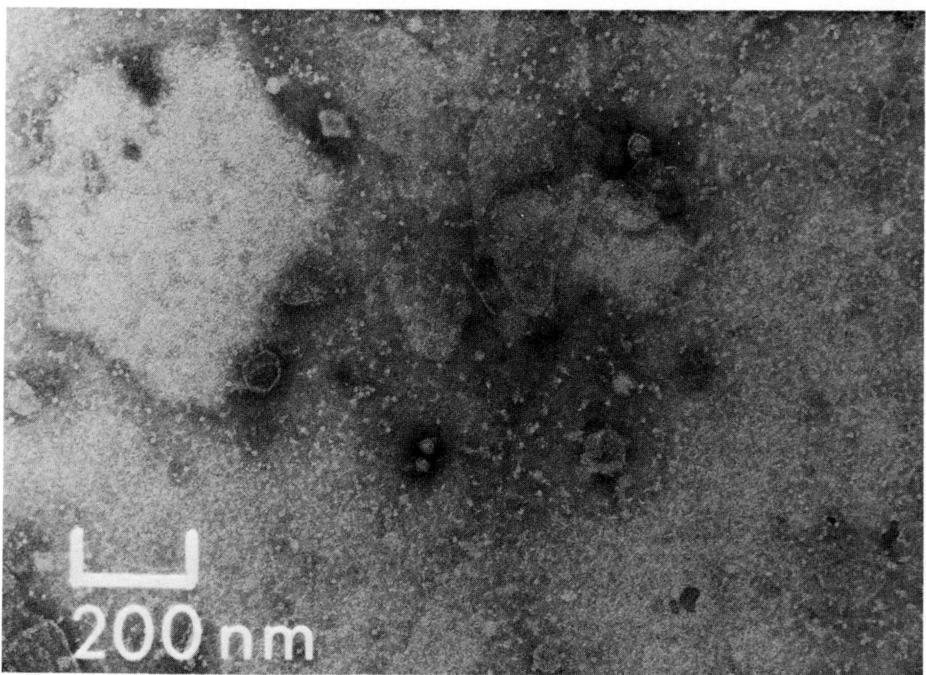

Fig. 5. Electron micrograph of ryegrass seed-borne virus using conventional quick preparation (Crown copyright).

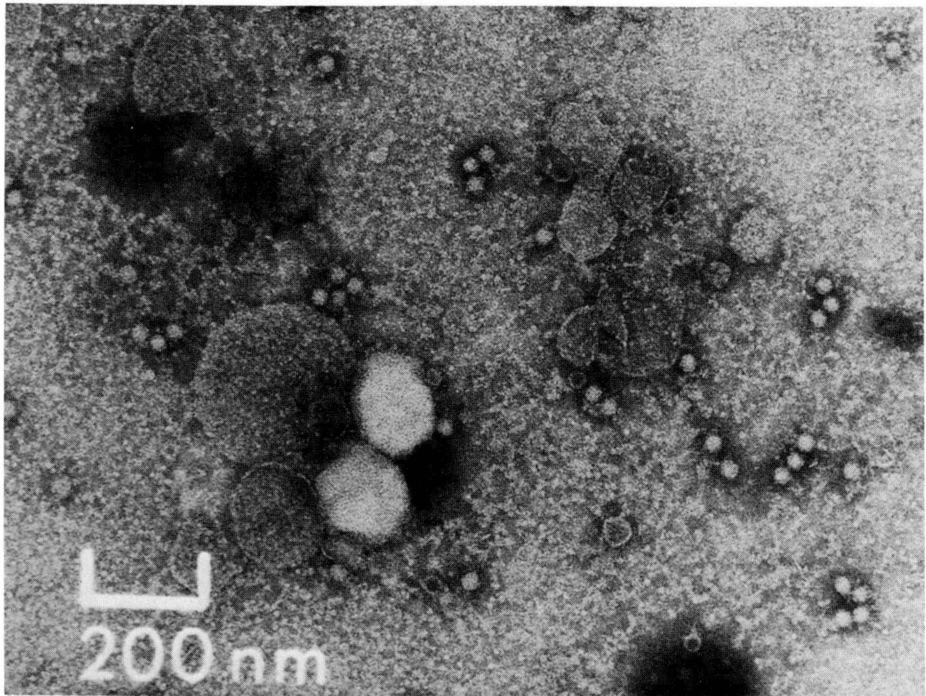

Fig. 6. Electron micrograph of ryegrass seed-borne virus particles using ISEM. Note the increase of virus particles present in comparison with Fig. 5 (Crown copyright).

4.1. Use of AVM

AVM was used to identify virus particles of 750 nm in length found in the pot plant *Dieffenbachia picta*. The affected plants were less vigorous than healthy ones, had

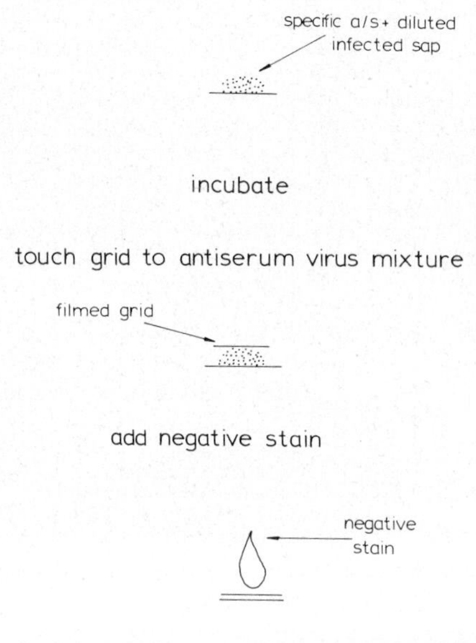

Fig. 7. Diagram of antiserum virus mixture procedure.

chlorotic leaves and were stunted. The younger leaves were distorted, with pale green streaks instead of the normal variegation and the unfurling leaf was often distorted (Fig. 8). Leaf squash preparations were examined in the electron microscope and found to contain 750 nm flexuous filamentous virus particles. Thus the virus could be a member of the potyvirus (*pot*ato *Y* virus) group. Symptoms and length of particle all suggested that the disease was caused by dasheen mosaic virus (DMV), non-indigenous in the United Kingdom. This was confirmed [19] by coating the particles with DMV antiserum. As proof of the specificity of the technique, in control tests particles of potato virus Y and potato virus X (*pot*ato *X* virus group) were included together with the *Dieffenbachia* extract but remained uncoated (Fig. 9). As with ISEM there are many variations to this technique but the basic procedure is described in the appendix (section 7.2).

4.2. Variations in AVM method

As little as 5 μl each of antiserum and virus extract can be mixed together and for periods as short as 15 min, to achieve clumping. Room temperature incubation

Fig. 8. *Dieffenbachia picta*, left—healthy plant, right—plant showing symptoms of dasheen mosaic virus (Crown copyright).

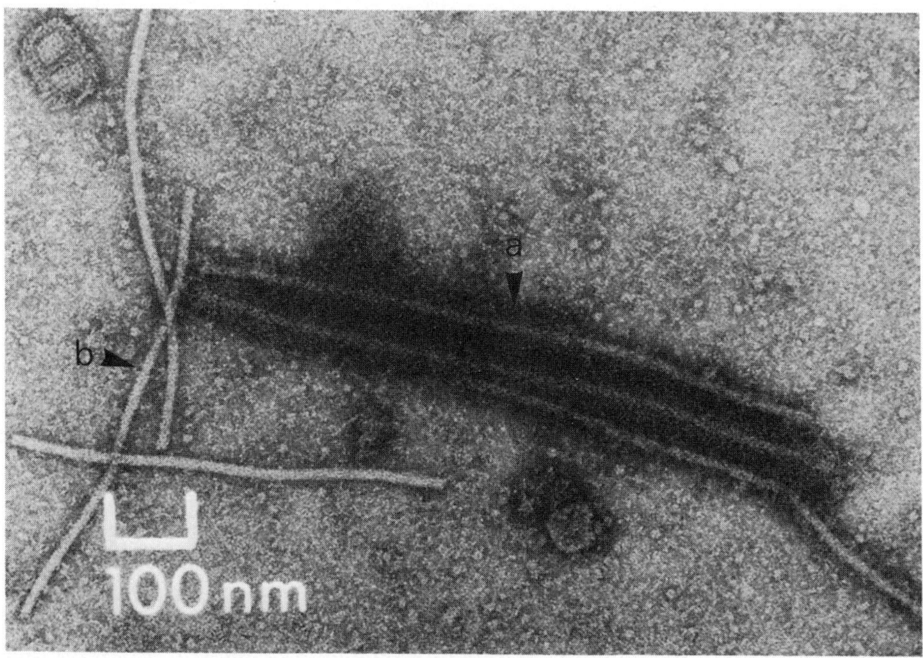

Fig. 9. Electron micrograph of a preparation using antiserum virus mixture showing (a) particles of dasheen mosaic virus coated with specific antibody and (b) uncoated particles of potato virus X (Crown copyright).

gives sufficiently rapid results but in certain circumstances incubation at 37°C may be better providing the virus is not affected by the high temperature and the preparation is protected from evaporation. Incubation at 4°C tends to give a more even coating of antibodies on the particles, avoids problems of evaporation and obviates the necessity for constant supervision. AVM can also be used for crude sap preparation (see section 7.3).

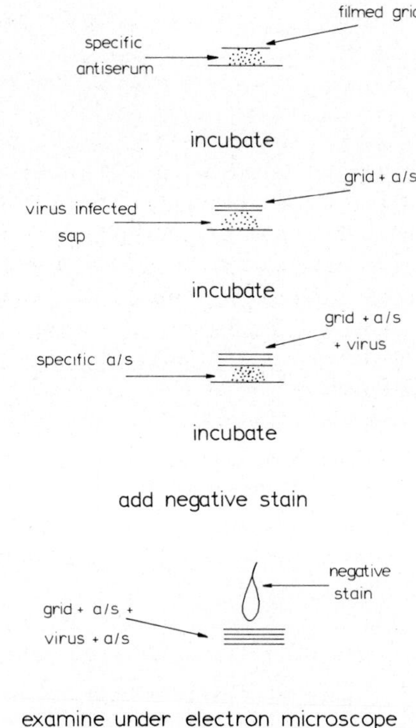

Fig. 10. Diagram of decoration procedure.

5. Decoration/antibody coating (Fig. 10)

The third IEM method and perhaps the most lengthy is used most regularly for specifically labelling virus particles and is called 'decoration' or 'antibody coating' (see section 7.4). As the name implies the technique involves decorating or coating virus particles which have already been attracted to the antiserum layering the grid. This is performed by exposing them to a second and lower dilution of the same specific antiserum which thus coats further available antigenic sites. The 'decoration' method may combine the two previous methods and may be considered the most versatile.

Such flexibility of technique is extremely valuable in diagnostic situations. The procedure can be altered so that as many virus particles as possible are attracted by the specific antiserum before introducing the lower dilution of antiserum to coat them. However, decoration can also be used for specifically labelling virus in crude sap extracts without the initial ISEM stage [16] (see section 7.5). Decoration can be used for titration of antisera by determining the highest dilution giving reliably detectable antibody coating. This is usually one or 2-fold higher than the gel diffusion or slide-precipitin titre. However, by far the greatest use is for confirmation of specific serological identification of unknown viruses or for differentiating morphologically identical viruses in mixed infections.

5.1. *Use of decoration*

Amongst viruses affecting field beans are two which are morphologically similar, broad bean stain (BBSV) and broad bean true mosaic (BBTMV). Since the symptoms caused by these viruses are also very similar it is difficult to know which is more important in a disease outbreak. In order to identify the virus responsible in a recent outbreak, the ISEM plus decoration method was employed. A mixture of antisera to both viruses was used for the initial grid coating to attract all possible particles and then the sap extract was added. One of the antisera, anti-BBTMV, at a dilution of 1:100 was used for the antibody coating with the result that almost all the virus particles present were coated and those left uncoated were of broad bean stain (Fig. 11). Thus it was comparatively easy to show that BBTMV was the predominant virus but that BBSV was present in small amounts. The method can also be used to detect these viruses in seed providing the seed is softened first to allow it to be ground to extract the virus.

Pelargoniums are susceptible to a number of viruses including Pelargonium leaf curl, flower break, ring pattern, leaf pattern and tomato ringspot, all of which are spherical in shape and approximately 30 nm in diameter. By combining the antisera of all likely candidates of infection and decorating with each antiserum in turn, conclusive identification of specific viruses in a mixture can be obtained.

5.2. *Variations in the decoration method*

Incubation times can vary from 1 min to 24 h at temperatures of 4°C to 37°C but 15 min at room temperature seems to give good and sufficient coating. Dilutions of antisera can be modified from 1:10 to 1:100 or even higher depending on the intensity of antibody coating required. The lower the dilution the more intense the antibody coating, but the chance of the particles being concealed as a result is also increased. Generally 1:50 to 1:100 should be enough to give an even coating but at the same time allow the particles themselves to be seen. Optimum dilution for each virus and antiserum combination should be determined by experiment.

6. Disadvantages of IEM methods

All three techniques are straightforward and can be adapted to suit individual need, but it is important to appreciate the possible difficulties. For example, the

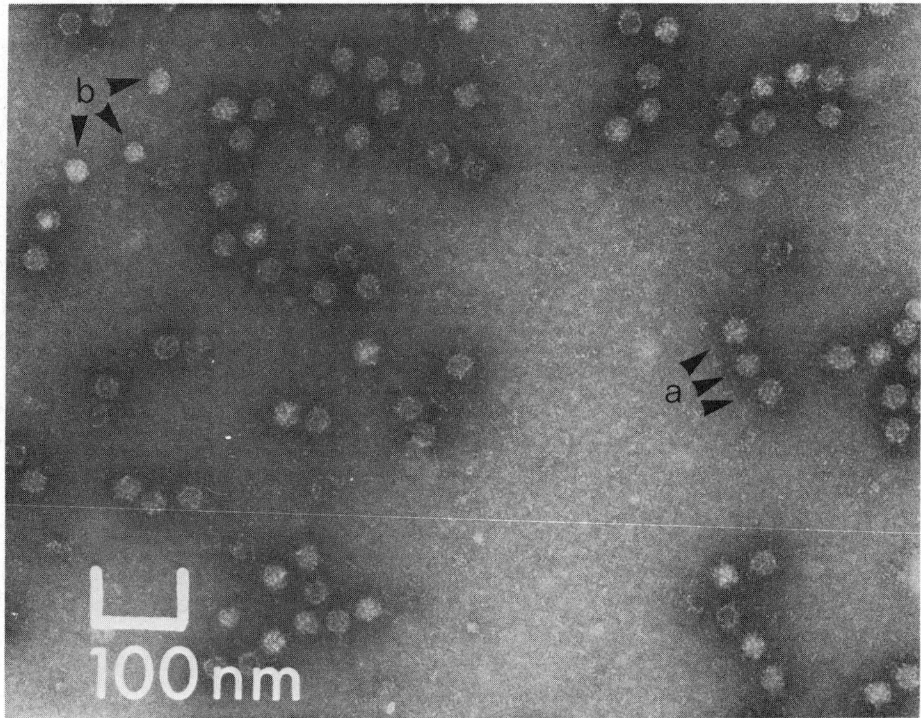

Fig. 11. Electron micrograph of a preparation using decoration showing (a) particles of broad bean true mosaic virus coated with specific antibody and (b) uncoated particles of broad bean stain virus (Crown copyright).

importance of electron microscopical examination for mushroom virus diagnosis has increased since it was found that the frequency of virus particles of one particular type in quick squash preparations corresponded with a noticeable drop in crop yield [20]. Thus electron microscopy provided convincing evidence that yield loss was due to viral agents and not other problems. However, by the time the virus problem was discovered the yield loss was already apparent and the diagnosis was in fact retrospective. With the knowledge that IEM could detect very low numbers of virus particles it was hoped that this method might provide early diagnosis of virus presence which would allow preventative steps to be taken to avoid the build up of high virus levels. However, preliminary investigations have raised further problems. Virus particles of the three main types—mushroom virus 1 (MV1) a 25 nm sphere; mushroom virus 3 (MV3) which has 50 × 19 nm bacilliform particles and mushroom virus 4 (MV4) a 35 nm sphere, are certainly attracted by the use of their specific antiserum (Fig. 12), but the numbers of each particle type attracted relates in part to the serum activity and not to the real particle frequency. MV4 is most commonly found and has been shown to cause the greatest yield loss when seen in large numbers in conventional electron microscopical preparations. MV1 and MV3 are not often seen in such conventional preparations and when they are

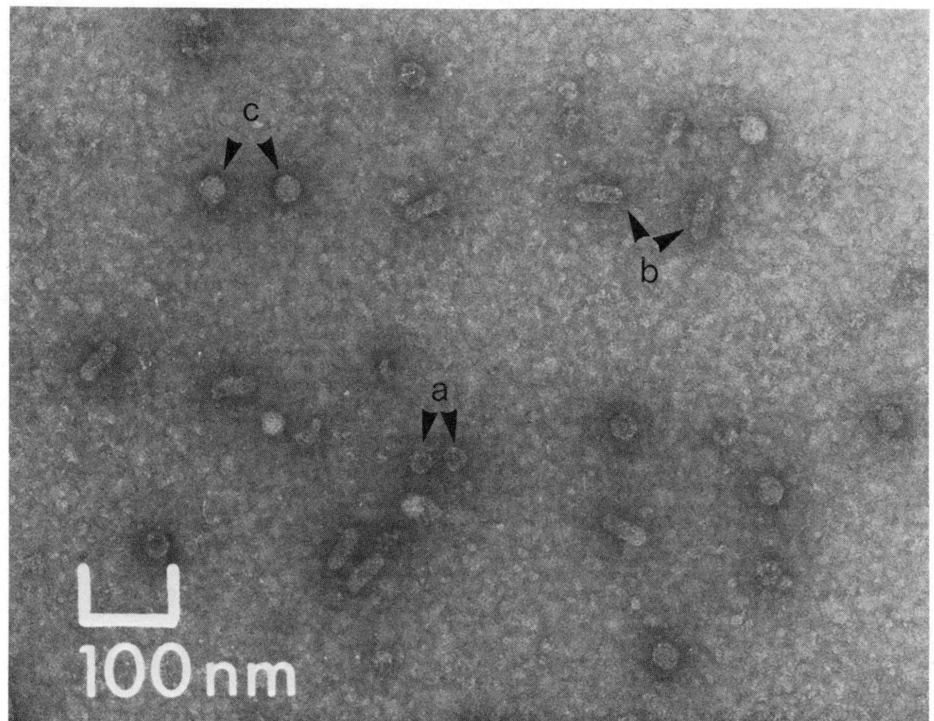

Fig. 12. Electron micrograph of a preparation showing three viruses affecting mushroom (a) MV1, (b) MV3, (c) MV4 (Crown copyright).

they occur in very low numbers. By using ISEM, MV1 and MV3 are seen much more frequently but only seem to be associated with yield loss when MV4 is present. A great deal more work remains to be done to resolve this difficulty.

This experience illustrates the need for caution when interpreting results obtained by the use of various IEM methods. The potential for these methods must not be under-estimated, but rather their limitations realised.

7. Appendix

7.1. *Immunosorbent electron microscopy (ISEM)*

Procedure

1) Line a glass Petri dish with Parafilm® or coat a microscope slide with wax and place in a Petri dish lined with moist filter paper.
2) Carefully place a number of 10–20 μl drops of diluted antiserum, 1:500–1:1000,* in 0.06 M phosphate buffer pH 6–7,* on the waxed slide or Parafilm® taking care not to 'spread' them. Place one grid on each antiserum drop, support film side down and incubate at either 37°C or room temperature for $\frac{1}{2}$–3 h*.

---

* ISEM gives results within the ranges stated but optimum conditions should be determined by experiment.

3) Prepare the sap extract by grinding a small amount of infected material in the 0.06 M phosphate buffer plus a small amount of 600 mesh carborundum, until liquid.
4) Centrifuge for 5–15* min at 8,000 g in a microangle centrifuge.
5) Wash antiserum coated grids either by floating on several millilitres of phosphate buffer twice for 10 min or by holding each grid in forceps and exposing the treated side to 20 consecutive drops of buffer from a Pasteur pipette. Drain grids.
6) Place each grid on to a separate 20 μl drop of the sap extract on Parafilm® (as for 1 above) and incubate at either 4°C for 24* h or room temperature for 1–5* h.
7) Remove the grid from the sap extract, drain and stain with one drop of negative stain (if using uranyl acetate, the grid must first be washed with 30 consecutive drops of distilled water, drained then stained with 5–6 drops of uranyl acetate and drained; uranyl acetate is incompatible with phosphate and with pH values above 5).

7.2. Antiserum virus mixture (AVM)

1) Dilute antiserum in phosphate buffer as for ISEM to not less than 1:16 (the optimum dilution is usually about the tube precipitin titre). Antiserum dilutions should be freshly made each time.
2) The virus preparation should be extracted as for ISEM and centrifuged to remove as much debris as possible. It should then be diluted with phosphate buffer until less than one rod shaped or filamentous particle or less than five spherical virus particles are seen per field of view at a magnification of 20,000 in a conventional preparation in the electron microscope.
3) Place 40 μl of diluted antiserum on to a waxed slide on damp filter paper in a Petri dish or on to Parafilm® lining a Petri dish.
4) Add 20 μl of diluted sap and mix together keeping the drop as small as possible (a tight meniscus). Cover Petri dish to avoid evaporation.
5) Incubate drops at 37°C for $\frac{1}{2}$–3† h or 4°C for 3–36† h.
6) When incubation is completed touch the surface of the drop with a filmed grid and stain with a few drops of negative stain, drain, dry and examine.

7.3. AVM using crude sap preparations [16]

1) Squash infected material in a small amount of diluted antiserum and incubate in a humid container for 15 min.
2) Touch a filmed grid to the mixed drop, wash with 30 drops of phosphate buffer, then 50 drops of water and, finally, 5 drops of 2% uranyl acetate.

---

* ISEM gives results within the ranges stated but optimum conditions should be determined by experiment.

† AVM gives results within the ranges stated but optimum conditions should be determined by experiment.

Different laboratories have preferences for different stains. If 2% sodium or potassium phosphotungstate, methylamine tungstate or ammonium molybdate is used, the treated grid can be stained immediately after touching the AVM. However, if uranyl acetate is used the grid must first be washed with 30 drops of water and sometimes fixed on 0.2% osmium tetroxide for 1–5 min before staining.

7.4. Decoration

1) Follow steps 1–6 of the ISEM procedure.
2) Dilute antiserum in 0.06 M phosphate buffer (dilutions from 1:10 to 1:100* may be used). Place 5–10 μl drops of dilute antiserum on to either a waxed slide on damp filter paper in a Petri dish, or a Parafilm® lined Petri dish. Cover to prevent evaporation.
3) Wash antiserum plus antigen treated grids with 20 consecutive drops of phosphate buffer.
4) Drain and place on to the antiserum drops prepared in 2 above.
5) Cover Petri dish and leave at room temperature for 15 min.
6) Remove the grids with forceps and wash the treated side with 30 consecutive drops of water, drain, then stain with 5–6 drops of uranyl acetate. If using potassium/sodium phosphotungstate, methylamine tungstate or ammonium molybdate as with the other methods it is unnecessary to wash the grid before staining.

7.5. Decoration using crude sap extracts [16]

1) Crush a small amount of virus infected leaf in 10–20 μl distilled water on a glass slide.
2) Touch a film-coated grid to the preparation and remove after a few seconds.
3) Rinse with 30 consecutive drops of phosphate buffer, drain but do not dry.
4) Follow steps 4–6 of the decoration procedure.

8. Acknowledgements

Flowers in a Vase by Jan van Huysum is reproduced by kind permission of the Trustees of the Wallace Collection. All other photographs and electron micrographs are reproduced by kind permission of the Ministry of Agriculture, Fisheries and Food.

My thanks go to Stephen A. Hill, National Virology Specialist, ADAS, Cambridge and my sister Valerie for all their help and encouragement in the preparation of this chapter.

* Decoration gives results within the ranges stated but optimum conditions should be determined by experiment.

9. References

1. Mayer, A. (1886) Landwirtsch. Vers. Stn. 32, 451 (Phytopathological Classic 7, 11).
2. Ivanowski, D. (1892) Concerning the mosaic disease of the tobacco plant. St. Petersb. Acad. Imp. Sci. Bull. 37, 67 (Phytopathological Classic 7, 27).
3. Beijerinck, M.W. (1898) Concerning a contagium vivum fluidum as cause of the spot disease of tobacco leaves. Vernand. Kon Akad. Weten. Amsterdam 6, 3 (Phytopathological Classic 7, 33).
4. Holmes, F.O. (1928) Accuracy in quantitative work with tobacco mosaic virus. Bot. Gaz. 86, 66.
5. Holmes, F.O. (1929) Local lesions in tobacco mosaic. Bot. Gaz. 87, 39.
6. Smith, K.M. (1937) A Text Book of Plant Virus Diseases, 1st Edn. Churchill, London.
7. Purdy, H.A. (1929) Immunologic reactions with tobacco mosaic virus. J. Exp. Med. 49, 919–935.
8. Hollings, M., Stone, O.M. and Last, F.T. (1966) Mushroom viruses. Rep. GCRI, 97.
9. Hill, S.A. and Wright, D.M. (1979) Mushroom diagnosis: the relationship between mycelial growth rate and infection with mushroom virus 4. Plant Pathol. 28, 123–127.
10. Anderson, T.F. and Stanley, W.M. (1941) A study by means of the electron microscope of the reaction between tobacco mosaic virus and its antiserum. J. Biol. Chem. 139, 339–344.
11. Derrick, K.S. (1973) Quantitative assay for plant viruses using serologically specific electron microscopy. Virology 56, 652–653.
12. Roberts, I.M. and Harrison, B.D. (1979) Detection of potato leaf roll and mop-top viruses by immunosorbent electron microscopy. Ann. Appl. Biol. 93, 289–297.
13. Roberts, I.M. and Brown, D.J.F. (1980) Detection of six nepoviruses in their nematode vectors by immunosorbent electron microscopy. Ann. Appl. Biol. 96, 187–192.
14. Paliwal, Y.C. (1981) Detection of barley yellow dwarf virus in aphids by serologically specific electron microscopy. Can. J. Bot. 60, 179–185.
15. Chester, I.B., Hill, S.A. and Wright, D.M. (1982) Serological detection of ryegrass mosaic virus and ryegrass seed-borne virus. Ann. Appl. Biol. 102, 325–329.
16. Milne, R.G. and Luisoni, E. (1975) Rapid high-resolution immune electron microscopy of plant viruses. Virology 68, 270–274.
17. Shukla, D.D. and Gough, K.H. (1979) The use of Protein A, from *Staphylococcus aureus*, in immune electron microscopy for detecting plant virus particles. J. Gen. Virol. 45, 533–536.
18. Gough, K.H. and Shukla, D.D. (1980) Further studies on the use of Protein A in immune electron microscopy for detecting virus particles. J. Gen. Virol. 51, 415–419.
19. Hill, S.A. and Wright, D.M. (1980) Identification of dasheen mosaic virus in *Dieffenbachia picta* and *Xanthosoma helliborifolium* by immune electron microscopy. Plant Pathol. 29, 143–144.
20. Gaze, R.H. and Hill, S.A. (1976) ADAS Mushroom virus survey. Mush. J. October, 1976.

Immunolabelling for Electron Microscopy (Polak/Varndell, eds.).

CHAPTER 23

Hybridisation histochemistry: In situ hybridisation at the electron microscope level

Nancy J. Hutchison

Division of Genetics, Fred Hutchinson Cancer Research Center, 1124 Columbia Street, Seattle, WA 98104, U.S.A.

CONTENTS

1. INTRODUCTION

This chapter will review briefly the history and rationale of electron microscope (EM) level in situ hybridisation; describe some recent modifications of previously reported procedures; suggest some areas where improvements are most needed, and present some potential applications for ultrastructural identification of nucleic acids. Recently we reported two methods for in situ hybridisation to whole mount chromosome preparations at the EM level [1–3]. One method employed tritium-labelled probes and autoradiographic detection in very much the standard style of light microscopical (LM) in situ hybridisation. The second method, based on the use of a biotinylated nucleotide analogue, provides a new, rapid, high resolution approach to in situ hybridisation at both LM and EM levels. The methods we devised are still fairly new and we are continually working to streamline them and improve their sensitivity. For either method the detailed protocols depend on the starting material, for example, sections, cells, chromosome whole mounts, etc. In addition, although less emphasis has been placed on this in the past, the target choice, i.e. DNA or RNA, also dictates procedural details. We hope that at least these methods will provide a starting point for others in the application to particular experimental situations.

The literature on in situ hybridisation, autoradiography, and EM labelling is extensive. In many cases, rather than listing all relevant publications, I have chosen recent references which themselves provide access into the earlier literature.

1.1. History and rationale

The development of in situ hybridisation in 1969 [4, 5] provided a dramatic new approach to localising DNA sequences within tissues, nuclei, and chromosomes at the LM level. Initially, many kinds of repetitive sequences were mapped [6]. Now recombinant DNA technology provides pure probes which are essential for mapping unique sequences. Nick-translation [7] and other enzymatic techniques allow nucleic acid probes to be labelled to 10^7–10^8 cpm/μg with tritium, giving the high specific activity and resolution needed for unique sequence mapping. Refinements in the hybridisation protocols [8–10] allowed improved probe access to target sequences while the generation of probe networks [11, 12] increased the delivery of probe signal to target sequences. These improvements have combined to allow the mapping of single copy sequences in cells and chromosomes at the LM level [13–16]. In addition, since 1975 avidin-biotin interactions [17] have been exploited to move in situ hybridisation away from the constraints of autoradiography at both the LM and EM levels. The application of this newer methodology at the EM level will be discussed in detail in section 3. Section 2 will consider the autoradiographic method and its application at the EM level.

2. Hybridisation to DNA and detection of radioactive probes

The introduction of in situ hybridisation at the LM level provided a unique tool for answering many kinds of questions. Thus it was no surprise that soon after its development, attempts were made to use this technique at the EM level for questions requiring greater resolution and analysis of ultrastructural detail [18–23]. Geuskens [24], in a very thorough review, summarised these attempts as mixed successes troubled primarily by variability in results and by low sensitivity. He noted that there is competition between the conditions necessary for efficient hybridisation and treatments necessary to preserve ultrastructural details.

Generally the EM level methods have been adaptations of the LM level procedures. Readers unfamiliar with the LM procedures should consult one or more review articles for grounding in the methods and philosophy of this approach [25–29]. Briefly, for hybridisation to DNA a squash or air-dried preparation is made on a slide using acid-fixed tissues or cells. The material is treated with ribonuclease to remove endogenous RNA which might compete in the hybridisation reaction. The DNA in the preparation is denatured with sodium hydroxide, or by heating usually in the presence of formamide. A high specific activity ^3H or ^{125}I-probe nucleic acid in an appropriate hybridisation solution is then applied to the slide followed by incubation with time and temperature selected from consideration of the particular nucleic acid probe, its guanine and cytosine content, nucleotide length, and the concentration of salt and formamide used. Frequently a heterologous unlabelled DNA or tRNA is included to reduce background due to non-specific sticking to the preparation. Following hybridisation, excess probe is washed away and the slides are treated with RNAse to remove probe RNA not involved in RNA/DNA hybrids, or rinsed extensively to remove unhybridised probe DNA in DNA/DNA hybridisations. The dried slides are coated with autoradiographic emulsion and stored in the dark for exposures ranging from days to months. Finally the emulsion is developed, the chromosomes are stained, and the slides are examined with the light microscope for analysis and karyotyping. Conceptually, the procedures are the same for EM level hybridisation. Geuskens [24] has already reviewed the variations in fixation, embedding, and sectioning which have been tried for tissues and cells.

We introduced a new protocol suitable for whole mount nuclei and chromosomes prepared by a modified Miller spreading procedure [30]. Full details of the procedure have been published [1, 2]; the hybridisation and autoradiographic procedures are quite similar to those described above. The development of this whole mount hybridisation procedure complements the previous methods which were designed for use with tissues and whole cells. The improved resolution of the EM method permits mapping sequences in systems where the metaphase chromosomes are very small with a sensitivity level that appears to be similar to the LM level procedures. In addition, it allows high resolution mapping for sequences within interphase nuclei as shown in Figure 1.

At present, the drawbacks include the long autoradiographic exposure times and variability in results reported in all the examples to date. Preservation of morphological detail has not been as good as one would hope. On the positive side, the sensitivity of this method is fairly good, and the resolution is improved over that available at the LM level. Continuing improvements in autoradiographic procedures and emulsions such as the new Kodak 129–01 promise that these methods will find great utility.

3. In situ hybridisation using biotin-labelled nucleotide analogues

3.1. Development of the concept

Until recently, radioactive probes and autoradiographic detection were implicit for in situ hybridisation. However, the long exposure times, the limited resolution, and technical expertise required made the prospect of alternative approaches worth

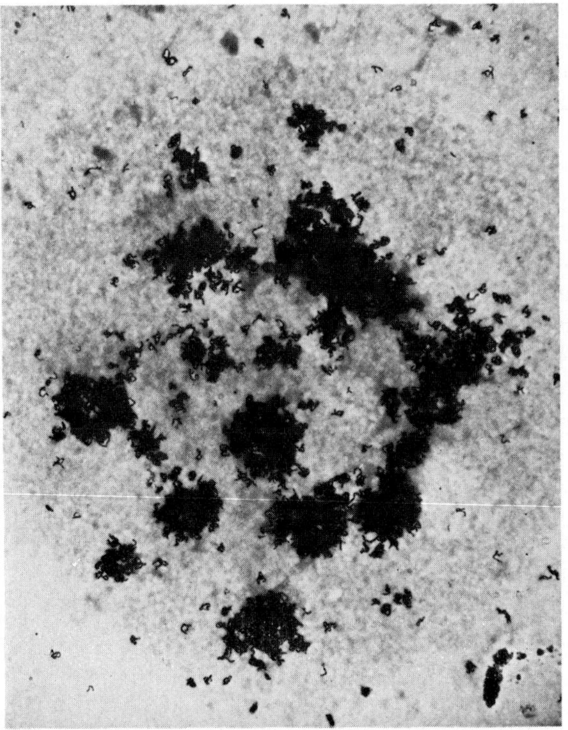

Fig. 1. Electron microscope autoradiograph of mouse L929 whole mount interphase nucleus hybridised with ^{3}H cRNA to mouse satellite DNA. Satellite sequences form clusters throughout the nucleus and associated with the nucleolus.

exploration. A new conceptual approach was pioneered by Manning et al. [31] and elegantly refined by Ward and his collaborators. This approach is based on the strong binding of the small molecule biotin by the protein avidin (Kd = 10^{-15} M) (see Chapter 8, pp. 95–111).

In the original work Manning et al. [31] labelled *Drosophila* ribosomal RNA with biotin through cytochrome c bridges. Following hybridisation and washes, hybrids were detected via binding by avidin covalently coupled to electron-dense polymethacrylate spheres. By combined LM and scanning EM analysis they were able to observe labelling over the nucleolus organiser region of polytene chromosome spreads. Ward and his collaborators [32] greatly simplified the nucleic acid labelling by synthesising a uridine nucleotide derivative with biotin on an alkylamine linker arm extending from the pyrimidine base. Now DNA can be rapidly and stably labelled with biotin by nick translation, or similarly biotinyl cRNA made using the ribonucleotide derivative and RNA polymerase. These biotin-labelled nucleic acids are relatively unchanged in denaturation and reassociation characteristics [32].

Although initial hybrid detection schemes were based on avidin-biotin binding, recognition of strong background binding between avidin and DNA led to alternatives, primarily streptavidin, a fungal analogue of egg avidin, and antibodies specific for biotin. Thus at the LM level, hybrid detection schemes now use fluorescence tagged secondary antibodies or enzymatic reaction products such as immunoperoxidase staining for visualisation [33, 34].

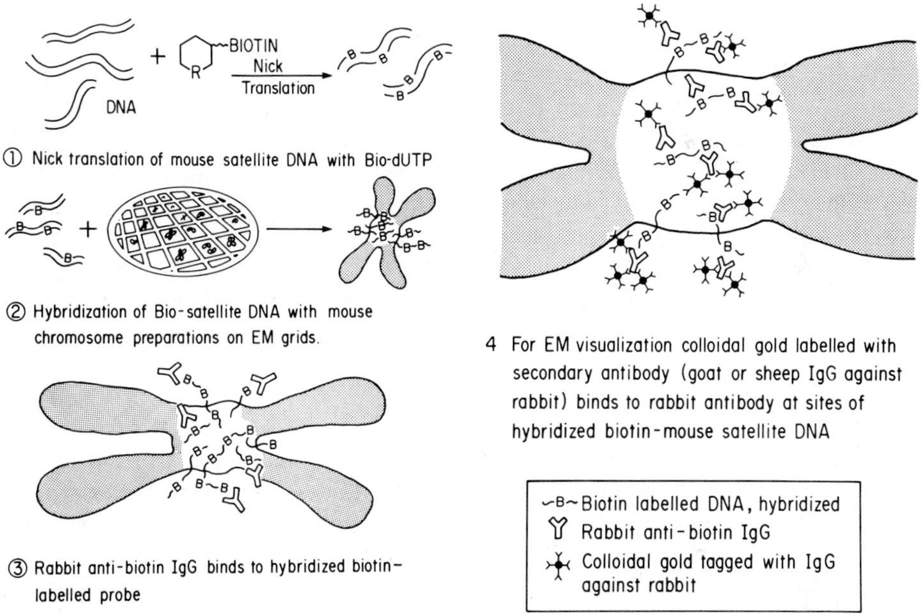

Fig. 2. Schematic diagram illustrating steps of electron microscopical in situ hybridisation using biotin nucleotide and colloidal gold detection.

Elaboration of this system for the EM level obviously requires the use of an electron-dense detection system. Although we knew peroxidase staining could work for EM visualisation, at the time we were developing these methods colloidal gold, which was very new as an EM marker, seemed even more suitable. As numerous chapters in this volume now confirm, gold particles are indeed excellent EM markers, easy to prepare and use whilst providing unambiguous recognition in the electron microscope.

Using mouse satellite DNA and mouse L929 tissue culture whole mount chromosomes as a test system, we developed an EM level hybridisation protocol which produced specific labelling of C-band positive heterochromatic regions, primarily centromeres [1–3]. This pattern is characteristic for mouse satellite DNA. Thus, the method allowed rapid EM level localisation with an estimated resolution of 260 Å or better. Figure 2 illustrates the steps involved in this hybridisation and detection scheme. The utility of the mouse satellite system is that it now provides a basis for testing potential improvements in the procedure; some of these are described below.

3.2. Technical comments

The original protocol is described fully [1]; space limitations prohibit repeating the full details here. Some of the procedural changes described here have been briefly reported [35].

The whole mount chromosome preparation remains unchanged: a modified Miller spreading procedure is used to obtain chromosomes on gold EM grids. Chromosomes are fixed in 0.1–0.5% glutaraldehyde, washed, then rinsed in 0.1% Photoflo solution, and dried on filter paper. The DNA is denatured in 2 × SSC (0.3 M NaCl, 0.03 M Na citrate, pH 7) carefully adjusted to pH 12 with NaOH. The hybridisation solution is as described [1] including 0.6 M NaCl, 10% dextran sulphate, 50% formamide, and the nick-translated biotin-labelled probe DNA. After hybridisation and washes, the grids are incubated with primary anti-biotin serum followed by rinses and incubation with colloidal gold tagged with the secondary antibody. Following washes to remove unbound colloidal gold the grids are rinsed in Photoflo, dried, stained, and examined in the EM.

3.2.1. Probe labelling

In our original experiments [1–3] the biotinylated nucleotide was constructed with a four atom ('short') linker arm joining biotin to the uridine base [32]. Since then Ward has synthesised and tested two biotin-nucleotides with longer linker arms of 11 and 16 atom lengths ([36]; Ward, personal communication; see also Chapter 8, pp. 95–111). By extending the biotin moiety further from the nucleotide on the 11 atom linker arm, antibody binding was optimised, and binding of avidin and streptavidin improved. Synthesis of the 16 atom linker arm further improved avidin and streptavidin binding to biotin-labelled hybrid molecules.

3.2.2. Colloidal gold markers

Our earlier experiments used only 20 nm colloidal gold particles prepared by the citrate reduction method [1, 2]. Recently, we have tested 5 nm colloidal gold markers (generously provided by J. De Mey). Following our standard hybridisation protocol with biotinylated mouse satellite DNA, the 5 nm gold particles give a greatly increased labelling intensity over satellite DNA-containing regions (S. Narayanswami, personal communication [35]). These smaller particles also provide a 4-fold improvement in resolution over the use of 20 nm particles. The increased labelling is probably attributable to decreased steric hindrance and/or reduced charge repulsion of the negatively charged gold particles.

In our first experiments with mouse satellite DNA hybridisation, we often observed gold labelling both over and around the centromeric regions of mouse chromosomes. In all our more recent experiments using the protocol just described, we no longer observe the halo of peripheral labelling. Instead the gold particles are directly over the satellite-containing centromeric regions as shown in Fig. 3. The reason for this improvement is unclear. We think it may result from better denaturation (careful adjustment of denaturing solution to pH 12) of chromosomal DNA. If so, it may be useful to try other means to assure complete

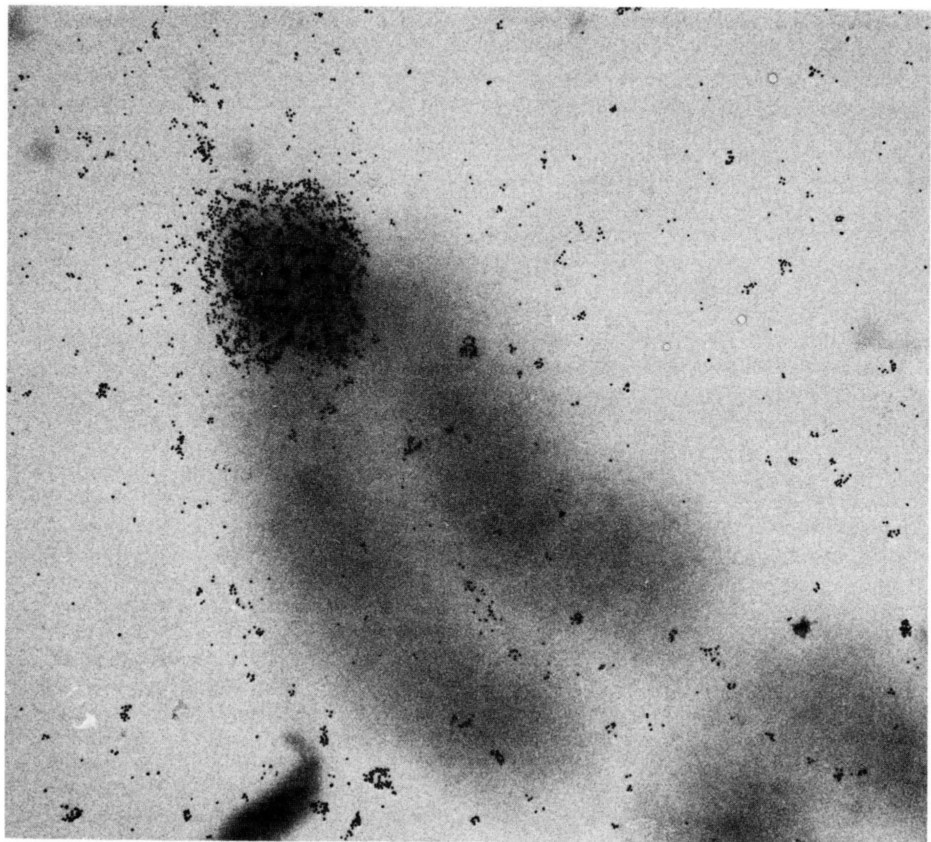

Fig. 3. Whole mount mouse L929 metaphase chromosome hybridised with mouse satellite DNA labelled with biotin (16 atom linker arm). Hybrid detection uses rabbit antibiotin followed by goat anti-rabbit colloidal gold (20 nm). The centromeric heterochromatin is intensely labelled. (Photograph courtesy of S. Narayanswami.)

denaturation of DNA, particularly for sequences with high guanine and cytosine content.

The introduction of the longer biotin linker arm and the 5 nm colloidal gold particles each improved the observed labelling intensity. In combination, the labelling intensity is now very high and specific to satellite DNA-containing regions of the chromosomes (S. Narayanswami, personal communication).

3.2.3. Sensitivity

As yet we have made no direct measurements of the sensitivity of this method. We initially chose mouse satellite DNA for this work because it represented a large portion (6%) of the mouse genome thereby increasing our chances for detection in our initial studies. In recent preliminary experiments this system has been used for the EM localisation of 5S genes on *Xenopus* chromosomes with good signal intensity (Narayanswami and Hamkalo, personal communication). Further testing will be necessary to determine the detection capability.

Numerous other chapters in this volume describe recent developments in the use of colloidal gold markers and other kinds of electron-dense tags. Hopefully, we and other investigators will also be testing these systems for improved EM level in situ hybridisation.

3.2.4. Commercially available reagents

Several of the reagents used in these experiments are now available through commercial suppliers. Both the biotinylated nucleotide (11 atom linker arm) and goat anti-biotin antibody are available from Enzo Biochem, New York, U.S.A.

Numerous companies supply secondary antibodies including those coupled to peroxidase which could be used for enzymatic detection at the EM level [34]. A peroxidase staining kit available from Vector Systems, Burlingame, CA, U.S.A., uses an avidin DH peroxidase complex. This product is made with a highly purified form of avidin which has no problem with background binding to DNA and gives good LM labelling [37].

Several companies now produce avidin and streptavidin as well as combinations of these with coupled antibodies or enzymatic complexes. Gold particles of 20 and 5 nm sizes with or without antibody tags are now available through Janssen Pharmaceutica Life Sciences, Beerse, Belgium.

3.3. *Areas of potential improvement*

The method described above is suitable for whole mount chromosomes. As yet, no comparable EM level method for tissues and sections exists. Presumably the LM procedures of Brigati and colleagues may be adapted to the EM level. For any kind of starting material, we will want to continue to improve the preservation of morphology and reduce the loss of material from samples. Additionally there is the potential for dual localisations if the biotin technology can be duplicated with other small molecules such as dinitro phenol (DNP).

4. Hybridisation to RNA at the electron microscope level

Hybridisation to cellular RNA at the LM level has recently been generating a great deal of interest [8, 10, 38–47]. No doubt similar attempts are currently in progress at the ultrastructural level although there seem to be no reports in the current literature. Certainly the lack of published reports on EM level hybridisation to cellular RNA points to an area where we must now concentrate our efforts. Particularly important will be developing methods that preserve ultrastructural detail. Also, in hybridising to RNA one must take into account the differences in lability of RNA and DNA as well as recognising that RNA, while generally single stranded, is still tightly complexed within ribonucleoprotein particles. Thus most workers have tried to develop conditions of fixation and hybridisation that promote retention of RNA while increasing accessibility to RNA to the probe for the LM level. It is very likely that similar steps will be useful for subcellular localisation of RNA at the EM level.

Recently there have been several biochemical analyses suggesting specific sub-cellular localisation of mRNAs, particularly the idea that mRNAs actively translated are attached to the cytoskeleton in regions where their translation products are to be used [48]. Testing these ideas will be a challenge, probably requiring the use of ultrastructural in situ hybridisation. Since the biotin nucleotide system has recently been used at the LM level for the localisation of actin mRNA [36], this technology may eventually provide a rapid method for EM level RNA detection.

5. Future applications

The ability to identify and map nucleic acid sequences with ultrastructural resolution is sure to be an important research tool for the future. With these methods we can perform fine structure mapping of genes on chromosomes especially in systems with micro-chromosomes and double minute chromosomes. In addition we can map the locations of specific sequences within interphase nuclei and begin to study the interactions of different sequences in the nucleus as well as examining compartmentalisation of active and inactive sequences.

We would like to be able to hybridise to RNA at the ultrastructural level to examine the relationship of mRNA and the cytoskeleton and to identify particular active cells within a tissue. Recently Gee and Roberts [10] elegantly combined LM hybridisation to mRNA and immunocytochemical localisation of corresponding peptide hormones by treating alternate sequential sections. This kind of approach will be very useful for future understanding of regulatory systems and tissue specialisation. Finally, if we can apply ultrastructural hybridisation to RNA in nascent transcripts of Miller spreads [35] we will have an elegant tool for continued studies of chromosome structure and gene expression.

6. Acknowledgements

I would like to thank Jan De Mey for providing colloidal gold samples and Sandya Narayanswami and Barbara Hamkalo for providing their results prior to publication. I appreciate the work of Helen Devitt in preparing the manuscript. This chapter is written in acknowledgement of the contributions to in situ hybridisation from them and my other colleagues, Dave Ward and Pennina Langer-Safer.

7. References

1. Hutchison, N., Langer-Safer, P., Ward, D. and Hamkalo, B. (1982) In situ hybridization at the electron microscope level: hybrid detection by autoradiography and colloidal gold. J. Cell Biol. 95, 609–618.

2. Hutchison, N.J. (1982) In situ hybridization at the electron microscope level. PhD Thesis, University of California, Irvine.
3. Hamkalo, B. and Hutchison, N. (1984) Electron microscope in situ hybridization. In: Research Prospectives in Cytogenetics (in press).
4. Pardue, M.L. and Gall, J.G. (1969) Molecular hybridization of radioactive DNA to the DNA of cytological preparations. Proc. Natl. Acad. Sci. U.S.A. 64, 600–604.
5. John, H.A., Birnstiel, M.L. and Jones, K.W. (1969) RNA-DNA hybrids at the cytological level. Nature (London) 223, 582–587.
6. Eckhardt, R.C. (1976) Cytological localization of repeated DNAs. In: R.C. King (Ed.) Handbook of Genetics, Vol. 5: Molecular Genetics. Plenum Press, New York, pp. 31–53.
7. Rigby, P.W., Dieckmann, M., Rhodes, C. and Berg, P. (1977) Labelling deoxyribonucleic acid to high specific activity in vitro by nick translation with DNA polymerase. I. J. Molec. Biol. 113, 237–251.
8. Brahic, M. and Haase, A.T. (1978) Detection of viral sequences of low reiteration frequency by in situ hybridization. Proc. Natl. Acad. Sci. U.S.A. 75, 6125–6129.
9. Haase, A.T., Strowring, L., Harris, J.D., Traynor, B., Ventura, P., Peluso, R. and Brahic, M. (1982) Visna DNA synthesis and the tempo of infection in vitro. Virology 119, 399–410.
10. Gee, C. and Roberts, J. (1983) In situ hybridization histochemistry: a technique for the study of gene expression in single cells. DNA 2, 157–163.
11. Wahl, G., Stern, M. and Stark, G. (1979) Efficient transfer of large DNA fragments from agarose gels to diazobenzyloxymethyl-paper and rapid hybridization using dextran sulfate. Proc. Natl. Acad. Sci. U.S.A. 76, 3683–3687.
12. Tereba, A., Lai, M. and Murti, K. (1979) Chromosome 1 contains the endogenous RAV-O retrovirus sequences in chicken cells. Proc. Natl. Acad. Sci. U.S.A. 76, 6486–6490.
13. Harper, M.E. and Saunders, G.F. (1981) Localization of single copy DNA sequences on G-banded human chromosomes by in situ hybridization. Chromosoma 83, 431–439.
14. Gerhard, D.S., Kawasaki, E.S., Bancroft, F.C. and Szabo, P. (1981) Localization of a unique gene by direct hybridization in situ. Proc. Natl. Acad. Sci. U.S.A. 78, 3755–3759.
15. Wahl, G., Vitto, L., Padgett, R. and Stark, G. (1982) Single-copy and amplified CAD genes in Syrian hamster chromosomes localized by a highly sensitive method for in situ hybridization. Molec. Cell. Biol. 2, 308–319.
16. Neel, B.G., Jhanwar, S., Chaganti, R. and Hayward, W. (1982) Two human c-onc genes are located on the long arm of chromosome 8. Proc. Natl. Acad. Sci. U.S.A. 79, 7842–7846.
17. Bayer, E.A. and Wilchek, M. (1980) The use of the avidin-biotin complex as a tool in molecular biology. Methods Biochem. Anal. 26, 1–45.
18. Jacob, J., Todd, K., Birnstiel, M.L. and Bird, A. (1971) Molecular hybridization of ^{3}H-labelled ribosomal RNA with DNA in ultrathin sections prepared for electron microscopy. Biochim. Biophys. Acta 228, 761–766.
19. Jacob, J., Gillies, K., MacLeod, D. and Jones, K. (1974) Molecular hybridization of mouse satellite DNA complementary RNA in ultrathin sections prepared for electron microscopy. J. Cell Sci. 14, 253–261.
20. Croissant, O., Dauguet, C., Jeanteur, P. and Orth, G. (1972) Application de la technique d'hybridation moleculaire in situ a la mise en evidence au microscope electronique de la replication vegetative de 1-Adn viral dans les papillomes provoques par le virus de Shope chez de lapin cottontail. C.R. Acad. Sci. Ser. D. 274, 614–617.
21. Gueskens, M. and May, E. (1974) Ultrastructural localization of SV40 viral DNA in cells, during lytic infection, by in situ molecular hybridization. Exp. Cell Res. 87, 175–185.
22. Rae, P.M. and Franke, W.W. (1972) The interphase distribution of satellite DNA-containing heterochromatin in mouse nuclei. Chromosoma 39, 443–456.
23. Steinert, G., Thomas, C. and Brachet, J. (1976) Localization by in situ hybridization of amplified ribosomal DNA during *Xenopus laevis* oocyte maturation (a light and electron microscopy study). Proc. Natl. Acad. Sci. U.S.A. 73, 833–836.
24. Geuskens, M. (1977) Autoradiographic localization of DNA in nonmetabolic conditions. In: M. Hayat (Ed.) Principles and Techniques of Electron Microscopy, Vol. 7. Van Nostrand Reinhold Co., New York, pp. 163–201.

25. Pardue, M.L. and Gall, J.G. (1975) Nucleic acid hybridization to the DNA of cytological preparations. In: D. Prescott (Ed.) Methods in Cell Biology, Vol. X. Academic Press, New York, pp. 1–16.
26. Jones, K.W. (1974) The method of in situ hybridization. In: R. Pain and B. Smith (Eds.) New Techniques in Biophysics and Cell Biology, Vol 1. Interscience, New York, pp. 29–62.
27. Wimber, D.E. and Steffensen, D. (1974) Localization of gene function. Ann. Rev. Genet. 7, 205–223.
28. Steffensen, D. (1977) Human gene localization by RNA: DNA hybridization in situ. In: J.J. Yunis (Ed.) Molecular Structures for Human Chromosomes. Academic Press, New York, pp. 59–88.
29. Szabo, P., Elder, R., Steffensen, D. and Uhlenbeck, O. (1977) Quantitative in situ hybridization of ribosomal RNA species to polytene chromosomes of *Drosophila melanogaster*. J. Molec. Biol. 115, 539–563.
30. Rattner, J.B., Branch, A. and Hamkalo, B.A. (1975) Electron microscopy of whole mount metaphase chromosomes. Chromosoma 52, 329–338.
31. Manning, J., Hershey, N., Broker, T., Pellegrini, M., Mitchell, H. and Davidson, N. (1975) A new method of in situ hybridization. Chromosoma 53, 107–117.
32. Langer, P., Waldrop, A. and Ward, D. (1981) Enzymatic synthesis of biotin-labelled polynucleotides: novel nucleic acid affinity probes. Proc. Natl. Acad. Sci. U.S.A. 78, 6633–6637.
33. Langer-Safer, P., Levine, M. and Ward, D.C. (1982) An immunological method for mapping genes on Drosophila polytene chromosomes. Proc. Natl. Acad. Sci. U.S.A. 79, 4381–4385.
34. Manuelidis, L., Langer-Safer, P.R. and Ward, D.C. (1982) High resolution mapping of satellite DNA using biotin-labelled DNA probes. J. Cell Biol. 95, 619–625.
35. Narayanswami, S., Hutchison, N.J., Ward, D.C. and Hamkalo, B.A. (1982) An improved method for in situ hybridization at the EM level. J. Cell Biol. 95, 74a.
36. Singer, R.H. and Ward, D. (1982) Actin gene expression visualized in chicken muscle tissue culture by using in situ hybridization with a biotinated nucleotide analog. Proc. Natl. Acad. Sci. U.S.A. 79, 7331–7335.
37. Brigati, D.J., Myerson, D., Leary, J.J., Spalholz, B., Travis, S.Z., Fong, C.K.Y., Hsiung, G.D. and Ward, D.C. (1983) Detection of viral genomes in cultured cells and paraffin-embedded tissue sections using biotin labelled hybridization probes. Virology 126, 32–50.
38. Venezky, D., Angerer, L. and Angerer, R. (1981) Accumulation of histone repeat transcripts in the sea urchin egg pronucleus. Cell 24, 385–391.
39. Angerer, L. and Angerer, R. (1981) Detection of poly A+ RNA in sea urchin eggs and embryos by quantitative in situ hybridization. Nucl. Acids Res. 9, 2819–2840.
40. Godard, C. and Jones, K.W. (1979) Detection of AKR MuLV-specific RNA in AKR mouse cells by in situ hybridization. Nucl. Acids Res. 6, 2849–2861.
41. Godard, C.M. and Jones, K.W. (1980) Improved method for detection of cellular transcripts by in situ hybridization: detection of poly(A) sequences in individual cells. Histochemistry 65, 291–300.
42. John, H., Patrinou-Georgoulas, M. and Jones, K.W. (1977) Detection of myosin heavy chain mRNA during myogenesis in tissue culture by in vitro and in situ hybridization. Cell 12, 501–508.
43. Gall, J.G., Stephenson, E.C., Erba, H.P., Diaz, M.O. and Barsacchi-Pilone, G. (1981) Histone genes are located at the sphere loci of newt lampbrush chromosomes. Chromosoma 84, 159–171.
44. Macgregor, H. (1980) Recent developments in the study of lampbrush chromosomes. Heredity 44, 3–35.
45. Hudson, P., Penshow, J., Shine, J., Ryan, G., Niall, H. and Coghlan, J. (1981) Hybridization histochemistry: use of recombinant DNA as a 'homing probe' for tissue localization of specific mRNA populations. Endocrinology 108, 353–356.
46. Hafen, E., Levine, M., Garber, R. and Gehring, W. (1983) An improved in situ hybridization method for the detection of cellular RNAs in Drosophila tissue sections and its application for localizing transcripts of the homeotic Antennapedia gene complex. EMBO J. 2, 617–623.
47. Edwards, M.K. and Wood, W.B. (1983) Location of specific messenger RNAs in *Caenorhabditis elegans* by cytological hybridization. Dev. Biol. 97, 375–390.
48. Fulton, A.B., Wan, K.M. and Penman, S. (1980) The spatial distribution of polyribosomes in 3T3 cells and the associated assembly of proteins into the skeletal framework. Cell 20, 849–857.

Subject index

T